Telephone Projects
for the
Evil Genius

Evil Genius Series

Telephone Projects for the Evil Genius

TOM PETRUZZELLIS

New York   Chicago   San Francisco   Lisbon   London   Madrid
Mexico City   Milan   New Delhi   San Juan   Seoul
Singapore   Sydney   Toronto

Library of Congress Cataloging-in-Publication Data

Petruzzellis, Thomas.
  Telephone projects for the evil genius/Tom Petruzzellis.
    p. cm. — (Evil genius)
  ISBN 978-0-07-154844-1 (alk. paper)
1. Telephone—Equipment and supplies—Design and construction—Amateurs' manuals. I. Title.
TK9951.P48 2008
621.385—dc22

2008023311

McGraw-Hill books are available at special quantity discounts to use as premiums and sales promotions, or for use in corporate training programs. To contact a representative please visit the Contact Us pages at www.mhprofessional.com.

Telphone Projects for the Evil Genius

1 2 3 4 5 6 7 8 9 0   QPD/QPD   0 1 3 2 1 0 9 8

ISBN 978-0-07-154844-1
MHID 0-07-154844-0

Sponsoring Editor	**Proofreader**
Judy Bass	Penny Richardson
Acquisitions Coordinator	**Indexer**
Rebecca Behrens	Golden Parodox Indexing
Editing Supervisor	**Production Supervisor**
David E. Fogarty	Pamela A. Pelton
Project Manager	**Composition**
Jeremy Toynbee	Keyword Group Ltd.
Copy Editor	**Art Director, Cover**
Pat Baxter	Jeff Weeks

This book is dedicated to my first grandson Giovanni.
May he dream big dreams and see them realized.
Wishing him a long, interesting, and enjoyable life!

About the Author

Tom Petruzzellis is an electronics engineer with 30 years' experience, currently working with the geophysical field equipment department at the State University of New York-Binghamton. He has written extensively for industry publications such as *Electronics Now* and *Modern Electronics*, and is the author of numerous McGraw-Hill bestsellers, including *Electronics Sensors for the Evil Genius*.

Contents

Contents

Introduction

Since its inception in 1876, the telephone has captured the imagination of people around the world. From its simple origins, the telephone has evolved from the humble device it once was to the modern cell phone or satellite phone. The telephone has fundamentally and instantaneously changed the human race's ability to communicate over long distances. Today, not only can we communicate via the human voice but we can rapidly send scientific and health data, pictures, and information between any two people or organizations on earth in just a few seconds by picking up the phone and dialing the party at the other end. The development of the telephone has progressed at such a rapid rate in the last 50 years alone that it is truly amazing what can be done with the telephone in our modern lives; and it is now mostly taken for granted.

Our new book titled *Telephone Projects for the Evil Genius* is a fun and informative project book which we hope will spark your imagination and creative abilities. If you are interested in learning about the history of the telephone, and how a telephone operates, or if you want to know how a telephone remote control functions and how to remotely control things around your home or office, or perhaps, if you would like to know how to build a remote telephone listening device, then you will definitely want to read on.

This telephone project book was written for the general electronics enthusiast as well as for telephone enthusiasts and experimenters alike.

This book will appeal to both electronics neophytes and seasoned circuit builders of all ages.

Once you open up this book you will realize that the common telephone and telephone lines can be used in a variety of ways to support many different types of projects that you perhaps never knew existed. You will explore the basics of the telephone, and learn how to build an electronic telephone and a telephone modem/line protector, as well as remote ringer circuits, line use indicators, a telephone intercom, speaker-phone, a telephone transmitter, and a telephone listening device, just to name a few projects.

The book is broken down into 32 chapters. Chapter 1 discusses a bit of telephone history, while Chapter 2 describes how a conventional telephone functions. Chapter 3 shows readers how to identify components; this chapter is particularly useful for newcomers to electronics or for younger readers. Chapter 4 delves into reading schematics, and you will learn the difference between a schematic and a pictorial diagram, as well as how components are connected and work with one another. Next, Chapter 5 shows the reader how to install components on circuit boards and how to solder. Every electronic hobbyist should know how to solder well before attempting electronic projects.

Chapter 6 begins the actual project section of the book and presents a device called the privacy guard. The privacy guard prevents other persons around the house from listening to your telephone calls.

You will notice that we begin with easier projects and escalate to more advanced projects as we go along. Next, Chapter 7 illustrates how to build a fax/modem/answering machine protector which can save your phone devices from voltage or lightning spikes and strikes. In Chapter 8 we present two different types of telephone amplifiers. One amplifier is a direct connect device which will allow you to amplify the telephone conversation for a group, or for hard-of-hearing elders. The direct amplifier could also be used to amplify a telephone ring remotely. The second amplifier uses a phone suction cup with a wireless connection to the phone to amplify a phone conversation.

Chapter 9 illustrates an automatic phone recorder switch device which will start and stop a tape or voice recorder, in order to record phone conversations around your home or office. Moving on, in Chapter 10, we take a look at a phone conferencer circuit which will combine two phone line conversations. Chapter 11 depicts a device called the frustrator, which will make annoying people go away and never call back!

Next we present four chapters which cover themes of circuits and projects. Chapter 12 illustrates four telephone hold circuits from simple manual on-hold and music-on-hold circuits to touch-tone-controlled hold circuits and an automatic music-on-hold project.

Chapter 13 discusses telephone ringer circuits. We start first with a silent cell phone ringer circuit and then remote ringer circuits, both wired and wireless types. Our final circuit is a ring generator circuit which could be used for intercoms or phone testing.

Chapter 14 presents phone status circuits and projects, from in-use detector, to cut phone line detector; then we move on to single and dual line status monitors. We also present a smart phone light which turns on a room light when the phone rings so you can take a note or move around the room for a period of time; and finally a phone line monitor circuit, called the phone line vigilant.

Chapter 15 covers telephone and telephone line testers, from the simple phone device tester to phone line tester and a phone line simulator. Next we move on to some intermediate telephone projects, such as the automatic telephone transmitter project in Chapter 16, which will automatically broadcast a two-way phone conversation over to an FM radio, using phone-line power. Chapter 17 shows the reader how to build a complete modern electronic telephone.

Chapter 18 features a fun voice/music telephone ringer, which you can program to announce the phone ringing in your own favorite way. The tollsaver in Chapter 19 is a phone dialer project which can be used to call a local access phone number for saving money or can be used for an older person to immediately call a loved one.

Chapter 20 presents the infinity bug, which can be placed on a phone line you wish to monitor. When called silently, the bug will allow you to listen-in to a remote site for as long as you like.

If you operate a CB or ham radio setup, then you will be interested in building the radio/telephone phone patch described in Chapter 21. This circuit will allow you to interface your radio equipment with the telephone line. The phone patch will permit a person on the phone to talk to a person on the radio.

The phone intercom in Chapter 22 will turn the telephones in your home into an intercom system throughout your whole house. The main intercom controller is attached to the phone line and power, while an alert module is placed at each phone to form a complete intercom system using your in-house phone wiring.

Chapter 23 presents a speaker-phone project which will allow you to carry on a phone conversation without using your handset, thus freeing up your hands for more important tasks. The speaker-phone project will also permit a group of people to converse on the phone with a person at a distance.

Have you ever wanted your phone conversation to remain private from possible nosy friends, family, or coworkers? Then you will appreciate the phone scrambler project in Chapter 24, which will keep your conversations unintelligible.

The tattletale project illustrated in Chapter 25 was designed to remotely monitor alarm conditions and report to you via a remote phone call. You will be able to monitor a remote room or cottage while you are away from home or the office.

Chapter 26 will show you how to build a DTMF remote phone controller. With this project you can dial your home phone and remotely control devices such as lights, fans, and air-conditioners.

Our next projects feature the use of a small microcontroller called the BASIC STAMP 2.

In the remaining projects you will program the controller to perform specific telephone controls. Chapter 27 is an introduction to the BASIC STAMP 2 controller. This chapter illustrates various types of BASIC STAMP 2 controllers and how they work; next we will move on to actual microcomputer projects using the BASIC STAMP 2.

Chapter 28 will show you how to build a touch-tone generator, as well as a touch-tone decoder/display unit which will display a phone number that was dialed on a phone line or on a radio frequency, e.g., via ham radio.

Chapter 29 presents a caller-ID project which will display the phone number of persons calling you, so you can decide if you want to answer the phone or not.

The Page-Alert project in Chapter 30 forms a small local alarm system which pages you in the event of an alarm condition. You can create your own multi-sensor alarm to monitor your home or office and let the page-alert call you to report an alarm condition.

The Tele-Alert project in Chapter 31 will permit you to monitor your home or office with alarm sensors and report an alarm condition to your cell phone.

Finally, Chapter 32 presents a dial-up temperature alarm system, which will allow you to monitor the remote temperature of your vacation home or cabin in the comfort of your own home. If your creative interest has been aroused, you may want to expand the system to monitor other devices.

Telephone Projects for the Evil Genius will open up a whole new area for electronics project building fun, projects that both you and your family will find useful and enjoy, and which you can build at a fraction of the cost of similar commercial products. A number of projects in this book are not commercially available and will never be available to the general consumer, but with this book you will be able to create any number of telephone related projects that will save you time and money as well. Take your time and enjoy the book; you may even find a project that you cannot live without!

We hope you will find the projects educational and interesting and that you will want to build one of these projects to enhance the telephone system in your home or office. Have fun and be safe!

Acknowledgments

A book is a complex entity and requires the help of many people to see it through to completion. I would like to thank the following people in making this book possible. First I would like to thank senior editor Judy Bass, her assistant Rebecca Behrens and all the folks at McGraw-Hill Professional who had a part in making this book possible. I would like to also like to thank the following people and companies listed below, who had a part in bringing this book to completion. Our hope is that the book will encourage readers to build the projects and that it might help inspire young electronic enthusiasts to enter the fields of engineering and electronics.

Colin Mitchell/Talking Electronics

Rebecca Lowery/Vellemen, Inc

Larry Steckler/Poptronix

Ken Gracey/Parallax, Inc

Thomson Publications

Arthur Seymour/Elenco Electronics

Ramsey Electronics

Cengage Learning

Frank Montegari/Glolab, Inc

Bill Bowden

Dave Johnson

Anthony Caristi

Tom Engdahl

Telephone Projects for the Evil Genius

Chapter 1

Telephone History

Telephone comes from the Greek word "tele," meaning from afar, and phone, meaning voice or voiced sound. Generally, a telephone is any device which conveys sound over a distance. Talking produces acoustic pressure. A telephone reproduces sound by electrical means.

The dictionary defines the telephone as "an apparatus for reproducing sound, especially that of the voice, at a great distance, by means of electricity; consisting of transmitting and receiving instruments connected by a line or wire which conveys the electric current." Electricity operates the telephone and it carries your voice.

Telephone history begins, perhaps, at the start of human history. Man has always wanted to communicate from afar. People have used smoke signals, mirrors, jungle drums, carrier pigeons, and semaphores to get a message from one point to another. But a phone was something new. A real telephone could not be invented until the electrical age began. The electrical principles required to build a telephone were known in 1831 but it was not until 1854 that Bourseul suggested transmitting speech electrically. And it was not until 22 years later in 1876 that the idea became a reality. Telephone development did not proceed in an organized line like powered flight, with one inventor after another working to realize a common goal, rather, it was a series of often disconnected events, mostly electrical, some accidental, that made the telephone possible.

Probably no means of communication has revolutionized the daily lives of ordinary people more than the telephone. The actual history of the telephone was long and arduous with many twists and turns and is a subject of complex dispute to this day.

The actual telephone was built upon the work of many people who preceded Alexander Graham Bell. In 1729, English chemist Stephen Gray is believed to be the first person to transmit electricity over a wire.

He sent charges nearly 300 feet over brass wire and moistened thread. An electrostatic generator powered his experiments, one charge at a time.

In 1800, Alessandro Volta produced the first battery. A major development, Volta's battery provided sustained low-powered electric current at high cost. Chemically based, as all batteries are, the battery improved quickly and became the electrical source for further experimenting. But while batteries got more reliable, they still could not produce the power needed to work machinery.

Then in 1820 Danish physicist Christian Oersted demonstrated electromagnetism, the critical idea needed to develop electrical power and to communicate. In a famous experiment at his University of Copenhagen classroom, Oersted pushed a compass under a live electric wire. This caused its needle to turn from pointing north, as if acted on by a larger magnet. Oersted discovered that an electric current creates a magnetic field.

Around 1821, Michael Faraday reversed Oersted's experiment. He got a weak current to flow in a wire revolving around a permanent magnet. In other words, a magnetic field caused or induced an electric current to flow in a nearby wire. In so doing, Faraday had built the world's first electric generator. Mechanical energy could now be converted to electrical energy. Faraday worked through different electrical problems in the next 10 years, eventually publishing his results on induction in 1831.

Then in 1830 the great American scientist Professor Joseph Henry transmitted the first practical electrical signal. A short time before, Henry had invented the first efficient electromagnet. He also concluded similar thoughts about induction before Faraday but he did not publish them first. Henry's place in electrical history however, has always been secure, in particular for showing that electromagnetism could do more than create current or pick up heavy weights—it could communicate.

In 1837, Samuel Morse invented the first workable telegraph, applied for its patent in 1838, and was finally granted it in 1848. Joseph Henry helped Morse build a telegraph relay or repeater that allowed long distance operation. The telegraph later helped unite the country and eventually the world. In 1832, he heard of Faraday's recently published work on inductance, and at the same time was given an electromagnet to ponder over. An idea came to him and Morse quickly worked out details for his telegraph. His system used a key or switch, to make or break the electrical circuit, a battery to produce power, a single line joining one telegraph station to another, and an electromagnetic receiver or sounder that upon being turned on and off produced a clicking noise. He completed the package by devising the Morse code system of dots and dashes. A quick key tap broke the circuit momentarily, transmitting a short pulse to a distant sounder, interpreted by an operator as a dot. A lengthier break produced a dash. Telegraphy was not accepted initially but eventually it became big business.

In the early 1870s the world still did not have a working telephone. Inventors focused on telegraph improvements since the telegraph itself already had a proven market. Developing a telephone, on the other hand, had no immediate market, if one at all. Elisha Gray, Alexander Graham Bell, as well as others such as Antonio Meucci, and Philip Reis trying to develop a multiplexing telegraph—a device to send several messages over one wire at once. Such an instrument would greatly increase traffic without the telegraph company having to build more lines. As it turned out, for both men, the desire to invent one thing turned into a race to invent something altogether different.

The telegraph and telephone are both wire-based electrical systems, and Alexander Graham Bell's success with the telephone came as a direct result of his attempts to improve the telegraph. When Bell began experimenting with electrical signals, the telegraph had been an established means of communication for some 30 years.

Bell developed new and original ideas but did so by building on older ideas and developments. He succeeded specifically because he understood acoustics, the study of sound, and something about electricity. Other inventors knew electricity well but little of acoustics. The telephone is a shared accomplishment among many pioneers, therefore, although the credit and rewards were not shared equally.

In the 1870s, two inventors, Elisha Gray and Alexander Graham Bell, both independently designed devices that could transmit speech electrically, the device destined to be called the telephone. Both men rushed their respective designs to the patent office within hours of each other; Alexander Graham Bell patented his telephone first. Elisha Gray and Alexander Graham Bell entered into a famous legal battle over the invention of the telephone, which Bell eventually won.

The principle of the telephone was uncovered in 1874, but it was the unique combination of electricity and voice that led to Bell's actual invention of the telephone in 1876. Bell's original telephone is shown in Figure 1-1. Convincing Bell's partners, Gardiner Greene Hubbard, a prominent lawyer from Boston, and Thomas Sanders, a leather merchant with capital from Salem, about the potential for voice transmittal was not an easy task, and they often threatened to pull Bell's funding. Nonetheless, agreement was finally reached and the trio received U.S. Patent No. 174,465, issued on March 3, 1876, for "Improvements in Telegraphy," which is now considered to be the most valuable patent ever issued. Bell's experiments with his assistant Thomas Watson finally proved successful on March 10, 1876, when the

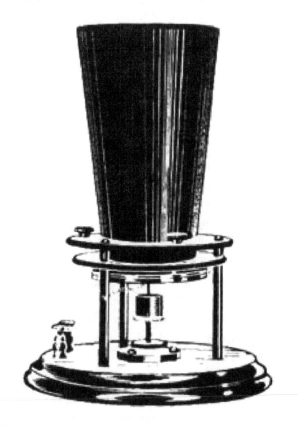

Figure 1-1 *Bell's first telephone instrument*

first complete sentence was transmitted: "Watson, come here; I want you."

Bell considered his invention's greatest advantage over every other form of electrical apparatus to be the fact that it could be used by anyone, as "all other telegraphic machines produce signals which require to be translated by experts, and such instruments are therefore extremely limited in their application, but the telephone actually speaks, and for this reason it can be utilized for nearly every purpose for which speech is employed."

Bell was nearly beaten to the patent office by Elisha Gray, who had independently developed a very similar invention. Gray arrived just hours after Bell at the Patent Office, filing a "caveat," a confidential report of an invention that was not yet perfected. Western Electric, co-founded by Gray, became one of the Bell System's major competitors. Western Union was another major competitor, already having established itself as a communications provider with the telegraph system.

In 1877, construction of the first regular telephone line from Boston to Somerville, Massachusetts, was completed, a distance of three miles. Commercial telephone service began in the United States in 1877. The workable exchange, developed in 1878, enabled calls to be switched among any number of subscribers rather than requiring direct lines. Exchanges were handled manually, first by boys, then by the now-famous women operators.

By the end of 1880, there were 47,900 telephones in the United States. The following year telephone service between Boston and Providence had been established. Service between New York and Chicago started in 1892, and between New York and Boston in 1894. Transcontinental service by overhead wire was not inaugurated until 1915. The first switchboard was set up in Boston in 1877. On January 17, 1882, Leroy Firman received the first patent for a telephone switchboard.

The first regular telephone exchange was established in New Haven in 1878. Early telephones were leased in pairs to subscribers. The subscriber was required to put up his own line to connect with another.

In 1889, Almon B. Strowger, a Kansas City undertaker, invented a switch that could connect one line to any of 100 lines by using relays and sliders. This step by step switch used to receive the dial pulses became known as "The Strowger Switch" after its

inventor and was still in use in some telephone offices well over 100 years later. Almon Strowger was issued a patent on March 11, 1891, for the first automatic telephone exchange. The first exchange using the Strowger switch was opened in La Porte, Indiana, in 1892 and initially subscribers had a button on their telephone to produce the required number of pulses by tapping.

Strowger installed his automatic exchanges in the United States and Europe. In 1924, the Bell Telephone System decided that using operators was *not* the way to go, and they licensed Strowger's technology.

An associate of Strowger invented the rotary dial in 1896 and this replaced the button. In 1943, Philadelphia was the last major area to give up dual service (rotary and button).

In 1879, telephone subscribers began to be designated by numbers rather than names—as a result of an epidemic of measles. A doctor from Lowell, Massachusetts, concerned about the inability of replacement exchange operators to put calls through because they would not be familiar with the names associated with all the jacks on the switchboards, suggested the alpha-numeric system of identifying customers by a two-letter and five-digit system.

Long-distance telephone service was established and grew in the 1880s using metallic circuits. The common-battery system, developed by Hammond V. Hayes in 1888, permitted a central battery to supply all telephones on an exchange with power, rather than relying upon each unit's own troublesome battery.

A young inventor, Dr. Lee De Forest, began work in 1906 on applying what was known as an "Audion," a three-element vacuum tube, which could amplify radio waves. He recognized the potential for installing Audions, which became the major component in what would be called repeaters on telephone lines, in order to amplify the sound waves at mid-points along the wires. The Bell System bought the rights to De Forest's patents in 1913. Long-distance telephone service was constructed on the New York to San Francisco circuit using loading coils and repeaters.

American Telephone and Telegraph (AT&T) took control of the Western Union Telegraph Company in a "hostile takeover," in 1911, having purchased the Western Union stocks through a subsidiary. The two eventually merged, sharing financial data and telephone lines. By 1918, ten million Bell System telephones were in service.

The next major improvement to the automatic switching of large numbers of calls was made possible in 1921, using "phantom circuits," which allowed three telephone conversations to be conducted on two pairs of wires. The "French" phone, with the transmitter and receiver in a single handset, was developed by the Bell System around 1904, but was not released on a widespread basis because it cost more than the desk sets. They ultimately became available to subscribers in 1927. The first transatlantic service, from New York to London, became operational in 1927. Research in electronic telephone exchanges began in 1936 in Bell Labs, and was ultimately perfected in the 1960s with its Electronic Switching System (ESS).

In 1938, the Bell System introduced crossbar switching to the central office. The first No. 1 crossbar was placed into service at the central office in Brooklyn, New York on February 13. AT&T improved on work done by the brilliant Swedish engineer Gotthilf Ansgarius Betulander. They even sent a team to Sweden to look at his crossbar switch. Western Electric's models earned a worldwide reputation for ruggedness and flexibility. Installed by the hundreds in medium to large cities, crossbar technology advanced in development and popularity until 1978, when over 28 million Bell System lines were connected to one.

In the mid-1940s, the first mobile wireless phone services appeared in the United States. These services used one tower in each metropolitan area. Since the technology was very expensive and the market small, one tower could handle all the phone calls. However, demand for mobile phone services began to grow, and technology improved so that phones could be smaller and less expensive. Engineers anticipated these trends, and in the 1960s began researching and developing what is now today's cellular phone service.

The Bell System benefited greatly from U.S. defense spending during World War II in its laboratories. Wartime experiments, innovations, and inventions brought Bell to the forefront of telecommunications in the post-war era. The first commercial mobile telephone service was put in service in 1946, linking moving vehicles to telephone networks by radio. The same year brought transmission via coaxial cables, resulting in a major improvement in service as they were less likely to be interrupted by other electrical interference. Microwave tube radio transmission was used for long-distance telephony in 1947. The transistor, a key to modern electronics, was invented at Bell Labs in 1947. A team consisting of William Schockley, Walter Brattain, and John Bardeen demonstrated the "transistor effect," using a germanium crystal that they had set up in contact with two wires two-thousandths of an inch apart. The development of the transistor not only made possible the advances to telephony but miniaturization of the transistor led to the development of small personal computer and laptops that we use today.

On August, 17, 1951, the first transcontinental microwave system began operating. One hundred and seven relay stations spaced about 30 miles apart formed a link from New York to San Francisco. It cost the Bell System $40,000,000; a milestone in their development of radio relay begun in 1947 between New York and Boston. In 1954, over 400 microwave stations were scattered across the country. By 1958, microwave carrier made up 13,000,000 miles of telephone circuits or one quarter of the nation's long distance lines. Six hundred conversations or two television programs could be sent at once over these radio routes.

Years of development led up to 1956 when the first transatlantic telephone cable system started carrying calls; this is an interesting story in itself. Two coaxial cables about 20 miles apart carried 36 two-way circuits. Nearly 50 sophisticated repeaters were spaced from 10 to 40 miles along the way. Each vacuum tube repeater contained 5,000 parts and cost almost $100,000. On the first day this system took 588 calls, 75% more than the previous 10 days' average with AT&T's transatlantic radio-telephone service.

The 1960s began a dizzying age of projects, improvements, and introductions. In 1961, the Bell System along with the help of the U.S. government started work on a classic cold war project, finally completed in 1965. It was the first coast to coast atomic bomb blast resistant cable network system.

The six major transcontinental cables were evenly distributed from the southern United States to the northern states. The trunk lines were about six inches in diameter with 22 coax cables and control and alarm wires within the major trunk cable. The telephone network also consisted of north–south links which were accomplished with microwave radio links interconnecting with the six major truck cables. The original trunk line cables were later replaced with half-inch diameter fiber optical cables. Over 950 buried concrete repeater stations were constructed, and stretched along the 19 state route were 11 manned test centers, buried 50 feet below ground, complete with air filtration, living quarters and food and water, for up to a month of operation.

The original repeater sites—every 20 miles along the transcontinental cables were highly disguised earthquake proof underground bunkers which could maintain communications in the event of a nuclear war or so it was believed. It was believed that a small crew could man each repeater site with food and fuel enough to power a jet engine turbine for a month of operation. Special air filter/scrubbers were installed in each location to allow the crew to be self contained in the event of a nuclear attack. AT&T contracted with many different companies which flew bi-weekly over the main transcontinental cables by overhead helicopter flights to monitor any activity or digging near the cables.

In 1963, the first modern touch-tone phone was introduced, the Western Electric 1500. It had only 10 buttons. Limited service tests had started in 1959. Also in 1963 digital carrier techniques were introduced. Previous multiplexing schemes used analog transmission, carrying different channels separated by frequency, much like those used by cable television. Transmission One, or T1, by comparison, reduced analog voice traffic to a series of electrical plots, binary

coordinates to represent sound. T1 quickly became the backbone of long distance toll service and then the primary handler of local transmission between central offices. The T1 system handles calls throughout the telephone system to this day.

In 1965, the first commercial communications satellite was launched, providing 240 two-way telephone circuits. The year 1965 also marked the debut of the No. 1ESS, the Bell System's first central office computerized switch. The product of at least 10 years of planning, 4,000 man years of research and development, as well as $500 million dollars in costs, the first Electronic Switching System was installed in Succasunna, NJ. Built by Western Electric, the 1ESS used 160,000 diodes, 55,000 transistors, and 226,000 resistors. These and other components were mounted on thousands of circuit boards. Not a true digital switch, the 1ESS did feature Stored Program Control, a fancy Bell System name for memory, enabling all sorts of new features like speed dialing and call forwarding.

Progress in miniaturization in the last 20 to 30 years has been astonishing. Few people realize that today's palm-sized phones originated from the "bag" or "briefcase phones" of the early and mid-1980s. This miniaturization of the modern cell phones would not have been possible without the cellular architecture, which uses low power cell phone towers or base station to hand-off calls between cell phone towers. The regularly spaced "cell" tower concept has permitted the rapid proliferation of the small pocket cell phone, since the pocket phone can utilize a low power, small size, stable UHF transmitter and receiver.

Chapter 2

How the Telephone Works

The telephone is so ubiquitous and transparent in our modern world that most often we take it for granted, yet most people really do not understand how the telephone works. In this chapter we will take a closer look at the telephone and its individual components as well as how the telephone companies' Central Office (CO) equipment controls our home telephone.

A telephone uses an electric current to convey sound information from your home to that of a friend. When the two of you are talking on the telephone, the telephone company is sending a steady electric current through your telephones from the telephone company's central office battery system. The two telephones, yours and your friend's, are sharing this steady current. But as you talk into your telephone's microphone, the current that your telephone draws from the telephone company fluctuates up and down. These fluctuations are directly related to the air pressure fluctuations that are the sound of your voice at the microphone.

Because the telephones are sharing the total current, any change in the current through your telephone causes a change in the current through your friend's telephone. Thus as you talk, the current through your friend's telephone fluctuates. A speaker or earphone in that telephone's handset responds to these current fluctuations by compressing and rarefying the air. The resulting air pressure fluctuations reproduce the sound of your voice. Although the nature of telephones and the circuits connecting them have changed radically in the past few decades, the telephone system still functions in a manner that at least simulates this behavior.

The current which powers your telephone is generated from the 48-V battery in the central office. The 48-V voltage is sent to the telephone line through some resistors and indicators (typically there are 2,000 to 4,000 ohms in series with the 48-V power source). The old ordinary offices had about 400 ohm line relay coils in series with the line.

When your telephone is in on-hook state the "TIP" is at about 0 V, while "RING" is about −48 V with respect to earth ground. When you go off hook, and current is drawn, TIP goes negative and RING goes positive (I mean less negative). A typical off-hook condition is TIP at about −20 V and ring at about −28 V. This means that there is about 8 V voltage between the wires going to telephone in normal operation condition. The DC-resistance of typical telephone equipment is in 200–300 ohm range and current flowing through the telephone is in 20–50 mA range.

The −48-V voltage was selected because it was enough to get through kilometers of thin telephone wire and still low enough to be safe (electrical safety regulations in many countries consider DC voltages lower than 50 V to be safe low voltage circuits). A voltage of 48 V is also easy to generate from normal lead acid batteries (4×12-V car battery in series). Batteries are needed in telephone central to make sure that it operates also when mains voltage is cut and they also give very stable output voltage which is needed for reliable operation of all the circuit in the central office. Typically, the CO actually runs off the battery chargers, with the batteries in parallel getting a floating charge.

The line feeding voltage was selected to be negative to make the electrochemical reactions on the wet telephone wiring to be less harmful. When the wires are at negative potential compared to the ground the metal ions go from the ground to the wire, replacing a situation where positive voltage would cause metal from the wire to leave, causing quick corrosion.

Some countries use other voltages in typically the 36 to 60 V range. PBXs may use as low as 24 V and can possibly use positive feeding voltage instead of the negative one used in normal telephone network. Positive voltage is more commonly used in many electronics circuits, so it is easier to generate; electrolysis in telecommunications wiring is not a problem in typical environments inside office buildings.

Ordinary telephones utilize only two wires, which carry both speaker and microphone signals. This is called full duplex operation in single wire pair. Full-Duplex is a term used to describe a communications channel which is capable of both receiving and sending information simultaneously.

The signal path between two telephones, involving a call other than a local one, requires amplification using a 4-wire circuit. The cost and cabling required ruled out the idea of running a 4-wire circuit out to the subscribers' premises from the local exchange and an alternative solution had to be found. Hence, the 4-wire trunk circuits were converted to 2-wire local cabling, using a device called a "hybrid." The hybrid device can send and receive audio signals at the same time; it is accomplished by designing the system so that there is a well-balanced circuit in both ends of the wire which are capable or separating incoming audio from outgoing signal. This function is done by telephone hybrid circuit contained in the network interface of the telephone.

A standard plain old telephone system or POTS telephone line has a bandwidth of 3 kHz. A normal POTS line can transfer the frequencies between 400 Hz and 3.4 kHz. The frequency response is limited by the telephone transmission system.

Telephone signaling

Ringing

When someone places a call to you, the CO is notified and the switch gear sends an AC ringing signal which will ring the bell in your telephone. Most of the world uses frequencies in 20 to 40 Hz range and voltage in the 50 to 150 V range. The ringer is built so that it will not pass any DC current when it is connected to telephone line (traditionally there has been a capacitor in series with the bell coil). So only the AC ring signal can go though the bell and make it ring. The bell circuit is either designed so that it has high impedance in audio frequencies or it is disconnected from the line when phone is picked off hook.

Dialing

There are two types of dials in use around the world: pulse dialing and tone dialing. The most common one is called pulse dialing (also called loop disconnect or rotary dialing). Pulse dialing is the oldest form of dialing—it has been with us since the 1920s. Pulse dialing is traditionally accomplished with a rotary dial, which is a speed governed wheel with a cam that opens and closes a switch in series with your phone and the line. It works by actually disconnecting or "hanging up" the telephone at specific intervals. The mostly used standard is one disconnect per digit (so if you dial a "1," your telephone is "disconnected" once and if you dial "2" your telephone is "disconnected" twice and for zero the line is "disconnected" ten times) but there are also other systems used in some countries.

Tone dialing is a more modern dialing method and is usually named Touch-tone, Dual Tone Multi-Frequency (DTMF) or Multi-Frequency (MF) in Europe. Touch-tone is fast and less prone to error than pulse dialing. Bell Labs developed DTMF in order to have a dialing system that could travel across microwave links and work rapidly with computer controlled exchanges. Touch-tone can therefore send signals around the world via the telephone lines, and can be used to control phone answering machines and computers (this is used in many automatic telephone services which you operate using your telephone keypad). Each transmitted digit consists of two separate audio tones that are mixed together (the four vertical columns on the keypad are known as the high group and the four horizontal rows as the low group). Standard DTMF dials will produce a tone as long as a key is depressed. No matter how long you press, the tone will be decoded as the appropriate digit. The shortest duration in which a digit can be sent and decoded is about 100 milliseconds (ms).

Other telephone signals

The telephone CO can send any different types of signals to the caller telling the status of a telephone call. Those signals are typically audio tones generated by the CO. Typical kinds of tones are dialing tone (typically constant tone of around 400 Hz), calling tone (tone telling that the telephone at the other end is ringing)

Figure 2-1 *Model 500C telephone*

or busy tone (usually like quickly on and off switched dialing tone). The exact tones used vary from country to country.

In the United States, the standard black telephone, which was provided by the Bell Telephone system for many years to the general public, was the classic black standard model 500C shown in Figure 2-1. The schematic or wiring diagram for the model 500C is shown in Figure 2-2; it included the switch-hook, the network device or hybrid, the microphone, the speaker, the bell, and the dial.

Telephone components

Switch hook

A "switch hook" is a manual control switch mechanism that answers and hangs up a call on a telephone. When you place the handset in the telephone cradle, it depresses the switch hook's button and hangs up (puts the phone "on hook"). When the phone is lifted from the cradle the switch hook is activated and the phone goes "off hook" and you will then get a dial tone.

The switch hook has two basic states: an "on-hook" state and an "off-hook" state. When the phone is on hook, both its microphone and speaker are disabled, as well as the network, while the bell is connected to the telephone line and awaits the ringing voltage when the phone is called. In the off-hook state, that is, when the phone is lifted from the cradle, the microphone and speaker are connected together through the network device and the bell is disconnected from the line. You will now hear a dial tone and you will be able to place or dial your call.

Network device

The standard telephone has a circuit called a voice network or telephone hybrid, which connects the

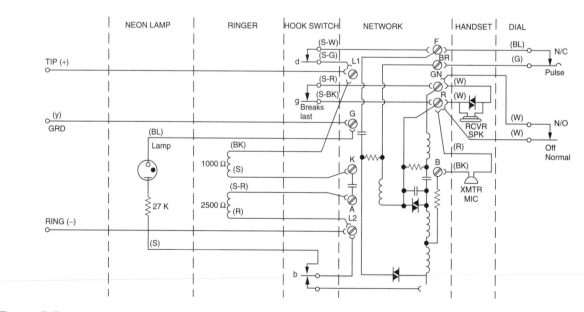

Figure 2-2 *Telephone schematic*

microphone and speaker to the telephone line (see telephone schematic). Network interface circuitry is designed so that it sends only the current changes the other telephone causes to the speaker. The current changes that the telephone's own microphone generates are not sent to the speaker. All this is accomplished using quite ingenious transformer circuitry. In theory the hybrid circuit can separate all incoming audio from the audio sent out at the same time if all the impedances in the circuitry (hybrids on both ends and the wire impedance in between) are well matched. Unfortunately, the hybrid is by its very nature a "leaky" device. As voice signals pass from the 4-wire to the 2-wire portion of the network, the higher energy level in the 4-wire section is also reflected back on itself, creating the echoed speech. Because the circuit does not work perfectly and you can still hear some of your own voice in the speaker, this is called side-tone or feedback, which is actually a desired effect if the volume level is kept low.

Rotary dial

The common rotary dial was 3″ in diameter, and had 10 finger holes that were cut through its outer perimeter, as seen in Figure 2-3. The dial is mounted via a shaft extending from inside the telephone or mounting and sits above a faceplate. In North America, traditional dials had letter codes displayed with the numbers under the finger holes: 1, 2 ABC, 3 DEF, 4 GHI, 5 JKL, 6 MNO, 7 PRS,

8 TUV, 9 WXY, and 0 Operator. A curved device called a finger stop sits above the dial at the 4 o'clock position.

Microphone

A basic telephone consists of a microphone, a speaker, and a simple electronic network that improves the behavior of the telephone. It also has a system for dialing and a bell to announce an incoming call. Modern telephones often contain sophisticated electronic devices, such as audio amplifiers, radio transmitters and receivers, lights, and audio recorders, but the basic concepts are still the same. When you talk into the microphone, it changes the amount of current flowing through the telephone. In older telephones, the microphone contains a small canister of carbon granules between two metal sheets; see Figure 2-4.

Since carbon conducts electricity somewhat, the microphone is a resistor. Current from one metal sheet flows to the other sheet over a circuitous path through the granules. The more tightly packed the carbon granules, the more they touch one another and the more direct the current path becomes. Compression causes the carbon microphone's electric resistance to decrease. Expansion causes it to increase. As you talk into the microphone, the air pressure fluctuations in your voice alternately compress and expand the granules and make the resistance of the microphone fluctuate up and down. Because the microphone is connected between the two telephone wires, this fluctuating resistance causes a

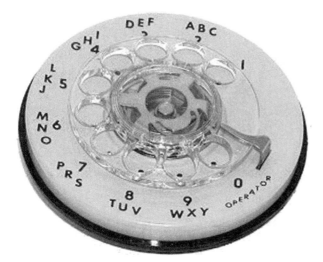

Figure 2-3 *Rotary dial*

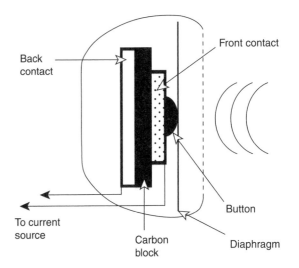

Figure 2-4 *Telephone microphone*

fluctuating current to flow through the telephone. Since all of the telephones in the parallel circuit share the same current, talking into the microphone of one telephone changes the currents flowing through each of the other telephones. Carbon microphones have poor frequency response and bad signal-to-noise ratios and they are only suitable for telephones and such communication applications.

In more modern telephones, more sophisticated electronic microphones and amplifiers have replaced the carbon microphone. There are literally dozens of types of microphones but the most popular one for telephones is the electret microphone. An electret is a thin insulating film that has charges permanently trapped in its surfaces. One surface is positively charged and the other surface is negatively charged—the film is electrically polarized. Although this charge separation slowly disappears, it takes hundreds or even thousands of years to vanish. In an electret microphone, the electret film is drawn taut like the head of a drum and is suspended just above a metal surface. As you talk into the microphone, pressure fluctuations in the air distort the electret film up and down, and it moves toward and away from the metal surface below it. Charges in the metal surface experience fluctuating forces as the polarized electret moves back and forth. As a result of these forces, current flows alternately toward and away from the metal surface through a wire that touches it.

In principle, talking into your own telephone microphone should cause the speaker of your telephone to reproduce your voice, too. However, this effect is undesirable because it would affect your speech. If you were to hear the full audio signal from your voice through your own speaker, you would think you were talking too loudly and you would unconsciously start to talk more softly. The sound you hear in your telephone speaker when you talk into your telephone microphone is called side-tone. Each telephone contains a balancing network that reduces side-tone. This network is usually a simple collection of electronic components that detects audio signals created by the telephone's microphone and keeps them from causing current changes in the telephone's speaker. The balancing network reduces the extent to which audio signals from your microphone affect your speaker so that you are not fooled into talking too quietly.

Telephone speaker

When the current through the telephone changes, the telephone's speaker or earphone creates sound. The speaker is a device that converts an electric current into pressure fluctuations in the air. A conventional speaker pushes and pulls on the air with a paper or plastic membrane, usually in the form of a cone; see Figure 2-5. The cone is driven in and out by the electric current through the telephone.

This sound generation process requires a conversion of electric power into mechanical power. The speaker or earphone unit performs this conversion using electromagnets. The speaker contains a permanent magnet that is fixed to the back of the speaker so that it is immobile. It also has a mobile coil of wire that becomes a magnet when current flows through it. The permanent magnet and the coil are arranged so that they attract one another if current flows in one direction through the coil and repel one another if the current is reversed. When an alternating current flows back and forth through the coil, the coil is alternately attracted to the permanent magnet and repelled from it. The speaker coil is attached to the speaker cone so that the two move together. The cone is loosely supported by the speaker's frame only at its periphery. The coil and cone have very little mass so that they accelerate in and out very easily. Their main resistance to motion is the air itself. As the cone moves out, it compresses the air in front of it and as it moves in, it rarefies the air in front of it. When it moves in and out rapidly,

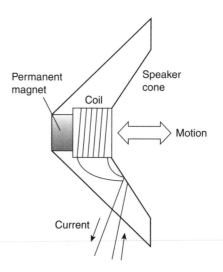

Figure 2-5 *Telephone earphone*

it produces sound. The speaker does a good job of converting an electric current representing sound into actual sound. The compressions and rarefactions of the air that it produces are very closely related to the current fluctuations passing through it.

The bell

When your telephone is not in use, it is "on-hook." When the handset is placed back into the cradle it electrically disconnects the microphone and the speaker from the two telephone wires. When the phone is on-hook it connects the bell equivalent ringer unit across the telephone phone line. The bell is a device that responds to an alternating current sent through the two telephone wires by the telephone company when a call is coming in. While the voltage on the telephone wires during a conversation is low and safe, the voltages used to drive current through the bell are large enough to give you a mild shock if you touch both wires while the bell is ringing.

A real telephone bell uses this alternating current to energize an electromagnet (see Figure 2-6).

One end of the electromagnet becomes a north pole and the other a south pole. Each time the current reverses, so do the poles of the electromagnet.

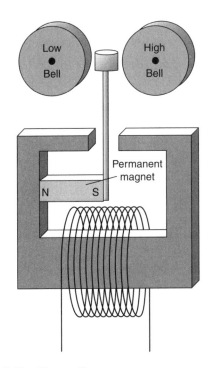

Figure 2-6 *Ringer diagram*

Situated between the two poles is an iron clapper. The clapper is magnetized by a small permanent magnet attached to its base so that it is attracted toward north poles and repelled by south poles. The electromagnet attracts the clapper first toward one pole and then toward the other. The clapper swings back and forth between the poles as the current reverses and its end strikes two metal bells in the process. The bells ring. The two bells are usually tuned to an octave interval. That means that the high-pitched bell rings at twice the frequency of the low-pitched bell, giving the telephone its characteristic sound.

Most modern telephones have replaced the bell with an electronic ringer unit. They use the ring current to power a tone-generating circuit and an amplifier, and produce the electronic sound with a speaker.

How a cell phone works

Inside your cell phone, there is a compact speaker, a microphone, a keyboard, a display screen, and a powerful circuit board with microprocessors that make each phone a miniature computer, along with a stable UHF transmitter and receiver pair. When connected to a wireless network, this technological wonder allows you to make phone calls or exchange data with other phones and computers around the world. The cell phone components operate so efficiently that a lightweight battery can power your phone for days.

A cell phone is really a radio—a very sophisticated computer and versatile radio. Because these radios connect into a digital network, cell phones offer much more than the ability to call any telephone anywhere in the world; modern cell phones can access the Internet and data services world wide.

Wireless networks operate on a type of grid network that divides cities or regions into smaller cells. One cell might cover a few city blocks or up to 250 square miles. Every cell uses a set of radio frequencies or channels to provide service in its specific area. The power of these radios is controlled in order to limit the signal's geographic range. Because of this, the same frequencies can be re-used in nearby cells. So, many people can hold conversations simultaneously in different cells throughout the city or region, even though they are on the same channel.

When you turn on your cell phone, it searches for a signal to confirm that service is available. Then the phone transmits certain identification numbers, so the network can verify your customer information, such as your wireless provider and phone number.

If you are calling from a cell phone to a wired phone, your call travels through a nearby wireless antenna and is switched by your wireless carrier to the traditional land-line phone system. The call then becomes like any other phone call and is directed over the traditional phone network, and to the person you are calling. If you are calling another cell phone, your call may go through the land-line network to the recipient's wireless carrier, or it might be routed within the wireless network to the cell site nearest the person you called, but if you are calling someone further away, your call will be routed to a long distance switching center, which relays the call across the country or around the world through fiber-optic cables.

Most cell phones use digital technology, which converts your voice into binary digits or zeros and ones, just like a computer. These small packets of data are relayed through wireless networks to the receiving phone. On the other end, the conversion process is reversed and the person you are calling hears your voice.

The key to the successful modern cell phone system is the wireless network that senses when your signal is getting weaker and hands over your call to an antenna with a stronger signal. Using smaller cells enables your phone to use less power and keep a clear signal as you move. Even when you are not talking, your cell phone communicates with the wireless antenna nearest to you. So it is ready to connect your call at any time.

If you travel outside your home area and make a call, another wireless carrier may provide service for your cell phone. That provider sends a signal back to your home network, so you can send and receive calls as you travel. This is called roaming. Roaming is key to mobile communications, as wireless providers cooperate to provide callers service wherever they go.

Chapter Two: How the Telephone Works

Chapter 3

Identifying Electronic Components

Electronic circuits are comprised of electronic components such as resistors and capacitors, diodes, semiconductors and LEDs. Each component has a specific purpose that it accomplishes in a particular circuit. In order to understand and construct electronic circuits it is necessary to be familiar with the different types of components, and how they are used. You should also know how to read resistor and capacitor color codes, and recognize physical components and their representative diagrams and pin-outs. You will also want to know the difference between a schematic and a pictorial diagram. First, we will discuss the actual components and their functions and then move on to reading schematics. Then we will help you to learn how to insert the components into the circuit board. In the next chapter we will discuss how to solder the components to the circuit board.

The diagrams shown in Figures 3-1 to 3-3 illustrate many of the electronic components that we will be using in the projects presented in this book.

Types of resistors

Resistors are used to regulate the amount of current flowing in a circuit. The higher the resistor's value or resistance, the less current flows, and conversely a lower resistor value will permit more current to flow in a circuit. Resistors are measured in ohms (Ω) and are identified by color bands on the resistor body. The first band at one end is the resistor's first digit, the second color band is the resistor's second digit and the third band is the resistor's multiplier value. A fourth color band on a resistor represents the resistor's tolerance value. A silver band denotes a 10% tolerance resistor, while a gold band notes a 5% resistor tolerance. No fourth band denotes that a resistor has a 20% tolerance. As an example, a resistor with a (brown) (black) and (red) band will represent the digit (1), the digit (0) with a multiplier value of (00) or one thousand, so that the resistor will have a value of 1 K or 1,000 ohms. There are a number of different styles and sizes of resistor. Small resistors can be carbon, thin film, or metal. Larger resistors are made to dissipate more power and they generally have an element wound from wire.

A Potentiometer or (pot) is basically a variable resistor. It generally has three terminals and it is fitted with a rotary control shaft that varies the resistance as it is rotated. A metal wiping contact rests against a circular carbon or wire-wound resistance track. As the wiper arms move about the circular resistance, the resistance to the output terminals changes. Potentiometers are commonly used as volume controls in amplifiers and radio receivers.

A Trimpot is a special type of potentiometer which while variable, is intended to be adjusted once or only occasionally. For this reason a control shaft is not provided but a small slot is provided in the center of the control arm. Trimpots are generally used on printed circuit boards.

A light dependent resistor (LDR) is a special type of resistor that varies its resistance value according to the amount of light falling on it. When it is in the dark, an LDR will typically have a very high resistance, i.e., millions of ohms. When light falls on the LDR the resistance drops to a few hundred ohms.

Types of capacitors

Capacitors block DC current while allowing varying or AC current signals to pass. They are commonly used for coupling signals from part of a circuit to another part of a circuit; they are also used in timing circuits.

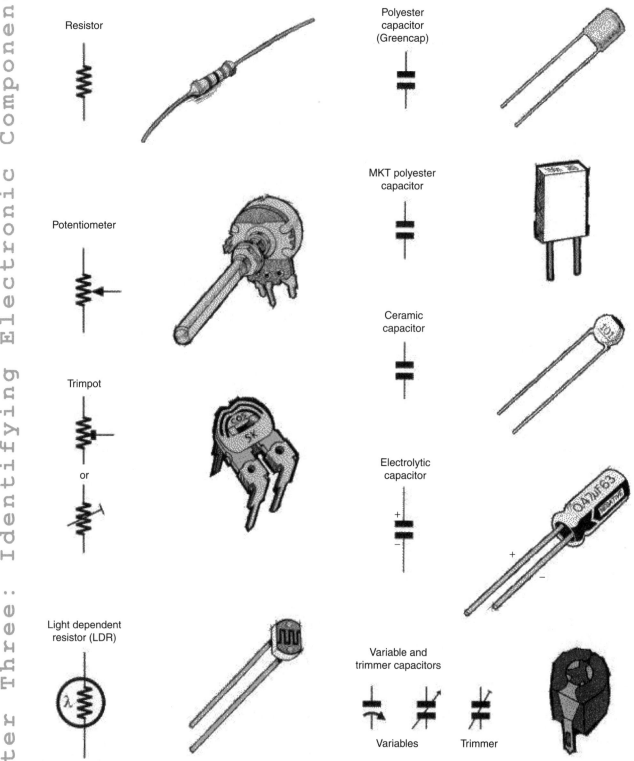

Figure 3-1 *Electronic components I*

Resistor

Potentiometer

Trimpot

or

Light dependent
resistor (LDR)

Polyester
capacitor
(Greencap)

MKT polyester
capacitor

Ceramic
capacitor

Electrolytic
capacitor

Variable and
trimmer capacitors

Variables   Trimmer

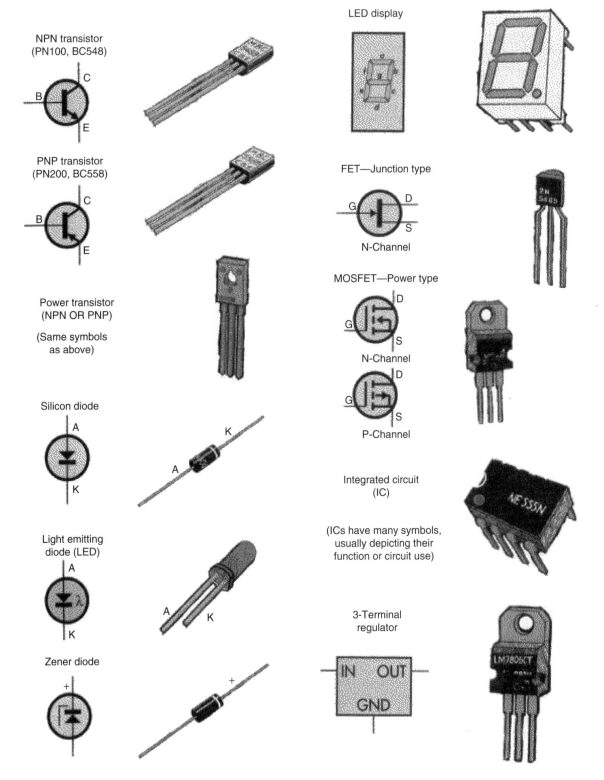

Figure 3-2   *Electronic components II*

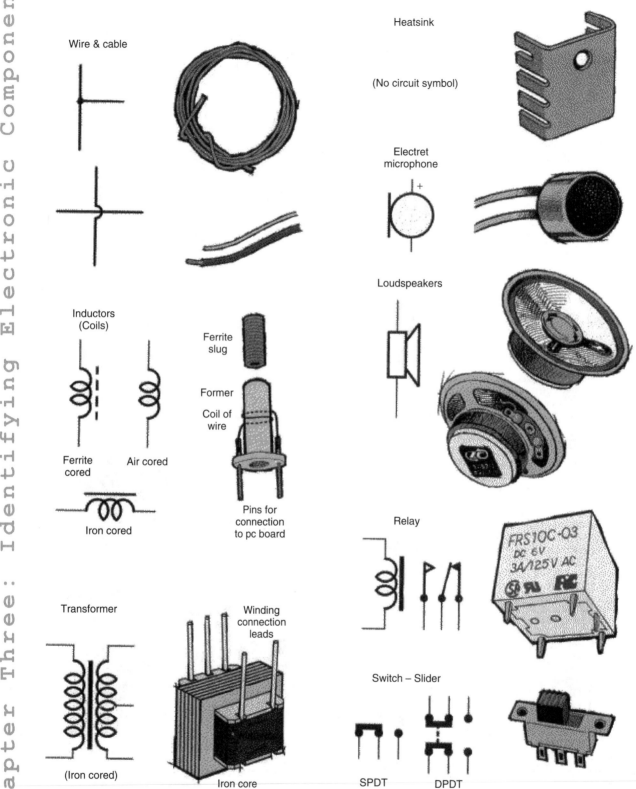

Wire & cable

Heatsink

(No circuit symbol)

Electret microphone

Loudspeakers

Inductors (Coils)

Ferrite slug

Former

Coil of wire

Ferrite cored   Air cored

Iron cored

Pins for connection to pc board

Relay

Transformer

Winding connection leads

(Iron cored)

Iron core

Switch – Slider

SPDT   DPDT

Figure 3-3 *Electronic components III*

There are a number of different types of capacitor as described below:

- Polyester capacitors use polyester plastic film as their insulating dielectric. Some polyester capacitors are called greencaps since they are coated with a green or brown color coating on the outside of the component. Their values are specified in microfarads (μF), or nanofarads (nF), or picofarads (pF) and range from 1 nF up to about 10 μF. These types of capacitors do not have polarity markings and can be installed in either direction.

- MKT capacitors are another type of capacitor, but they are rectangular or (block) in shape and are usually yellow in color. One of the major advantages of these capacitors is a more standardized lead spacing, making them more useful for PC board projects. Their components can generally be substituted for polyester types.

- Ceramic capacitors use a tiny disk of ceramic or porcelain material in their construction for a dielectric and they range in value from 1 pF up to 2.2 μF. Those with values above 1 nF are often made with multiple layers of metal electrodes and dielectric, to allow higher capacitance values in smaller bodies. Their capacitors are usually called "multilayer monolithics" and are distinguished from lower value disc ceramic types. Ceramic capacitors are often used in Rf radio circuits and filter circuits.

- Electrolytic capacitors use very thin film of metal oxide as their dielectric, which allows them to provide a large amount of capacitance in a very small volume. They range in value from 100 nF up to hundreds and thousands of microfarads (μF). They are commonly used to filter power supply circuits, coupling audio circuits, and in timing circuits. Electrolytic capacitors have polarity and must be installed with respect to these polarity markings. The capacitor will have either a white or black band denoting polarity with a plus (+) or minus (−) marking next to the color band.

- Variable capacitors are used in circuits for (trimming) or adjustment, i.e., for setting a frequency. A variable capacitor has one set of fixed plates and one set of plates which can be moved by turning a knob. The dielectric between the plates is usually a thin plastic film. Most variable capacitors have low values up to

a few tens of picofarads (pF) and a few hundreds of microfarads for larger variable capacitors.

Diodes

A diode is a semiconductor device which can pass current in one direction only. In order for current to flow, the anode (A) must be positive with respect to the cathode (K). In this condition, the diode is said to be forward biased and a voltage drop of about .6 V appears across its terminals. If the anode is less than .6 V positive with respect to the cathode, negligible current will flow and the diode behaves as an open circuit.

Types of transistors

Transistors are semiconductor devices that can either be used as electronic switches or to amplify signals. They have three leads, called the collector, base, and emitter. A small current flowing between base and emitter (junction) causes a much larger current to flow between the emitter and collector (junction). There are two basic types of transistors: PNP and NPN styles.

A Field Effect Transistor or FET is a different type of transistor, which usually still has three terminals but works in a different way. Here the control element is the "gate" rather than the base, and it is the gate voltage which controls the current flowing in the "channel" between the other terminals—the "source" and the "drain." Like ordinary transistors FETs can be used either as electronic switches or amplifiers; they also come in P-channel and N-channel types, and are available in small signal types as well as power FETs.

Power transistors are usually larger than the smaller signal type transistors. Power transistors are capable of handling larger currents and voltages. Often metal tabs and heat sinks are used to remove excess heat from the part. These devices are usually bolted to the chassis and are used for amplifying RF or audio energy.

Integrated Circuits

Integrated circuits or ICs contain in one package all or most of the components necessary for a particular circuit function. Integrated circuits contain as few as 10 transistors or many millions of transistors, plus many resistors, diodes and other components. There are many shapes, styles, and sizes of integrated circuits; in this book we will use the dual-in-line style IC either 8, 14, or 16 pin devices.

Three-terminal regulators are special types of integrated circuits, which supply a regulated, or constant and accurate, voltage from their output regardless (within limits) of the voltage applied to their input. They are most often used in power supplies. Most regulators are designed to give specific output voltages, so that an "LM7805" regulator provides a 5-V output, but some IC regulators can provide adjustable output based on an external potentiometer which can vary the output voltage.

Heat sinks

Many electronic components generate heat when they are operating. Generally, heat sinks are used on semiconductors such as transistors to remove heat. Overheating can damage a particular component or the entire circuit. The heat sink cools the transistor and ensures a long circuit life by removing the excess heat from the circuit area.

Light-emitting diodes

A light-emitting diode or LED, is a special diode which has a plastic translucent body (usually clear, red, yellow, green, or blue in color) and a small semiconductor element which emits light when the diode passes a small current. Unlike an incandescent lamp, an LED does not need to get hot to produce light. LEDs must always be forward biased to operate. Special LEDs can also produce infrared light.

LED displays consist of a number of LEDs together in a single package. The most common type has seven elongated LEDs arranged in an "8" pattern. By choosing which combinations of LEDs are lit, any number of digits from "0" through "9"can be displayed. Most of these "7-segment" displays also contain another small round LED which is used as a decimal point.

Types of inductors

An inductor or "coil" is basically a length of wire, wound into a cylindrical spiral (or layers of spirals) in order to increase its inductance. Inductance is the ability to store energy in a magnetic field. Many coils are wound on a former of insulating material, which may also have connection pins to act as the coils' terminals. The former may also be internally threaded to accept a small core or "slug" of ferrite, which can be adjusted in position relative to the coil itself to vary the coil inductance.

A transformer consists of a number of coils or windings or wire wound on a common former, which is also inside a core of iron alloy, ferrite, or other magnetic material. When an alternating current is passed through one of the windings (primary), it produces an alternating magnetic field in the core and this in turn induces AC voltages in the other (secondary) windings. The voltages produced in the other winding depends on the number of turns in those windings, compared with the turns in the primary winding. If a secondary winding has fewer turns than the primary, it will produce a lower voltage, and be called a step-down transformer. If the secondary winding has more windings than the primary then the transformer will produce a higher voltage and it will be a step-up transformer. Transformers can be used to change the voltage levels of AC power and they are available in many different sizes and power-handling capabilities.

Microphones

A microphone converts audible sound waves into electrical signals which can then be amplified.

In an electret microphone, the sound waves vibrate a circular diaphragm made from very thin plastic material containing a permanent charge. Metal films coated on each side form a capacitor, which produces a very small AC voltage when the diaphragm vibrates. All electret microphones also contain FET which amplifies the very small AC signals. To power an FET amplifier the microphone must be supplied with a small DC voltage.

Loudspeakers

A loudspeaker converts electrical signals into sound waves that we can hear. It has two terminals which go to a voice coil, attached to a circular cone made of either cardboard or thin plastic. When electrical signals are applied to the voice coil, this creates a varying magnetic field from a permanent magnet at the back of the speaker. As a result the cone vibrates in sympathy with the applied signal to produce sound waves.

Relays

Many electronic components are not capable of switching higher currents or voltages so a device called a relay is used. A relay has a coil which forms an electromagnet, attracting a steel "armature," which itself pushes on one or more sets of switching contacts. When a current is passed through the coil to energize it, the moving contacts disconnect from one set of contacts to another, and when the coil is de-energized the contacts go back to their original position. In most cases a relay needs a diode across the coil to prevent damage to the semiconductor driving the coil.

Switches

A switch is a device with one or more sets of switching contacts, which are used to control the flow of current in a circuit. The switch allows the contacts to be controlled by a physical actuator of some kind, such as a press-button toggle lever, rotary, or knob. As the name denotes, this type of switch has an actuator bar which slides back and forth between the various contact positions. In a single-pole, double throw (SPDT) slider switch, a moving contact links the center contact to either of the two end contacts. In contrast a double-pole, double throw (DPDT) slider switch has two of these sets of contacts, with their moving contacts operating in tandem when the slider is actuated.

Wire

A wire is simply a length of metal conductor, usually made from copper since its conductivity is good, which means its resistance is low. When there is a risk of a wire touching another wire and causing a short, the copper wire is insulated or covered with a plastic coating which acts as an insulating material. Plain copper wire is not usually used since it will quickly

oxidize or tarnish in the presence of air. A thin metal alloy coating is often applied to the copper wire; usually an alloy of tin or lead is used.

Single or multi-strand wire is covered in colored PVC plastic insulation and is used quite often in electronic applications to connect circuits or components together. This wire is often called a "hook-up" wire. On a circuit diagram, a solid dot indicates that the wires or PC board tracks are connected together or joined, while a "loop-over" indicates that they are not joined and must be insulated. A number of insulated wires enclosed in outer jackets are called electrical cables. Some electrical cables can have many insulated wires in them.

Semiconductor substitution

Often, when building an electronic circuit, it is difficult or impossible to find or locate the original transistor or integrated circuit. There are a number of circuits featured in this book which feature transistors, SCRs and UJT and FETs that are specified but cannot be found in this country. Where possible, many of these foreign components are converted to substitute values, either with a direct replacement or close substitution. Many foreign parts can be easily converted directly to a commonly used transistor or component. Occasionally an outdated component has no direct common replacement, so the closest specifications of that component are attempted.

In some instances we have specified replacement components with substitution components from the NTE brand or replacements. Most of the components for the project used in this book are quite common and easily located or substituted without difficulty.

When substituting components in the circuit, make sure that the pin-outs match the original components. Sometimes, for example, a transistor may have a bottom view drawing, while the substituted value may have a drawing with a top view. Also be sure to check the pin-outs of the original components versus the replacement. As an example, some transistors will have EBC versus ECB pin-outs, so be sure to look closely at possible differences which may occur.

Chapter 4

Reading Schematics

Before we take a look at some electronics schematics, it is important that you understand the relationship between voltage and current and resistance. This relationship is called Ohm's Law and it is the backbone of simple electronics formulas for electronics. In its simplest form it is shown as the equation $V = I * R$, which says voltage (V) is equal to current (I) multiplied by resistance (R). Voltage is represented as electromotive force where V is used instead of E. Current represented as (I) is in amperes or amps; in typical electronic circuits milliamps may be used, and this is shown as 1/1,000 or an ampere. One milliamp = .001 amp, abbreviated as mA or ma. The three most important Ohm's Law equations are $V = I * R$; $I = V/R$; and $R = V/I$. The most used Ohm's Law power equation is shown as $P = I * V$, where P is power in watts or wattage, which is equal to the current of a circuit multiplied by the voltage in the circuit. Thus a voltage of 100 V multiplied by a current consumption of two amps of current would equate to a power consumption of 200 watts.

The diagram shown in Figure 4-1 depicts a very common circuit called a voltage divider.

The basic voltage divider consists of two resistors, usually connected in series as shown. The total resistance is simply the sum of the two. In this example, it would be 22 K ohms plus 33 ohms = 22,033 ohms.

If a volt signal is applied to the input end of R1 the 22 K ohm resistor, the current through the whole circuit would be $I = V/R = 1/22,033$ or .000,045,386,4 amps or about .05 milliamps. Voltage dividers are used to reduce or step down voltages or signal levels to voltage converter circuits, amplifiers, etc. Voltage dividers are often set up using resistor ratios or 1,000:1, 100:1, or 10:1, for example a 10 to 1 resistor divider would have a 10 ohms and a 1 ohm resistor for R1 and R2 values.

The introduction of the operational amplifier or Op-Amp in the late 1960s and early 1970s transformed the electronics industry like no device before it. Op-Amps began life as large integrated amplifier modules, some measuring up to 3″ in length. Early Op-Amps were large and noisy but nonetheless a monumental development in electronics. The Op-Amp permitted a simple building block approach to electronic circuit design. Early Op-Amps were primarily used for pre-amps and instrumentation amplifiers. As Op-Amp quality progressed and their sizes and power consumption reduced over time, Op-Amps were used in all types of electronic circuit designs. Op-Amps are now utilized in audio amplifiers, filter design, oscillators, comparators, regulators, etc. Since Op-Amps are now found in most all electronic circuits, we will spend some time identifying different Op-Amp configurations and applications.

Integrated circuits (ICs) are used throughout much of modern electronics, so understanding how they are represented is important. ICs generally contain many individual circuits or components, and are shown schematically as functional blocks. Op-amps can be configured in many different ways, including non-inverting amplifiers.

The ubiquitous operational amplifier or Op-Amp is shown in the diagram of Figure 4-2 as a triangle with a plus (+) and minus (−) input and a single output at the point of the triangle.

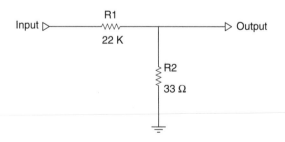

Figure 4-1 *Voltage divider circuit*

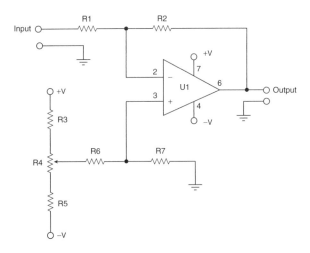

Figure 4-2 *Inverting Op-Amp amplifier*

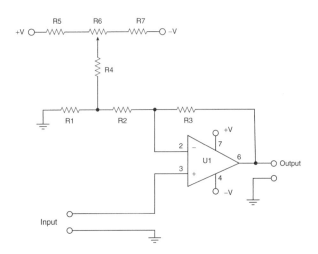

Figure 4-3 *Non-inverting Op-Amp amplifier*

This type of Op-Amp configuration is called an inverting input amplifier circuit, since the input is fed through the minus (–) input of the Op-Amp. Note that the signal input is applied through the input resistor at R1. Also notice that a second resistor R2 is shown from the minus input across to the output of the Op-Amp. Resistor R2 is called the feedback resistor. The plus (+) input of the Op-Amp is shown connected to a resistor network used to balance the input through a bias control at resistor R4. Both plus and minus voltage sources are connected to the bias network.

Simulate driving a current through the inverting input by placing a 1 volt on the input at R1 and assume that the right end has 0 volts on it. The current will be $I = V/R = 1/1 \text{ K} = 1 \text{ ma}$. The voltage output will try to counter this by driving a current of the opposite polarity through the feedback resistor into the inverting input. The required voltage to do that will be $V = -(I * R) = -(1 \text{ ma} * 10 \text{ K}) = -10 \text{ V}$. Thus we get a voltage to current conversion, a current to voltage conversion, a polarity inversion and, most importantly, amplification. G = –(Feedback Resistor/Input Resistor) In this example, it is shown as Gain or $G = -(R2/R1)$.

This inverting input Op-Amp configuration is commonly used in modern amplifiers and filters, etc. Op-amps generally require both plus (+) and minus (–) voltages for operation; this allows a voltage swing either side of zero.

The diagram illustrated in Figure 4-3 shows a non-inverting Op-Amp configuration.

This configuration is used when you wish to maintain the same phase or polarity from the input, through to

the output. Notice that in this configuration the signal input is applied to the plus (+) input of the Op-Amp. The balance and trim capability is performed via the bias network formed by resistors R4 through R7, which connect to both the plus and minus power supply.

The circuit shown in Figure 4-4 depicts a basic differential Op-Amp instrumentation amplifier.

In this circuit, notice that there are two separate inputs and they are referenced against ground connection, which is separate from the inputs. This type of circuit is used in high gain instrument amplifiers, when you wish to keep common modes noise from reaching the Op-Amp. This configuration is commonly used for high input impedance sensor inputs, where low leads may be used and low noise is required. Note the balance or trim for Op-Amp is provided through the resistor network composed of R4 through R7, which allows bias from both the plus and minus power sources.

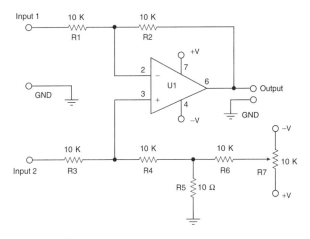

Figure 4-4 *Differential Op-Amp amplifier*

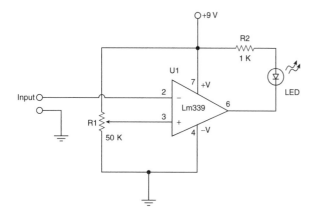

Figure 4-5   *Differential Op-Amp comparator*

Figure 4-5 illustrates how an Op-Amp can be used as a comparator to trigger an LED when a certain voltage threshold is reached. In this configuration, the plus (+) input of the Op-Amp is connected to a 50 K potentiometer which is connected between the plus 9-V source and ground.

Note in this configuration, only a single voltage is required. The minus (–) input lead of the Op-Amp is connected to the voltage source you wish to monitor. In operation, you would apply a voltage to the minus input pin of the Op-Amp and adjust the R1, so the LED is not lit. Raising the voltage at the input would now offset the comparator and its trip point and the LED would become lit.

Op-Amps can also be used in oscillators and function generators as shown in Figure 4-6.

In this circuit the Lm339 comparator Op-Amp is used to generate a square-wave signal. A resistor and the capacitor at C1 form a timing network which is used to establish the frequency of the square-wave generator.

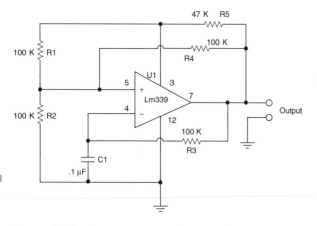

Figure 4-6   *Square-wave Op-Amp oscillator*

The duty cycle is set up through resistors to produce a symmetrical square-wave output.

As you can see from these examples, the Op-Amp is a very powerful electronic building block tool for greatly simplifying the design of electronic circuits. We have only covered a small number of Op-Amp applications. Op-Amps are also used for filters, power regulators, integrators, voltage to current and current to voltage convertors. For more information on Linear Op-Amps theory and applications look for the IC Op-Amp Cookbook by Walter Jung or point your browser to w1.859.telia.com/~u85920178/begin/opamp00.htm for a discussion on *Operational Basics* by Harry Lythall.

The above examples illustrate linear Op-Amps, and they represent nearly a half of the spectrum of integrated circuits. On the other side of the spectrum of integrated circuits are the digital integrated circuits which generally contain multiple digital gates, switches, and memory functions all in one package. One of the more common digital integrated circuits is the Quad 2-input AND gate, shown in Figure 4-7.

This diagram illustrates a fourteen pin 74LS08 Quad 2-input AND gate, and its truth table. The 74LS08 contains four AND gates. The inputs to the first AND gate are at 1A and 1B represented by pins 1 and 2, while the output is at pin3. The second AND gate has inputs 2A and 2B on pins 4 and 5 with its output on pin 6. The common ground connection is shown on pin 7, while 5-volt power is supplied to pin 14. The third AND gate has its inputs at 3A and 3B, on pins 9 and 10, with its output on pin8. The 4th AND gate shown as 4A and 4B are on pins 12 and 13, with its output on pin 11. The truth table or function table for the 74LS08 describes the input vs. the output condition for each of the gates. If inputs 1A and 1B are HIGH then the resulting output at 1Y will be HIGH. If input 1A is LOW and input at 1B is either HIGH or LOW, then the output at 1Y will be LOW. If input 1B is LOW and input 1A is either HIGH or LOW the output at 1Y will be LOW once again.

In working with electronic circuits, you will come across many schematics as well as pictorial diagrams. Schematics are the electronic representations on paper of an electronics circuit. A pictorial diagram is a generally a physical layout diagram of the same circuit. In electronics work you will come across both types of diagrams; they look quite different but they are really

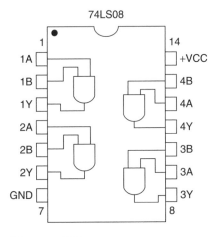

74LS08

1A	1	14 +VCC
1B		4B
1Y		4A
2A		4Y
2B		3B
2Y		3A
GND	7	8 3Y

Function table

Inputs		Output
A	B	T
H	H	H
L	X	L
X	L	L

Figure 4-7 *Quad 2-input AND gate*

the same circuit shown differently. Experience in looking and comparing these two types of diagrams will help you enormously in your electronics projects.

The diagram shown in Figure 4-8 illustrates a schematic of a two transistor audio amplifier circuit.

The input jack at J1 is coupled to an input capacitor at C1. Capacitor C1 couples the microphone of input source to the amplifier. It acts as a coupling device as well as a blocking device which can keep any constant DC component or voltage from reaching the input of the amplifier. The resistors R1 and R2 form the input impedance and frequency response characteristics of the input to the amplifier. The NPN Transistor at Q1 forms the first stage of amplification of the amplifier. Resistors R4 and R5 form the bias or voltage supply for Q1. The output of the first amplifier stage at Q1 is connected to capacitor C3, which is used to couple the first amplifier stage to the next amplifier stage. Capacitor C3 is next fed to variable resistor R6, which acts as the systems volume control. Capacitor C5 couples the audio signal

to the second amplifier stage at Q2. Resistor R10 is used to bias the second stage amplifier stage. Capacitor C7 is used to connect the output of transistor Q2 to the final amplifier stage or to the input of a transmitter, at J2. Note the ground bus or common connection for ground at the bottom of the circuit, and the power bus connection at the top of the schematic diagram.

The pictorial diagram shown in Figure 4-9 serves to illustrate the same two transistor amplifier circuits shown in Figure 4-8. Both diagrams are electrically the same, but look quite different. The pictorial diagram shows how the circuit would look, and how it may look wired and laid out in a chassis box. Compare the two diagrams so that you can see and understand how they differ from each other.

An Op-Amp audio amplifier is shown schematically in Figure 4-10.

The input to the amplifier is shown at input jack at J1. Resistor R1 serves to control the input signal or current flowing into the amplifier circuit. Resistor R1

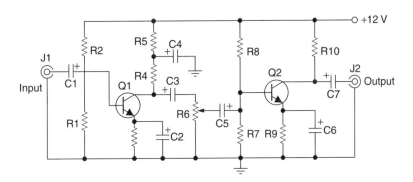

Figure 4-8 *Transistor amplifier schematic diagram*

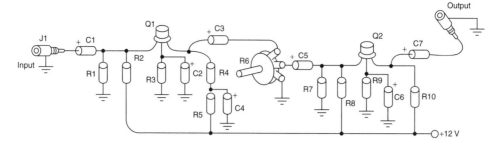

Figure 4-9 *Transistor amplifier pictorial diagram*

establishes the input impedance to the amplifier circuit. R1 is next fed to the .1-µF capacitor at C1. Capacitor C1 is used to block low frequency signals and couple the input to the next stage as well as keep any constant DC voltage from the guitar away from the input to the Op-Amps. The low frequencies to be blocked are dependent upon resistors R1 and R2. The triangle symbol shown at U1 represents the first Op-Amp amplifier in the circuit. The Op-Amp has two inputs, an inverting input and a non-inverting input. The input signal from resistor R2 is fed to the non-inverting input of the Op-Amp. The connection from the input to the output pin of the Op-Amp forms the feedback network which generally determines the gain of the amplifier. In this example, there is no resistor but a direct connection between the input and the output. The direct connection from the input to the output of the amplifier establishes a unity gain, or no change in the signal input. The purpose of the first stage in this example is to reduce the amount of current the guitar must supply to the

amplifier. Resistor R3, represents a variable resistor which is as a volume control for the amplifier. The center tap on the variable resistor is fed to resistor R4, which is used to couple the first stage of the amplifier to the second stage. Resistors R4 and R5 establish the gain of the second amplification stage. This gain stage forms the real muscle of the amplifier. The output of the second Op-Amp at pin 6 is fed directly to a speaker through resistor R7, which is used to protect the output of the amplifier as well as protecting the speaker from too much current.

The diagram shown in Figure 4-11 depicts the same audio amplifier circuit but it is now shown pictorially instead of schematically.

The pictorial diagram shows the circuit as it might appear physically in its enclosure of circuit board. The diagram looks quite different but it is really the same circuit drawn slightly differently. You will need to recognize and become familiar with the differences

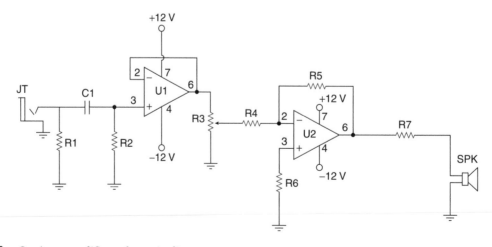

Figure 4-10 *Op-Amp amplifier schematic diagram*

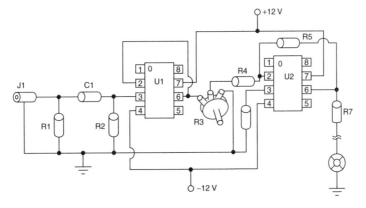

Figure 4-11 *Op-Amp amplifier pictorial diagram*

between these types of drawings. The best way to familiarize yourself with schematics and pictorial diagrams is to see and compare a number of them over time. Once you get some practice it will become second nature to you. There are a number of good books on how to read schematics, which cover the topic in more depth.

Before we discuss installing electronic components to printed circuit boards, we will take a few minutes to discuss the resistor and capacitor codes which are used to help identify these components. Table 4-1 lists the resistor color code information.

Resistors will generally have at least three or four color bands which help identify the resistor values. The color band will start from one edge of the

resistor body. This is the first color code, which represents the first digit; the second color band depicts the second digit, while the third band is the resistor multiplier value. The fourth color band represents the resistor tolerance value. If there is no fourth band then the resistor has a 20% tolerance, if the fourth band is silver, then the resistor has a 10% tolerance value, and if the fourth band is gold then the resistor has a 5% tolerance value. Therefore, if you have a resistor with the first band having a brown color, the second band with a black color and a third band with a red color, then this resistor would have a value of 1 K or 1,000 ohms.

Capacitor types are quite varied; there are ceramic, mylar, electrolytic, polyester, paper, tantalum, mica,

Table 4-1
Resistor color code chart

Color Band	1st Digit	2nd Digit	Multiplier	Tolerance
Black	0	0	1	
Brown	1	1	10	1%
Red	2	2	100	2%
Orange	3	3	1,000 (K)	3%
Yellow	4	4	10,000	4%
Green	5	5	100,000	
Blue	6	6	1,000,000 (M)	
Violet	7	7	10,000,000	
Gray	8	8	100,000,000	
White	9	9	1,000,000,000	
Gold			0.1	5%
Silver			0.01	10%
No color				20%

polyproplene, etc. When installing the small non-polarized capacitors such as the ceramic or mylar types, you will need to refer to the capacitor chart in Table 4-2, which illustrates the three-digit capacitor code values.

Most large capacitors, such as electrolytic types, will have their actual value printed on the body of the capacitor. Often some small capacitors are just too small to have their actual capacitor values printed on them, so an abbreviated code was devised to mark the capacitor using a three-digit code to save space as shown. If you refer to the chart you will notice that a code marking of (104) would denote a small capacitor with a value of .1 mF (microfarads) or 100 nF (nanofarads) value. Once you understand how these component codes work, you can easily identify the resistors and capacitors and proceed to mount them on the printed circuit board.

Table 4-2

Three-digit capacitor codes

pF	nF	µF	CODE	pF	nF	µF	CODE
1.0			**1R0**	3,900	3.9	.0039	**392**
1.2			**1R2**	4,700	4.7	.0047	**472**
1.5			**1R5**	5,600	5.6	.0056	**562**
1.8			**1R8**	6,800	6.8	.0068	**682**
2.2			**2R2**	8,200	8.2	.0082	**822**
2.7			**2R7**	10,000	10	.01	**103**
3.3			**3R3**	12,000	12	.012	**123**
3.9			**3R9**	15,000	15	.015	**153**
4.7			**4R7**	18,000	18	.018	**183**
5.6			**5R6**	22,000	22	.022	**223**
6.8			**6R8**	27,000	27	.027	**273**
8.2			**8R2**	33,000	33	.033	**333**
10			**100**	39,000	39	.039	**393**
12			**120**	47,000	47	.047	**473**
15			**150**	56,000	56	.056	**563**
18			**180**	68,000	68	.068	**683**
22			**220**	82,000	82	.082	**823**
27			**270**	100,000	100	.1	**104**
33			**330**	120,000	120	.12	**124**
39			**390**	150,000	150	.15	**154**
47			**470**	180,000	180	.18	**184**
56			**560**	220,000	220	.22	**224**
68			**680**	270,000	270	.27	**274**
82			**820**	330,000	330	.33	**334**
100		.0001	**101**	390,000	390	.39	**394**
120		.00012	**121**	470,000	470	.47	**474**
150		.00015	**151**	560,000	560	.56	**564**
180		.00018	**181**	680,000	680	.68	**684**
220		.00022	**221**	820,000	820	.82	**824**
270		.00027	**271**		1,000	1	**105**
330		.00033	**331**		1,500	1.5	**155**

Table 4-2—cont'd

Three-digit capacitor codes

pF	nF	μF	CODE	pF	nF	μF	CODE
390		.00039	**391**	2,200	2.2		**225**
470		.00047	**471**	2,700	2.7		**275**
560		.00056	**561**	3,300	3.3		**335**
680		.00068	**681**	4,700	4.7		**475**
820		.00082	**821**		6.8		**685**
1,000	1.0	.001	**102**		10		**106**
1,200	1.2	.0012	**122**		22		**226**
1,500	1.5	.0015	**152**		33		**336**
1,800	1.8	.0018	**182**		47		**476**
2,200	2.2	.0022	**222**		68		**686**
2,700	2.7	.0027	**272**		100		**107**
3,300	3.3	.0033	**332**		157		**157**

Code = 2 significant digits of the capacitor value + number of zeros to follow

For values below 10 pF, use "R" in place of decimal e.g. 8.2 pF = 8R2

10 pF = 100

100 pF = 101

1,000 pF = 102

22,000 pF = 223

330,000 pF = 334

1 μF = 105

Chapter 5

Electronic Parts Installation and Soldering

Before we begin constructing our telephone projects, you will need to understand how to prepare the electronic components before installing them onto the circuit board. First, you will need to know how to "dress" or prepare resistor leads, which is shown in Figure 5-1.

Resistor leads are first bent to fit the PC board component holes, then the resistor is mounted to the circuit board. Figure 5-2 depicts how capacitor leads are bent to fit the PC board holes.

Once the capacitor leads are prepared, the capacitor can be installed on to the circuit board. The diagram shown in Figure 5-3 illustrates how to prepare diode leads before mounting them to the PB board.

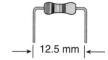

(a) Grip leads with pliers near body and bend down free ends

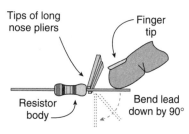

(b) Leads bent down and parallel, spaced to match board holes

(c) Leads passed through PC board holes, bent and soldered to pads

(d) Excess leads trimmed off with sidecutters

Figure 5-1   *Preparing resistor leads*

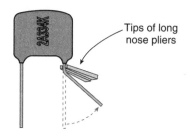

(a) Grip leads with pliers near body and bend out or in by 45°

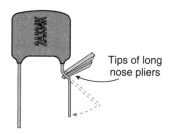

(b) Grip now about 2 mm down and bend down by 45°, parallel again

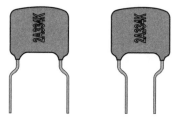

(c) Both leads cranked in (or out) if necessary to match PC board hole spacing

Figure 5-2   *Dressing capacitor leads*

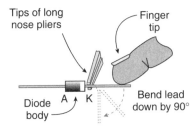

Tips of long nose pliers

Finger tip

Diode body A K

Bend lead down by 90°

(a) Grip leads with pliers near body and bend down free ends

|← 12.5 mm →|

(b) Leads bent down and parallel, spaced to match board holes

(c) Leads passed through PC board holes, bent and soldered to pads

(d) Excess leads trimmed off with sidecutters

Figure 5-3 *Preparing diode leads*

Finally, Figure 5-4 depicts transistor lead preparation, while Figure 5-5 shows how integrated circuit leads are prepared for installation.

Learning how to solder

Everyone working in electronics needs to know how to solder well. Before you begin working on a circuit, carefully read this chapter on soldering. In this section you will learn how to make good solder joints when soldering point-to-point wiring connections as well as for PC board soldering connections.

In all electronics work the wiring connections must be absolutely secure. A loose connection in a radio results in noise, scratching sounds, or no sound at all. In a TV, poor connections can disrupt the sound or picture. The safe operation of airplanes and the lives of astronauts in flight depend on secure electronics connections.

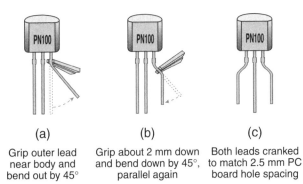

(a) Grip outer lead near body and bend out by 45°

(b) Grip about 2 mm down and bend down by 45°, parallel again

(c) Both leads cranked to match 2.5 mm PC board hole spacing

Figure 5-4 *Dressing transistor leads*

Soldering joins two pieces of metal, such as electrical wires, by melting them together with another metal to form a strong, chemical bond. Done correctly, it unites the metals so that electrically they act as one piece of metal. Soldering is not just gluing metals together. Soldering is tricky and intimidating in practice, but easy to understand in theory. Basic supplies include a soldering iron, which is a prong of metal that heats to a specific temperature through electricity, like a regular iron. The solder, or soldering wire, often an alloy of aluminum and lead, needs a lower melting point than the metal you are joining. Finally, you need a cleaning resin called flux that ensures the joining pieces are incredibly clean. Flux removes all the oxides on the surface of the metal that would interfere with the molecular bonding, allowing the solder to flow into the joint smoothly. You also need two things to solder together.

The first step in soldering is cleaning the surfaces, initially with sandpaper or steel wool and then by

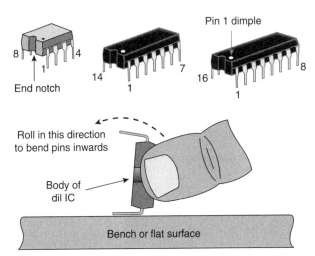

Pin 1 dimple

8   4

1

End notch

14   7

1

16   8

1

Roll in this direction to bend pins inwards

Body of dil IC

Bench or flat surface

Figure 5-5 *Preparing IC leads*

melting flux onto the parts. Sometimes, flux is part of the alloy of the soldering wire, in an easy to use mixture. Then, the pieces are both heated above the melting point of the solder (but below their own melting point) with the soldering iron. When touched to the joint, this precise heating causes the solder to "flow" to the place of highest temperature and makes a chemical bond. The solder should not drip or blob, but spread smoothly, coating the entire joint. When the solder cools, you should have a clean, sturdy connection.

Many people use soldering in their field, from electrical engineering and plumbing to jewelry and crafts. In a delicate procedure, a special material called solder, flows over two pre-heated pieces and attaches them through a process similar to welding or brazing. Various metals can be soldered together, such as gold and sterling silver in jewelry, brass in watches and clocks, copper in water pipes, or iron in leaded glass stained windows. All these metals have different melting points, and therefore use different solder. Some "soft" solder, with a low melting point, is perfect for wiring a circuit board. Other "hard" solder, such as for making a bracelet, needs a torch rather than a soldering iron to get a hot enough temperature. Electrical engineers and hobbyists alike can benefit from learning the art and science of soldering.

Solder

The best solder for electronics work is 60/40 rosin-core solder. It is made of 60% tin and 40% lead. This mixture melts at a lower temperature than either lead or tin alone. It makes soldering easy and provides good connections. The rosin keeps the joint clean as it is being soldered. The heat of the iron often causes a tarnish or oxide to form on the surface. The rosin dissolves the tarnish to make the solder cling tightly. Solders have different melting points, depending on the ratio of tin to lead. Tin melts at 450°F and lead at 621°F. Solder made from 63% tin and 37% lead melts at 361°F, the lowest melting point for a tin and lead mixture. Called 63-37 (or eutectic), this type of solder also provides the most rapid solid-to-liquid transition and the best stress resistance. Solders made with different lead/tin ratios have a plastic state at some temperatures.

If the solder is deformed while it is in the plastic state, the deformation remains when the solder freezes into the solid state. Any stress or motion applied to "plastic solder" causes a poor solder joint.

The 60/40 solder has the best wetting qualities. Wetting is the ability to spread rapidly and bond materials uniformly. This solder also has a low melting point. These factors make it the most commonly used solder in electronics.

Some connections that carry high current cannot be made with ordinary tin–lead solder because the heat generated by the current would melt the solder. Automotive starter brushes and transmitter tank circuits are two examples. Silver-bearing solders have higher melting points, and so prevent this problem. High-temperature silver alloys become liquid in the 1100 to 1200°F range, and a silver–manganese (85–15) alloy requires almost 1800°F.

Because silver dissolves easily in tin, tin bearing solders can leach silver plating from components. This problem can be greatly reduced by partially saturating the tin in the solder with silver or by eliminating the tin. Tin-silver or tin-lead-silver alloys become liquid at temperatures from 430°F for 96.5–3.5 (tin-silver), to 588°F for 1.0–97.5–1.5 (tin–lead–silver). A 15.0–80.0–5.0 alloy of lead–indium–silver melts at 314°F.

Never use acid-core solder for electrical work. It should be used only for plumbing or chassis work. For circuit construction, use only fluxes or solder-flux combinations that are labeled for electronic soldering.

The resin or the acid is a flux. Flux removes oxide by suspending it in solution and floating it to the top. Flux is not a cleaning agent! Always clean the work before soldering. Flux is not part of a soldered connection— it merely aids the soldering process. After soldering, remove any remaining flux. Resin flux can be removed with isopropyl or denatured alcohol. A cotton swab is a good tool for applying the alcohol and scrubbing the excess flux away. Commercial flux-removal sprays are available at most electronic part distributors.

The soldering iron

Soldering is used in nearly every phase of electronic construction so you will need soldering tools. A soldering tool must be hot enough to do the job and sufficiently lightweight for agility and comfort. A temperature controlled iron works well, although the cost is not justified for occasional projects. Get an iron with a small conical or chisel tip. Soldering is not like gluing; solder does more than bind metal together and provide an electrically conductive path between them. Soldered metals and the solder combine to form an alloy.

You may need an assortment of soldering irons to do a wide variety of soldering tasks. They range in size from a small 25-W iron for delicate printed-circuit work to larger 100- to 300-W sizes used to solder large surfaces. If you could only afford a single soldering tool when initially setting up your electronics workbench, then an inexpensive to moderately priced pencil-type soldering iron with between 25- and 40-W capacity is best for PC board electronics work. A 100-watt soldering gun is overkill for printed-circuit work, since it often gets too hot, cooking solder into a brittle mess or damaging small parts of a circuit. Soldering guns are best used for point-to-point soldering jobs, for large mass soldering joints, or large components. Small "pencil" butane torches are also available, with optional soldering-iron tips. A small butane torch is available from the Solder-It Company. Butane soldering irons are ideal for field service problems and will allow you to solder where there is no 110-V power source. This company also sells a soldering kit that contains paste solders (in syringes) for electronics, pot metal and plumbing. See the Appendix for the address information.

Keep soldering tools in good condition by keeping the tips well tinned with solder. Do not run them at full temperature for long periods when not in use. After each period of use, remove the tip and clean off any scale that may have accumulated. Clean an oxidized tip by dipping the hot tip in Sal ammoniac (ammonium chloride) and then wiping it clean with a rag. Sal ammoniac is somewhat corrosive, so if you do not wipe the tip thoroughly, it can contaminate electronic soldering. You can also purchase a small jar of Tip Tinner, a soldering iron tip dresser from your local RadioShack store. Place the tip of the soldering into the Tip Tinner after every few solder joints.

If a copper tip becomes pitted, file it smooth and bright and then tin it immediately with solder. Modern soldering iron tips are nickel or iron clad and should not be filed. The secret of good soldering is to use the right amount of heat. Many people who have not soldered before use too little heat, dabbing at the joint to be soldered, and making little solder blobs that cause unintended short circuits. Always use caution when soldering. A hot soldering iron can burn your hand badly or ruin a tabletop. It is a good idea to buy or make a soldering iron holder.

Soldering station

Often when building or repairing a circuit, your soldering iron is kept switched on for unnecessarily long periods, consuming energy and allowing the soldering iron tip to burn, and develop a buildup of oxide. By using this soldering-iron temperature controller, you will avoid destroying sensitive components when soldering.

Buying a lower wattage iron may solve some of the problems, but new problems arise when you want to solder some heavy-duty component, setting the stage for creating a "cold" connection. If you have ever tried to troubleshoot some instrument in which a cold solder joint was at the root of the problem, you know how difficult such defects are to locate. Therefore, the best way to satisfy all your needs is to buy a temperature controlled electronics workbench.

A soldering station usually consists of a temperature controlled soldering iron with an adjustable heat or temperature control and a soldering iron holder and cleaning pad. If you are serious about your electronics hobby or if you have been involved with electronics building and repair for any length of time you will eventually want to invest in a soldering station. There are real low-cost soldering stations for hobbyists, for under $30, but it makes more sense to purchase a moderately-priced soldering station such as the quality Weller series. A typical soldering station is shown in Figure 5-6.

Soldering gun

An electronics workbench would be incomplete without a soldering gun. Soldering guns are useful for soldering

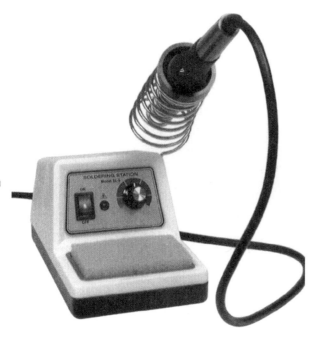

Figure 5-6 *Temperature controlled soldering station*

large components to terminal strips, or splicing wires together, or when putting connectors on coax cable. There are many instances where more heat is needed than a soldering iron can supply. For example, a large connector mass cannot be heated with a small soldering iron, so you would never be able to tin a connector with a small wattage soldering iron. A soldering gun is a heavy-duty soldering device which does in fact look like a gun. Numerous tips are available for a soldering gun and they are easily replaceable using two small nuts on the side arm of the soldering gun. Soldering guns are available in two main heat ranges. Most soldering guns have a two-step "trigger" switch which enables you to select two heat ranges for different soldering jobs. The most common soldering gun provides a 100-W setting when the trigger switch is pressed to its first setting. As the trigger switch is advanced to the next step, the soldering gun will provide 150 W when more heat is needed. A larger or heavy-duty soldering gun is also available, but a little harder to locate is the 200- to 250-W solder gun. The first trigger switch position provides 200 W, while the second switch position provides the 250-W heat setting. When splicing wires together either using the "Western Union" or parallel splice or the end splice, a soldering gun should be used, especially if the wire gauge is below size 22 ga. Otherwise the solder may not melt properly and the connections may reflect a "cold" solder joint and

therefore a poor or noisy splice. Soldering wires to binding post connections should be performed with a soldering gun to ensure proper heating to the connection. Most larger connectors should be soldered or pre-tinned using a soldering gun for even solder flow.

Preparing the soldering iron

If your iron is new, read the instructions about preparing it for use. If there are no instructions, follow this procedure. It should be hot enough to melt solder applied quickly to its tip (half a second when dry, instantly when wet with solder). Apply a little solder directly to the tip so that the surface is shiny. This process is called "tinning" the tool. The solder coating helps conduct heat from the tip to the solder joint. When soldering, bring the soldering tip in contact with one side of the solder joint. If you can place the tip on the underside of the joint, do so. With the tool below the joint, convection helps transfer heat to the joint. Place the solder against the joint directly opposite the soldering tool. It should melt within a second for normal PC connections, within two seconds for most other connections. If it takes longer to melt, there is not enough heat for the job at hand. Keep the tool against the joint until the solder flows freely throughout the joint. When it flows freely, solder tends to form concave shapes between the conductors. With insufficient heat solder does not flow freely; it forms convex shapes, or blobs. Once the solder shape changes from convex to concave, remove the tool from the joint. Let the joint cool without movement at room temperature. It usually takes no more than a few seconds. If the joint is moved before it is cool, it may take on a dull, satin look that is characteristic of a "cold" solder joint. Reheat cold joints until the solder flows freely and hold them still until cool. When the iron is set aside, or if it loses its shiny appearance, wipe away any dirt with a wet cloth or sponge. If it remains dull after cleaning, tin it again.

Overheating a transistor or diode while soldering can cause permanent damage. Use a small heat sink when you solder transistors, diodes, or components with plastic parts that can melt. Grip the component lead with a pair of pliers up close to the unit so that the heat is conducted away. You will need to be careful, since it is easy to damage delicate component leads.

A small alligator clip also makes a good heat sink to dissipate from the component.

Mechanical stress can damage components, too. Mount components so there is no appreciable mechanical strain on the leads.

Preparing work for soldering

If you use old junk parts, be sure to completely clean all wires or surfaces before applying solder. Remove all enamel, dirt, scale, or oxidation by sanding or scraping the parts down to bare metal. Use fine sandpaper or emery paper to clean flat surfaces or wire. (Note, no amount of cleaning will allow you to solder to aluminum. When making a connection to a sheet of aluminum, you must connect the wire by a solder lug or a screw.)

When preparing wires, remove the insulation with wire strippers or a pocket knife. If using a knife, do not cut straight into the insulation; you might nick the wire and weaken it. Instead, hold the knife as if you were sharpening a pencil, taking care not to nick the wire as you remove the insulation. For enameled wire, use the back of the knife blade to scrape the wire until it is clean and bright. Next, tin the clean end of the wire. Now, hold the heated soldering-iron tip against the under surface of the wire and place the end of the rosin-core solder against the upper surface. As the solder melts, it flows on the clean end of the wire. Hold the hot tip of the soldering iron against the under surface of the tinned wire and remove the excess solder by letting it flow down on the tip. When properly tinned, the exposed surface of the wire should be covered with a thin, even coating of solder.

How to solder

The two key factors in quality soldering are time and temperature. Generally, rapid heating is desired, although most unsuccessful solder jobs fail because insufficient heat has been applied. Be careful; if heat is applied too long, the components or PC board can be damaged, the flux may be used up, and surface oxidation can become a problem. The soldering-iron tips should be hot enough to readily melt the solder without burning, charring, or discoloring components, PC boards, or wires. Usually, a tip temperature about 100°F above

the solder melting point is about right for mounting components on PC boards. Also, use solder that is sized appropriately for the job. As the cross section of the solder decreases, so does the amount of heat required to melt it. Diameters from 0.025 to 0.040″ are good for nearly all circuit wiring.

Always use a good quality multi-core solder. A standard 60% tin–40% lead alloy solder with cores of non-corrosive flux will be found easiest to use. The flux contained in the longitudinal cores of multi-core solder is a chemical designed to clean the surfaces to be joined of deposited oxides, and to exclude air during the soldering process, which would otherwise prevent these metals coming together. Consequently, do not expect to be able to complete a joint by using the application of the tip of the iron loaded with molten solder alone, as this usually will not work. Having said that, there is a process called tinning where conductors are first coated in fresh, new solder prior to joining by a hot iron. Solder comes in gauges like wire. The two most common types of solder are 18 ga, used for general work, and the thinner 22 ga, used for fine work on printed circuit boards.

A well-soldered joint depends on:

1. Soldering with a clean, well-tinned tip
2. Cleaning the wires or parts to be soldered
3. Making a good mechanical joint before soldering
4. Allowing the joint to get hot enough before applying solder
5. Allowing the solder to set before handling or moving soldered parts.

Making a good mechanical joint

Unless you are creating a temporary joint, the next step is to make a good mechanical connection between the parts to be soldered. For instance, wrap the wire carefully and tightly around a soldering terminal or soldering lug, as shown in Figure 5-7.

Bend wire and make connections with long-nosed pliers. When connecting two wires together, make a tight splice before soldering. Once you have made a good mechanical contact, you are ready for the actual soldering.

The next step is to apply the soldering iron to the connection, soldering the connection as shown.

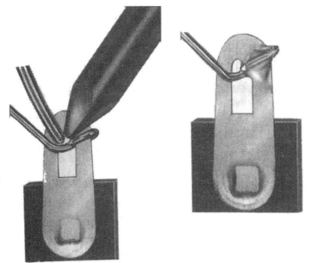

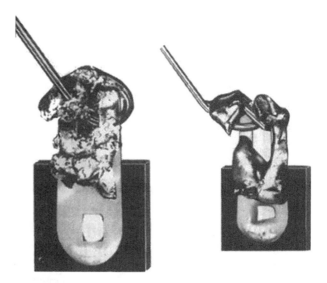

Figure 5-7 *Proper technique for soldering leads to lugs*

Figure 5-8 *Bad solder technique leads to lugs*

In soldering a wire splice, hold the iron below the splice and apply solder to the top of the splice. If the tip of the iron has a bit of melted solder on the side held against the splice, heat is transferred more readily to the splice and the soldering is done more easily. Don't try to solder by applying solder to the joint and then pressing down on it with the iron. Be sure not to disturb the soldered joint until the solder has set. It may take a few seconds for the solder to set, depending upon the amount of solder used in making the joint. Now take a good look at the joint. It should have a shiny, smooth appearance, not pitted or grainy. If it does have a pitted, granular appearance as seen in Figure 5-8, reheat the joint, scrape off the solder, and clean the connection.

Then start over again. After the solder is well set, pull on the wire to see if it is a good, tight connection. If you find that you made a poor soldering job do not get upset, be thankful you found it and do it over again. A quick reference solder checklist is shown in the listing in Table 5-1.

Table 5-1
Soldering checklist

1. Prepare the joint. Clean all surfaces and conductors thoroughly with fine steel wool. First, clean the circuit traces, then clean the component leads.

2. Prepare the soldering iron or gun. The soldering device should be hot enough to melt solder applied to the tip. Apply a small amount of solder directly to the tip, so that the surface is shiny.

3. Place the tip in contact with one side of the joint; if possible place the tip below the joint.

4. Place the solder against the joint directly opposite the soldering tool. The solder should melt within two seconds; if it takes longer use a larger iron.

5. Keep the soldering tool against the joint until the solder flows freely throughout the joint. When it flows freely the joint should form a concave shape; insufficient heat will form a convex shape.

6. Let the joint cool without any movement; the joint should cool and set up within a few seconds. If the joint is moved before it cools the joint will look dull instead of shiny and you will likely have a cold solder joint. Re-heat the joint and begin anew.

7. Once the iron is set aside, or if it loses its shiny appearance, wipe away any dirt with a wet cloth or sponge. When the iron is clean the tip should look clean and shiny. After cleaning the tip apply some solder.

Soldering printed circuit boards

Most electronic devices use one or more printed circuit (PC) boards. A PC board is a thin sheet of fiberglass or phenolic resin that has a pattern of foil conductors "printed" on it. You insert component leads into holes in the board and solder the leads to the foil pattern. This method of assembly is widely used and you will probably encounter it if you choose to build from a kit. Printed circuit boards make assembly easy. First, insert component leads through the correct holes in the circuit board. Mount parts tightly against the circuit board unless otherwise directed. After inserting a lead into the board, bend it slightly outward to hold the part in place.

When the iron is hot, apply some solder to the flattened working end at the end of the bit, and wipe it on a piece of damp cloth or sponge so that the solder forms a thin film on the bit. This is tinning the bit.

Melt a little more solder on to the tip of the soldering iron, and put the tip so it contacts both parts of the joint. It is the molten solder on the tip of the iron that allows the heat to flow quickly from the iron into both parts of the joint. If the iron has the right amount of solder on it and is positioned correctly, then the two parts to be joined will reach the solder's melting temperature in a couple of seconds. Now apply the end of the solder to the point where both parts of the joint and the soldering iron are all touching one another. The solder will melt immediately and flow around all the parts that are at, or over, the melting part temperature. After a few seconds, remove the iron from the joint. Make sure that no parts of the joint move after the soldering iron is removed until the solder is completely hard. This can take quite a few seconds with large joints. If the joint is disturbed during this cooling period it may become seriously weakened.

The most important point in soldering is that both parts of the joint to be made must be at the same temperature. The solder will flow evenly and make a good electrical and mechanical joint only if both parts of the joint are at an equal high temperature. Even though it appears that there is a metal to metal contact in a joint to be made, very often there exists a film of oxide on the surface that insulates the two parts. For this reason it is no good applying the soldering iron tip to one half of the joint only and expecting this to heat the other half of the joint as well.

It is important to use the right amount of solder, both on the iron and on the joint. Too little solder on the iron will result in poor heat transfer to the joint, too much and you will suffer from the solder forming strings as the iron is removed, causing splashes and bridges to other contacts. Too little solder applied to the joint will give the joint a half-finished appearance: a good bond where the soldering iron has been, and no solder at all on the other part of the joint.

The hard, cold solder on a properly made joint should have a smooth shiny appearance and if the wire is pulled it should not pull out of the joint. In a properly made joint the solder will bond the components very strongly indeed, since the process of soldering is similar to brazing, and to a lesser degree welding, in that the solder actually forms a molecular bond with the surfaces of the joint. Remember it is much more difficult to correct a poorly made joint than it is to make the joint properly in the first place. Anyone can learn to solder, it just takes practice.

The diagram in Figure 5-9 shows how to solder a component lead to a PC board pad.

The tip of the soldering iron heats both the lead and the copper pad, so the end of the solder wire melts when it is pushed into the contact. The diagram illustrated in Figure 5-10 show how a good solder joint is obtained.

Notice that it has a smooth and shiny "fillet" of solder metal, bonding all around to both the component lead and the copper pad of the PC board. This joint provides a reliable electrical connection.

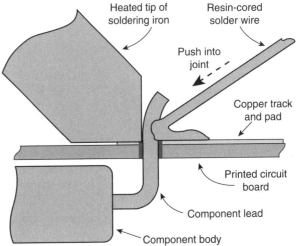

Figure 5-9 *Proper application of soldering iron to joint*

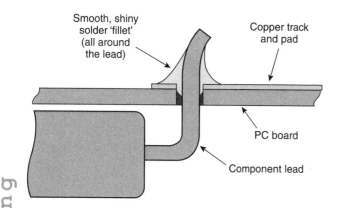

Figure 5-10 *Proper PC board soldering*

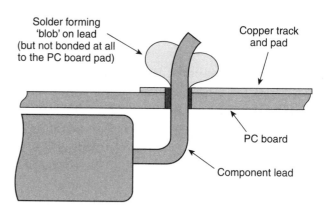

Figure 5-12 *Bad PC board solder joint II*

Try to make the solder joint as quickly as possible because the longer you take, the higher the risk that the component itself and the printed circuit board pad and track will overheat and be damaged. But don't work so quickly that you cannot make a good solder joint. Having to solder the joint over again always increases the risk of applying too much heat to the PCB.

As the solder solidifies, take a careful look at the joint you have made, to make sure there is a smooth and fairly shiny metal "fillet" around it. This should be broadly concave in shape, showing that the solder has formed a good bond to both metal surfaces. If it has a rough and dull surface or just forms a "ball" on the component lead, or a "volcano" on the PCB pad with the lead emerging from the crater, you have a "dry joint." If your solder joint looks like the picture shown in Figure 5-11, you will have to resolder the joint over again.

Figure 5-12 shows another type of dry solder joint which would have to be resoldered. These types of "dry" solder joints if now redone will cause the circuit to be unreliable and intermittent.

For projects that use one or more integrated circuits, with their leads closely spaced pins, you may find it easier to use a finer gauge solder, i.e. less than 1 mm in diameter. This reduces the risk of applying too much solder to each joint, and accidentally forming "bridges" between pads to PC "tracks."

The finished connection should be smooth and bright. Reheat any cloudy or grainy-looking connections. Finally, clip off the excess wire length, as shown in Figure 5-13.

Occasionally a solder "bridge" will form between two adjacent foil conductors. You must remove these bridges, otherwise a short circuit will exist between the two conductors. Remove a solder bridge by heating the bridge and quickly wiping away the melted solder with a soft cloth. Often you will find a hole on the board

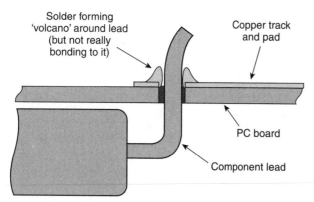

Figure 5-11 *Bad PC board solder joint I*

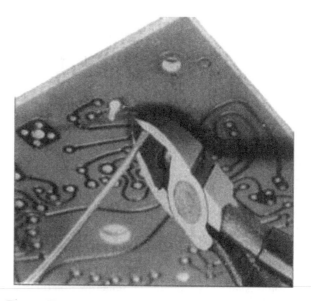

Figure 5-13 *Cutting excess components leads from PC board*

plugged by solder from a previous connection. Clear the hole by heating the solder while pushing a component lead through the hole from the other side of the board. Good soldering is a skill that is learnt by practice.

How to unsolder

In order to remove components, you need to learn the art of desoldering. You might accidentally make a wrong connection or have to move a component that you put in an incorrect location. Take great care while unsoldering to avoid breaking or destroying good parts. The leads on components such as resistors or transistors and the lugs on other parts may sometimes break off when you are unsoldering a good, tight joint. To avoid heat damage, you must use as much care in unsoldering delicate parts as you do in soldering them. There are three basic ways of unsoldering. The first method is to heat the joint and "flick" the wet solder off. The second method is to use a metal wick or braid to remove the melted solder. This braid is available at most electronics parts stores; use commercially made wicking material (braid). Place the braid against the joint that you want to unsolder. Use the heated soldering iron to gently press the braid against the joint. As the solder melts, it is pulled into the braid. By repeating this process, you can remove virtually all the solder from the joint. Then reheat the joint and lift off the component leads.

Another useful tool is an air-suction solder remover. Most electronics parts stores have these devices. Before using a desolder squeeze bulb, use your soldering iron to heat the joint you want to unsolder until the solder melts. Then squeeze the bulb to create a vacuum inside. Touch the tip of the bulb against the melted solder. Release the bulb to suck up the molten solder. Repeat the process until you have removed most of the solder from the joint. Then reheat the joint and gently pry off the wires. This third method is easy, and is the preferred method, since it is fast and clean. You can use a vacuum device to suck up molten solder. There are many new styles of solder vacuum devices on the market that are much better than the older squeeze bulb types. The new vacuum desoldering tools are about 8 to 12″ long with a hollow teflon tip. You draw the vacuum with a push handle and set it. As you reheat the solder around the component to be removed, you push a button

on the device to suck the solder into the chamber of the desoldering tool.

Desoldering station

A desoldering station is a very useful addition to an electronics workshop or workbench but in most cases, they just cost too much for most hobbyists. desoldering stations are often used in production environments or as re-work stations, when production changes warrant changes to many circuit boards in production. Some repair shops use desoldering stations to remove components quickly and efficiently.

Another useful desoldering tool is one made specifically for removing integrated circuits. The specially designed desoldering tip is made the same size as the integrated circuit, so that all IC pins can be desoldered at once. This tool is often combined with a vacuum suction device to remove the solder as all the IC pins are heated. The IC desoldering tips are made in various sizes: there are 8-pin, 14-pin, and 16-pin versions, which are used to uniformly desolder all IC pins quickly and evenly, so as not to destroy the circuit board. The specialized soldering tips are often used in conjunction with vacuum systems to remove the solder at the same time.

Remember these things when unsoldering:

1. Be sure there is a little melted solder on the tip of your iron so that the joint will heat quickly.

2. Work quickly and carefully to avoid heat damage to parts. Use long-nosed pliers to hold the leads of components just as you did while soldering.

3. When loosening a wire lead, be careful not to bend the lug or tie point to which it is attached. Use pieces of wire or some old radio parts and wire. Practice until you can solder joints that are smooth, shiny, and tight. Then practice unsoldering connections until you are satisfied that you can do them quickly and without breaking wires or lugs.

Caring for your soldering iron

To get the best service from your soldering iron, keep it cleaned and well tinned. Keep a damp cloth on the bench as you work. Before soldering a connection, wipe the tip of the iron across the cloth, then touch some fresh solder to the tip. The tip will eventually become worn or pitted. You can repair minor wear by filing the tip back into shape. Be sure to tin the tip immediately after filing it. If the tip is badly worn or pitted, replace it. Replacement tips can be found at most electronics parts stores. Remember that oxidation develops more rapidly when the iron is hot. Therefore, do not keep the iron heated for long periods unless you are using it. Do not try to cool an iron rapidly with ice or water. If you do, the heating element may be damaged and need to be replaced or water might get into the barrel and cause rust. Take care of your soldering iron and it will give you many years of useful service.

Remember, soldering equipment gets hot! Be careful. Treat a soldering burn as you would any other. Handling lead or breathing soldering fumes is also hazardous. Observe these precautions to protect yourself and others!

Ventilation

Properly ventilate the work area where you will be soldering. If you can smell fumes, you are breathing them. Often when building a new circuit or repairing a "vintage" circuit you may be soldering continuously for a few hours at a time. This can mean you will be breathing solder fumes for many hours and the fumes can cause you to get dizzy or light-headed. This is dangerous because you could fall down and possibly hurt yourself in the process. Many people who are highly allergic are also allergic to the smell of solder fumes. Solder fumes can cause sensitive people to get sinus infections from smelling solder fumes. So ventilating solder fumes is important.

There a few different ways to handle this problem. One method is to purchase a small fan unit housed with a carbon filter which sucks the solder fumes into the carbon filter to eliminate them. This is the simplest method of reducing or eliminating solder fumes from the immediate area. If there is a window near your soldering area, be sure to open it to reduce the exposure to solder fumes. Another method of reducing or eliminating solder fumes is to buy or build a solder smoke removal system. You can purchase one of these systems but they tend to be quite expensive. You can create you own solder smoke removal system by locating or purchasing an 8–10′ piece of 2″ diameter flexible hose, similar to your vacuum cleaner hose. At the solder station end of the hose, you can affix the hose to a wooden stand in front of your work area. The other end of the hose is funneled into a small square "muffin" type fan placed near a window. Be sure to wash your hands after soldering, especially before handling food, since solder contains lead; also try to minimize direct contact with flux and flux solvents.

Chapter 6

Telephone Privacy Guard

Parts list

Parts Bin

Telephone Privacy Guard

Q1   BR100 DIAC
     semiconductor

Q2   2N6073A TRIAC
     semiconductor

Misc PC board, wire,
     connectors, etc.

The Telephone Privacy Guard is designed to prevent telephone calls from being heard by other telephones in the same subscriber telephone line in your home or office. When one telephone is picked up, the circuit prevents other telephones that may be picked up from listening in on your conversation, since this circuit "starves" the other phones on the line from obtaining telephone line current. The Privacy Guard circuit is installed between telephone line and telephone as shown in Figure 6-1.

You need to install a Privacy Guard circuit in each telephone on the same subscriber line in order for the system to work properly. The Privacy Guard circuit can be installed inside each telephone base unit, or it could be installed in a small plastic box next to each phone.

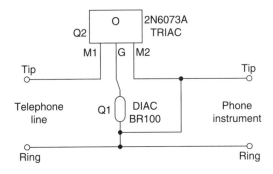

Figure 6-1 *Privacy group pictorial. Courtesy of Tom Engdahl*

The circuit is simple to build, it only consumes a small current from the phone line, and only consists of two components as shown in the circuit diagram.

The circuit can be built on a small circuit board or on a small piece of prototype board in just a few minutes. The Privacy Guard circuit not only prevents others from listening-in on your telephone call, but it also prevents others from interrupting modem communication between your computer and a distant computer or server.

Circuit description

The basic idea of the Privacy Guard circuit is to sense the voltage in the telephone line when the telephone is picked up. If that voltage is higher than about 30 V (normal on-hook voltage is about 48 V) then the circuit lets the telephone work normally. If the voltage is lower than 30 V it prevents the current from going from the telephone line to telephone (normally the voltage in line is about 6–10 V when one telephone is off-hook). The circuit is designed so that is passes the ring voltage to all telephones without problems.

The circuit is a very simple circuit built from one DIAC and one TRIAC. When a telephone headset is picked up, it prevents any operating current from flowing thorough the TRIAC. The triggering of the TRIAC is accomplished through DIAC at Q1, which will trigger the TRIAC if there is more than about 30 V present at the circuit. The TRIAC leads at M1 and M2 are connected in series with the telephone TIP line wires. When TRIAC Q2 starts to conduct, it will conduct as long as there is any current flowing through it. So TRIAC Q2 conducts until the telephone handset is put on-hook (call has ended).

The circuit is very similar to the operation of commercial adapters. But remember that this adapter

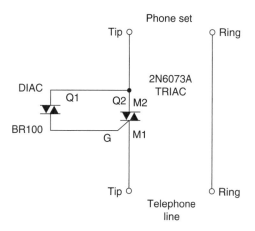

Phone set

Tip ○ ── ── ○ Ring

DIAC
Q1 Q2 M2
BR100
G M1

2N6073A
TRIAC

Tip ○ ── ── ○ Ring

Telephone
line

Figure 6-2 *Privacy guard schematic. Courtesy of Tom Engdahl*

is not the type approved for connection to the public telephone network. The component values used in the prototype could be replaced with similar types of components, such that a TRIAC with nearly any type which will handle at least 200 V can be triggered easily and keeps on conducting at currents as low as 15 mA.

The drawing in Figure 6-2 represents a pictorial diagram of the major components used in the prototype unit and how it is connected between the telephone line and the telephone unit.

Circuit assembly

Before we begin building the Telephone Privacy Guard project, let us first find a clean, well lit worktable or workbench. Next we will gather a small 20–30 W pencil tipped soldering iron, a length of 60/40 tin–lead solder, a small canister of "Tip Tinner," available at your local RadioShack store, along with a few small hand tools. Try and locate small flat blade and small Phillips screwdrivers. A pair of needle nose pliers, a pair of tweezers along with a magnifying glass, and a pair of end-cutters and we will begin constructing the project. Now locate your schematic diagram and parts layout diagram along with all of the components to build the project. Once all the components are in front of you, you can check them off against the project parts list to make sure you are ready to start.

So, finally we are ready to begin assembling the project, so let's get started! First, you will need to decide

if your are going to build the Privacy Guard circuits on small printed circuit boards, or small perf-boards, or proto-boards available from RadioShack. Once you have decided which route you are going to take, and you proceed to construct the Privacy Guard circuits, remember that you will need to build one for each phone in your home or office. Go ahead and locate both the DIAC and the TRAIC. Note that the DIAC only has two leads, while the TRIAC has three leads. The DIAC has no polarity; however, the TRIAC does. The TRIAC has an M1 and an M2 lead and a gate lead, as shown in both the schematic and pictorial diagrams. Go ahead and install the DIAC and TRIAC in their respective locations on the PC, then take your end-cutters and cut the excess component leads flush to the edge of the circuit board.

After a short break, we will inspect the circuit board for any "cold" solder joints or possible "short" circuits. Pick up the circuit board with the foil side facing upwards toward you. Carefully inspect the circuit board against the parts placement diagram and the schematic to make sure that the proper component was inserted in the correct PC hole. It is possible that you may have inserted a semiconductor backwards.

Once this preliminary inspection is over, you can inspect the circuit board for any possible cold solder joints or short circuits. Take a careful look at each of the PC solder joints; they should all look clean and shiny. If any of the solder joints look dull, dark, or blobby, then you unsolder the joint, remove the solder and then resolder the joint all over again. Next, examine the PC board for any short circuits. Short circuits can be caused from two circuit traces touching each other due to a stray component lead that stuck to the PC board from solder residue or from solder blobs bridging the solder traces. Once you have fully examined the circuit board, you can reconnect the circuit board back to the phone line and try out the phone transmitter once again.

Once the inspection has been completed, you can move on to installing the circuit in each of the telephone locations in your home or office. You may wish to install a Privacy Guard circuit inside of each telephone base or you could elect to place each Privacy Guard circuit inside a small plastic box with a phone jack at both ends, so that you could plug a telephone set in on one connector and then place short telephone extension cable between the Privacy Guard and the telephone line. If the Privacy Guard is placed inside your telephone base then you will likely want to mount a 4-position terminal strip on each PC board to facilitate connecting the phone wire inside the phone base to the Privacy Guard circuit. After wiring the telephone Privacy Guard circuits in or at each telephone then you are ready to test out the system.

Testing the privacy guard

Testing the Privacy Guard circuit is simple. Have someone call your home phone from a remote phone line or from a cell phone. Pick up one of the home phones and talk over the phone. Now have a second person pick up an extension phone in your home or office. Once a second extension phone is picked up, they should not be able to hear your conversation with the remote phone or cell phone. If the extension phone does hear the conversation, then you may have wired the Privacy Guard incorrectly. In that event you may want to carefully re-inspect each Privacy Guard circuit for an error.

Chapter 7

Modem Protector

Parts list

Parts Bin

Modem/fax Protector

SVP1, SVP2 gas surge
    arrestors 250 V
    CG-230L (C.P. Claire
    or equiv)

D1, D2, D3, D4 1N5386B
    Zener diodes –
    180 V, 5 W

Misc 1N5386B Zener
    diodes – 180 V,
    5 W

Protect your modem and fax machine or answering machine against errant transients with this easy to build circuit. Shield your modem, fax, or answering machine against harmful transients arriving over the phone line by building this fast-acting Modem/fax Protector, shown in Figure 7-1. Then you will be able to relax when electric storms bear down on your home or office.

It is an easy, inexpensive project, and low-cost insurance for your expensive telephone-dependent equipment. Even people who plug their computers and entertainment electronics into power-line surge protectors (and would not dream of leaving them unprotected) are likely to forget about protecting modems, faxes, and answering machines from transients arriving over the phone line. Then comes the first thunderstorm of the season. Before they remember to do anything about it, the innards of that equipment could disappear in a puff of smoke. You could, of course, disconnect your telephone gadgets with a switch when they are not in use. But will you remember to do it in time? Needless to say, the switch won't help much if the devices are running when the lightning starts to flash.

When a telephone handset is "on the hook" there are 48 V DC across the two wires designated "tip" and "ring." The green tip wire is at ground potential, or zero volts, and the red ring wire is at negative 48 V. Approximately 21 to 35 mA of current flow in this condition. When the handset is picked up the line voltage drops to 6 V DC. The 20- or 30-Hz telephone-ringing signal can be from about 100 V to about 120 V AC. It is superimposed across the normal 48-V DC signal. That "ringing" voltage determines the voltage ratings of the protective devices.

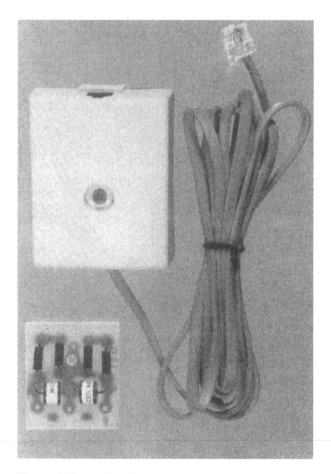

Figure 7-1 *Modem/fax protector*

Telephone protection

Although the effects of lightning are well known, many people believe that only a direct strike on a nearby phone or power line will cause damage. However, most damage to electronic equipment is caused by voltages induced in those conductors by direct strikes elsewhere. Harmful voltages can be caused by strikes as far away as 15 miles! Telephone circuits generally have resistive elements in each wire to protect the telephone handsets from transient spikes in excess of about 500 V. However, this rather crude passive protection is inadequate for protecting more vulnerable electronics.

Spark-Gap tubes or surge voltage protectors (SVP) have been used for many years to protect electronic circuits from man-made and natural surges arriving over either power or phone lines. They provide low resistance paths for excessive voltage transients but appear open to normal voltages. The devices are hermetically sealed gas discharge tubes. Typically made of ceramic with properly spaced electrodes, they are filled with a rare gas.

The main purpose of the SVP is to provide a conductive path for unwanted and excessive transients, thereby preventing the transient energy and associated voltages from damaging equipment and components and harming people. They are designed to switch current at a pre-established breakdown voltage.

The breakdown voltage causes the internal gas to ionize and change from a non-conducting to a conducting state, thus permitting an arc to form and short the connected wires to ground. During conduction, the SVP can momentarily carry high currents. After the voltage transient has been discharged, the gas de-ionizes and the SVP is ready for another voltage transient. The SVP is bipolar and has a symmetrical characteristic. In the restored or extinguished condition, it causes very little loss because of its high impedance. This characteristic contrasts with those of transient absorbtion Zener diodes and metal-oxide varistors that exhibit leakage. However, both of those devices have faster response times than SVPs.

The metal-oxide varistor (MOV or SIOV) has also been said to protect electronics connected to telephone lines. It is made from finely powdered zinc oxide mixed with binders and pressed into a disk. After firing, the disk becomes a matrix of conductive zinc-oxide grains separated by highly resistive boundaries. This property gives them a symmetrical electrical characteristic similar to the SVP (or two back-to-back Zener diodes). However, the MOV's breakdown time is too slow to protect connected electronics against the fastest voltage spikes caused by lightning. As a result, faster Zener diodes and SI/Pb have been combined in our Modem/fax Protector.

Circuit details

The Modem/fax Protector is shown schematically in Figure 7-2. It has a typical response time of about 10 ns—fast enough to protect your equipment against the speediest voltage spikes. Zener diodes D1 and D2 are connected back-to-back in parallel with surge voltage protector SVP1 between the ring and the PC board earth ground connection. Similarly, Zener diodes D3 and D4 are connected in parallel with SVP2 between the tip wire and the PC board earth ground. The SVPs are rated for a nominal DC voltage of 230 V with breakdown voltage from 195 to 265 V DC. The Zener diodes break down at 180 V DC within about 10 ns, to protect the telephone circuits from the fastest initial voltage spikes, and then the SVPs ionize to ground the over current. Note that the circuit has no batteries; all it really needs to protect your equipment is an effective ground.

Circuit assembly

Before we begin building the Modem/fax Protector, first you will need to locate a clean, well-lit worktable

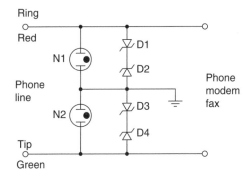

Figure 7-2 *Phone-modem-fax-answering machine protector*

or workbench. Next we will gather a small 25–30 W pencil tipped soldering iron, a length of 60/40 tin–lead solder, and a small canister of "Tip Tinner," available at your local RadioShack store; "Tip Tinner" is used to condition the soldering tip between solder joints. You drive the soldering tip into the compound and it cleans and prepares the soldering tip. Next grab a few small hand tools, try and locate small flat blade and Phillips screwdrivers. A pair of needle nose pliers, a pair of tweezers along with a magnifying glass and a pair of end-cutters and we will begin constructing the project. Now locate your schematic diagram and parts layout diagram along with all of the components needed to build the project. Once all the components are in front of you can check them off against the project parts list to make sure you are ready to start building the project. The prototype Modem/fax Protector was built on a small, square PC board measuring 1½″ per side. However, point-to-point wiring techniques can also be used.

The Modem/fax Protector is quite easily constructed using four Zener diodes and two gas surge arrestors. The gas surge arrestors are small cylindrical tubes with a lead at both ends. The gas surge arrestors do not have polarity and so they can be installed in either direction on the PC board. Place the gas surge arrestors in their proper locations on the circuit board and solder them in place. Once soldered in place, trim the excess lead lengths with a pair of end-cutters. Cut the excess leads flush to the edge of the circuit board.

The Modem/fax Protector project contains four Zener diodes at D1 through D4. These diodes also have polarity, which must be observed if the circuit is going to work correctly. You will notice that each diode will have a black or white colored band at one edge of the diode's body. The colored band represents the diode's cathode lead. Check your schematic and parts layout diagram when installing these diodes on the circuit board. Place all four Zener diodes on the circuit board and solder them in place on the PC board. Then cut the excess components lead with your end-cutter.

Take a short break and inspect the circuit board for any possible "cold" solder joints or "short" circuits. Take a careful look at each of the PC solder joints, they should all look clean and shiny. If any of the solder joints look dull, dark or blobby, then you unsolder the joint, remove the solder and then resolder the joint all over again. Next, examine the PC board for any short circuits.

Short circuits can be caused from two circuit traces touching each other due to a stray component lead that stuck to the PC board from solder residue or from solder blobs bridging the solder traces. Once you have fully examined the circuit board, you can now connect the circuit board between the phone line and the item that you wish to protect. The Modem/fax Protector itself does have a certain polarity in that the gas surge protectors are placed on the phone line side of the overall hookup, and the Zener diode side of the circuit is connected to the modem, fax, or answering machine.

Obtain a small modular plastic phone jack cover with an included jack. The jack should have a short section of 4-wire telephone cable attached. Cut off the black and yellow wires and connect the red ring and green tip wire pigtails within the box. Cut a bottom plate from a sheet of phenolic or other suitable thin but rigid insulating material slightly larger (about ⅟₁₆″) than the outer dimensions of the jack box. Carefully mark the location on the cover plate for a hole to accommodate the central screw so that it is opposite the hole in the jack box, and drill a hole of the same diameter through the cover.

Determine a suitable length for the 4-wire telephone cable between your telephone outlet or junction box and the Modem/fax Protector and attach a plug to one end of that cable. Carefully form a hole in the side wall of the jack box, making it large enough to permit the bare end of the telephone cable to be pulled through a distance of 3 to 4″. Clamp the cable with a plastic cable tie.

Strip the cable jacket back to about ⅛ inch from the cable tie, select out the red ring wire and green tip wire and strip their ends. Next, cut off the yellow and black wires close to the cable tie. Then strip the ends of the red and green wires from the jack. Solder both red and green wires to the circuit board. Cut a short length of insulated 14 or 16 AWG wire, strip both ends, and solder one end to the ground pad on the PC board.

Apply four drops of silicone RTV adhesive to the inside as shown, align the hole in the circuit board over the hole in the jack box, and bed the board down in the adhesive. Allow sufficient time for the adhesive to set up before proceeding. Determine a suitable length for a 14 to 16 AWG solid copper ground wire based on the proximity of your telephone apparatus to a suitable location for a ground rod (to be discussed later).

Form a loop in one end of the heavy ground wire to accommodate the central screw in the Modem/fax Protector. Then insert the loop in the ground wire, jack box, and circuit board. Tightly wrap the bare copper end of the ground wire on the circuit board several times around the central screw to complete the ground connection. Then apply solder to the outside of the turns.

Plug your modem, fax, or answering machine into the jack, and plug the length of cable into your telephone wall outlet. The ring and tip wires of the telephone line must remain consistent throughout. The tip lead must be positive with respect to the ring lead.

After making sure that all connections have been made correctly apply a thin layer of adhesive to the rim of the jack box, assemble the cover over the screw and clamp it in position with a washer and two nuts.

Good grounding

The necessity for a good ground in protective circuits cannot be overstated; a bad ground is no ground at all! The most effective ground is achieved with a metal rod, preferably copper, at least four feet long, and driven into moist soil. Connect the ground rod to the Modem/fax Protector with # 18 gauge insulated wire. The Modem Protector can be connected to the ground strap through a window opening or a hole drilled through the wall. Suitable grounding rods with wire connecting clamps are available from electronics supply stores. The next best grounding method is to connect the 14 to 16 AWG wire to a cold-water pipe.

Troubleshooting

If your Modem/fax Protector does not appear to function correctly or it shorts out the phone line, then you will need to disconnect the circuit from the phone line and carefully inspect the circuit board. It is likely that you made a soldering or parts placement mistake during construction. Pick up the circuit board with the foil side facing upwards toward you. Carefully inspect the circuit board against the parts placement diagram and the schematic to make sure that the proper component was inserted in the correct PC hole. You may have placed the wrong end at a particular location; or you may have inserted a Zener diode backwards. Once you have corrected the problem, reconnect the Modem/fax Protector between your modem and the phone line. The Modem/fax Protector should give you many years of trouble-free service. You may want to build a couple of the Modem/fax Protectors for other devices around your home or office.

Chapter 8

Telephone Amplifiers and Projects

Parts list

Magnetic Phone Amplifier I

R1, R7   270 K, $\frac{1}{4}$ W, 5% resistor

R2, R4, R5, R6   10 K, $\frac{1}{4}$ W, 5% resistor

R3   100 K $\frac{1}{4}$ W, 5% resistor

R8   10 K Trimpot

C1, C9   100 µF, 35 V electrolytic capacitor

C3, C10   220 pF, 35 V ceramic capacitor

C2   4.7 µF, 35 V MPE capacitor

C4, C5, C7   100 nF, 35 V [104]

C6, C8   10 µF, 35 volt electrolytic capacitor

U1   LM358 IC

U2   LM386 IC

J1   audio jack

M1   mini magnetic telephone suction pickup coil

SPKR   8 ohm mini speaker

Misc   IC socket, PC board, battery clip, wire, terminal block, etc

CK601 Amplifier kit

Kit available from:
www.electronickits.com

Telephone Amplifier II—Direct Connect

R1, R2, R3   10 K ohm, $\frac{1}{4}$ W, 5% resistor

R4, R5   1 K ohm, $\frac{1}{4}$ W, 5% resistor

R6   10 ohm

R7   3.3 K, $\frac{1}{4}$ W, 5% resistor

R8   560 ohm, $\frac{1}{4}$ W, 5% resistor

R9   10 K potentiometer (PC mount)

C1, C2   22 nF MKM, 35 V

C3   18 nF (183) or MKM, 35 V

C4, C5   47 nF (473) or MKM

C6   100 nF, 35 V

C7   1 µF, 35 V electrolytic capacitor

C8   47 µF, 35 V electrolytic capacitor

C9   100 µF, 35 V electrolytic capacitor

C10   470 µF, 35 V electrolytic capacitor

D1, D2   1N4148 silicon diode

D3, D4 1N4000 silicon
diode

T1 600 to 600 ohm mini
matching transformer
(PC type)

SPKR 8 ohm mini speaker
or headphone

Misc IC socket, PC
board, hookup wire,
etc

PWR supply 7-9 V or 9
to 12 V DC "wall
wart" power supply

K4900 Amplifier kit is
available from:
Velleman Kits

Project 1: Telephone amplifier I—Magnetic phone pickup

It is often very convenient for other people in the room to listen to both sides of a phone conversation. The telephone amplifier is ideal for older people who are hard of hearing. This circuit will allow them to better hear the phone conversation. A specially designed magnetic telephone pickup with a suction cup attaches onto the earpiece of your phone, available

from RadioShack with an ⅛" plug. A plug at the opposite end of the pickup's line cord connects into the amplifier circuit. The design is low cost and there is no direct electrical connection to the phone system, so it is completely isolated. This telephone amplifier could also be used to amplify the bell ringer as a remote ringer to notify you of an incoming phone call.

Circuit description

The telephone pickup is really a magnetic field fluctuation detector. It picks up the oscillating magnetic field from the receiver of your telephone when someone is speaking to you. But it will also collect any other oscillating magnetic fields which happen to be floating around in the air, for example, low- and high-frequency noise from your TV set or computer monitor, or the characteristic mains hum from power lines. High-frequency filters have been built into the circuit to reduce some of this unwanted noise.

The telephone amplifier circuit shown in Figure 8-1 consists of two high gain pre-amplifier stages in the Lm358 followed by a power amplifier to drive the speaker.

The ICs are low cost and easily available. Both pre-amplifiers in the Lm358 are biased to half the supply voltage by R4 and R5. This allows maximum voltage swing at the outputs before hitting either supply rail. C6 bypasses any AC signal to ground, stabilizing the DC bias voltage. R3 is necessary to couple the DC bias voltage to IC1:B while also providing a high impedance

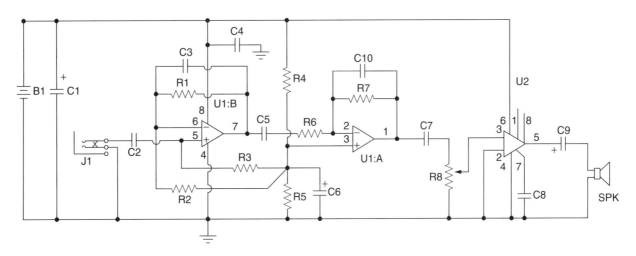

Figure 8-1 *Magnetic-pickup telephone amplifier. Courtesy of DIY Electronics*

to the input signal. The RC feedback circuit on both pre-amps will, like any RC circuit, have a cut-off frequency. Or to think of it another way, the capacitance starts to act as a short circuit as the frequency increases and the gain decreases. The cut-off frequency is given by the formula: $f = 1/(6.28 \times RC)$. This gives a cut-off frequency of 2.7 kHz, which has the effect of limiting the amount of high frequency noise, or "hiss." This high frequency roll-off does not greatly affect voice frequency signals since voice frequency is nominally in the range 300 Hz to 3.0 kHz.

The output of the second pre-amp is fed to the power amplifier stage via C7 and R8. Capacitor C7 removes any DC component from the amplified signal while R8 acts as a volume control. The Lm386 is very easy to use and requires a minimum of external components. C8 provides filtering and bypassing for the internal bias network. C9 removes any DC component from the output signal. The gain of the Lm386 may be set according to the combination of resistors and capacitors across pins 1 and 8. If no components are placed between pins 1 and 8 the overall gain would be 20.

Circuit Assembly

Locate a clean, well-lit worktable or workbench, a small pencil tip soldering iron, some 60/40 tin–lead solder and a small can of Tip Tinner from your local RadioShack store. Tip Tinner is used to clean and prep the soldering tip. You will also want to locate some small tools such as a pair of small end cutters, a few small screwdrivers, and a small Exacto type knife. Heat up your soldering iron and place all the project parts in front of you, along with your printed circuit board. When building electronic circuit boards with integrated circuits it is a good idea to install IC sockets on the PC board. In the event of an IC failure it is then a simple matter to simply remove the defective IC and just install a new one. Integrated circuit sockets are inexpensive as well as a good insurance policy.

Place your single-sided PC board in front of you along with the schematic and any parts layout diagrams. Next, place all of the parts in front of you and create a check-off list to make sure you have all the components you will need for this project. It is usually a good idea to begin installing the lowest profile components such as resistors first. Now, locate Tables 8-1 and 8-2, and place them in front of you. Table 8-1 illustrates the resistor code chart and how to read resistors, while Table 8-2 illustrates the capacitor code chart which will aid you in constructing the project.

Refer now to Table 8-1, which introduces the resistor color chart; this will help you identify resistors. Each resistor will have either three or four color bands, which

Table 8-1

Resistor color code chart

Color Band	1st Digit	2nd Digit	Multiplier	Tolerance
Black	0	0	1	
Brown	1	1	10	1%
Red	2	2	100	2%
Orange	3	3	1,000 (K)	3%
Yellow	4	4	10,000	4%
Green	5	5	100,000	
Blue	6	6	1,000,000 (M)	
Violet	7	7	10,000,000	
Gray	8	8	100,000,000	
White	9	9	1,000,000,000	
Gold			0.1	5%
Silver			0.01	10%
No color				20%

begin at one end of the resistor's body. The first colored band represents the resistor's first digit, while the second color represents the second digit of the resistor's value. The third colored band represents the resistor's multiplier value, and the fourth color band represents the resistor's tolerance value. If there is no fourth colored band then the resistor has a 20% tolerance value. If the resistor's fourth band is silver then the resistor has a 10% tolerance value and if the fourth band is gold then the resistor's tolerance value will be 5%.

Finally, we are ready to begin assembling the project, so let's get started. The prototype was constructed on a single-sided printed circuit board or PCB. With your PC board in front of you, we can now begin populating the circuit board. We are going to place and solder the lowest height components first—the resistors and the diodes. First locate resistor R1. Resistor R1 has a value of 270 K ohms and is a 5% tolerance type. Now take a look for a resistor whose first color band is red, violet, yellow, and gold. From the chart you will notice that red is represented by the digit (2), the second color band is violet, which is represented by (7). Notice that the third color is yellow and the multiplier is (10,000), so (2) (7) × 10,000 = 270,000 or 270 K ohms.

Go ahead and locate where resistor R1 goes on the PC and install it; next solder it in place on the PC board, and then with a pair of end-cutters trim the excess component leads flush to the edge of the circuit board. Next locate the remaining resistors and install them in their respective locations on the main controller PC board. Solder the resistors to the board, and remember to cut the extra lead with your end-cutter.

Now, let us begin installing the capacitors. Capacitors will usually have their actual values printed on them. Often small capacitors will not have enough room for the entire value to be printed so they will have a three digit code (see Table 8-2). Let us begin with C4, C5 and C7, these are 100 nF capacitors, which may be marked (104). Go ahead and place them in their respective locations on the PC board. Next locate C3 and C10, these are 220 pF capacitors, which may be marked (221) Install these capacitors in their proper locations and solder them in place on the PC board. Next, take your end-cutters and trim the excess capacitor leads flush to the edge of the PC board. The remaining capacitors are all electrolytic types and will have a plus or minus marking at one end of the capacitor.

Table 8-2
Three-digit capacitor codes

pF	nF	µF	CODE	pF	nF	µF	CODE
1.0			1R0	3,900	3.9	.0039	392
1.2			1R2	4,700	4.7	.0047	472
1.5			1R5	5,600	5.6	.0056	562
1.8			1R8	6,800	6.8	.0068	682
2.2			2R2	8,200	8.2	.0082	822
2.7			2R7	10,000	10	.01	103
3.3			3R3	12,000	12	.012	123
3.9			3R9	15,000	15	.015	153
4.7			4R7	18,000	18	.018	183
5.6			5R6	22,000	22	.022	223
6.8			6R8	27,000	27	.027	273
8.2			8R2	33,000	33	.033	333
10			100	39,000	39	.039	393
12			120	47,000	47	.047	473
15			150	56,000	56	.056	563
18			180	68,000	68	.068	683
22			220	82,000	82	.082	823

(Continued)

Table 8-2—cont'd

Three-digit capacitor codes

pF	nF	µF	CODE	pF	nF	µF	CODE
27			**270**	100,000	100	.1	**104**
33			**330**	120,000	120	.12	**124**
39			**390**	150,000	150	.15	**154**
47			**470**	180,000	180	.18	**184**
56			**560**	220,000	220	.22	**224**
68			**680**	270,000	270	.27	**274**
82			**820**	330,000	330	.33	**334**
100		.0001	**101**	390,000	390	.39	**394**
120		.00012	**121**	470,000	470	.47	**474**
150		.00015	**151**	560,000	560	.56	**564**
180		.00018	**181**	680,000	680	.68	**684**
220		.00022	**221**	820,000	820	.82	**824**
270		.00027	**271**		1,000	1	**105**
330		.00033	**331**		1,500	1.5	**155**
390		.00039	**391**		2,200	2.2	**225**
470		.00047	**471**		2,700	2.7	**275**
560		.00056	**561**		3,300	3.3	**335**
680		.00068	**681**		4,700	4.7	**475**
820		.00082	**821**			6.8	**685**
1,000	1.0	.001	**102**			10	**106**
1,200	1.2	.0012	**122**			22	**226**
1,500	1.5	.0015	**152**			33	**336**
1,800	1.8	.0018	**182**			47	**476**
2,200	2.2	.0022	**222**			68	**686**
2,700	2.7	.0027	**272**			100	**107**
3,300	3.3	.0033	**332**			157	**157**

Code = 2 significant digits of the capacitor value + number of zeros to follow

For values below 10 pF, use "R" in place of decimal e.g. 8.2 pF = 8R2

10 pF = 100

100 pF = 101

1,000 pF = 102

22,000 pF = 223

330,000 pF = 334

1 µF = 105

The capacitors have polarity, and this must be observed in order for the circuit to operate correctly. When installing the electrolytic capacitors pay close attention to the marking and install them correctly, referring to both the schematic and parts layout diagrams.

Once you have installed all of the electrolytic capacitors you can solder them in place on the circuit board. Remember to trim the excess leads; use your end-cutter to trim the leads flush to the edge of the circuit board.

Locate the potentiometer R8, a 10 K trim-pot and install it on the circuit board and then solder it in place.

The Telephone Amplifier circuit contains two integrated circuits (ICs). ICs are often static sensitive, so they must be handled with care. Use a grounded anti-static wriststrap and stay seated in one location when handling the integrated circuits. Take out a cheap insurance policy by installing IC sockets for each of the ICs. In the event of a possible circuit failure, it is much easier to simply unplug a defective IC than trying to unsolder 14 or 16 pins from a PC board without damaging the board. Integrated circuits have to be installed correctly if the circuit is going to work properly. IC packages will have some sort of markings which will help you orient them on the PC board. An IC will have either a small indented circle, a cutout, or notch at the top end of the IC package. Pin 1 of the IC will be just to the left of the notch, or cutout. Refer to the manufacturer's pin-out diagram, as well as the schematic when installing these parts. If you doubt your ability to correctly orient the ICs, seek the help of a knowledgeable electronics enthusiast. Go ahead and install the two IC sockets in their respective locations on the printed circuit board.

Now is a good time to take a short rest and when we come back, we will inspect the circuit board for any "cold" solder joints and any "shorts." Pick up the circuit board with the foil side facing upwards toward you. Examine the circuit board carefully, looking for any solder joins that look dark or blobby. If you see a solder joint that looks bad, use a desoldering tool and remove the solder and then resolder the joint all over again. Also inspect the foil side of the board for any short circuits which may be caused by stray or cut component leads, as they can often cause a bridge across the circuit traces.

Once you have finished building the amplifier circuit, you can install it in a small plastic or metal chassis box or enclosure. Install the integrated circuits into their respective sockets, paying close attention to the proper orientation of the IC in its socket. Mount the printed circuit board in the center of the chassis box using ¼″ plastic standoffs. You will most likely want to install a power on-off switch on the front of the enclosure. A 2- to 3″ speaker can be mounted at the top of the enclosure, but be sure to drill holes for the sound to exit the enclosure box. Install a plastic battery

holder at the bottom of the chassis box. Finally, install an ⅛″ phone jack for the telephone pickup coil on the front panel of the chassis box. When finished, apply power to the circuit and plug in your telephone pickup coil and you will be ready to listen to a phone conversation.

Your telephone amplifier is now ready to serve you. Simply plug the telephone pickup into the ⅛″ phone jack on the front panel, apply power to the circuit, and you are ready to amplify any phone conversation. Your older relatives will be quite pleased to use this device, as now they can hear the phone conversations loud and clear. The direct phone amplifier is also good in a party or group conversation when a number of people wish to contribute to the conversation.

When using the telephone amplifier, be sure to keep the speaker away from the telephone handset to prevent any feedback. The pickup will be affected by strong magnetic fields—mains wiring, a computer monitor, and TV set. Low roll-off filters (to reduce the 50-Hz hum) could have been included in the circuit just as the high roll-off filters have been.

If the telephone amplifier does not work immediately, it is likely that you inserted a component incorrectly or you have made some poor solder joints. First you will need to inspect the PC board for possible cold solder joints and short circuits. Pick up the PC board with the foil side facing upwards and facing you. Carefully inspect the solder joints, which should look smooth, clean, and bright. If any of the solder joints look dull, dark, or blobby, then you should remove the solder with a solder sucker or a solder wick and then resolder the joint all over again. Now we will inspect the PC board for possible shortcircuits. Shortcircuits can be caused by small solder blobs bridging the circuit traces or from cut component leads which can often stick to the circuit board once they have been cut. A sticky residue from solder will often trap component leads across the circuit traces. Look carefully for any bridging wires across circuit traces.

Now reapply power to the circuit and check to see if the amplifier is working; if not you may want to check the circuit board to make sure that you have not installed a particular component at the wrong location. Next, check the polarity of the capacitors, with respect to its installation, and finally check to make sure that the integrated circuits have been installed correctly.

Figure 8-2 *Telephone amplifier board. Courtesy of Velleman Kits*

Project 2: Telephone amplifier II—Direct connect

The second telephone amplifier in this project is a Direct Connect audio amplifier, as shown in Figure 8-2.

The circuit features high performance and small size, with both speaker and line outputs. The circuit is not really a speaker-phone, but a high performance isolated input amplifier that connects directly to a telephone line. Unlike many types of amplifiers, this one provides both speaker output and line output. Speaker output is perfect for "group listen" applications with standard telephones

for group gatherings without the inconvenience and interruption of vox operated speaker-phones.

This telephone amplifier also features line level output designed to be connected to the auxillary input of any standard audio mixer or amplifier. This makes it perfect for radio stations, home broadcasters, and larger applications that need to connect their phone line into their systems. It features a standard isolation transformer input for zero loading of the phone line. Speaker output is a half watt into 8 ohms. Line output is 0 dB (.775 Vrms). The circuit runs on 7–10 VAC or 9–12 VDC at 150 mA. Board measures 2.5 × 2.1 inches. Designed to be built into existing equipment, but the amplifier can also be used as a stand-alone amplifier by placing the circuit in a small enclosure. (Group listen is great, no speaker-phone interruptions, yet everyone in my office can hear!)

Circuit description

The telephone amplifier schematic is depicted in Figure 8-3. The amplifier circuit is connected directly to the telephone line through the resistor and diode network formed by resistors R1 and R2 placed across the phone line. Resistors R1 and R2 are placed in series with two capacitors and fed to resistor R4. Two silicon diodes are placed back-to-back across the input to the isolation transformer at T1. Transformer T1 is a 600 to 600 ohm isolation/matching transformer. The secondary

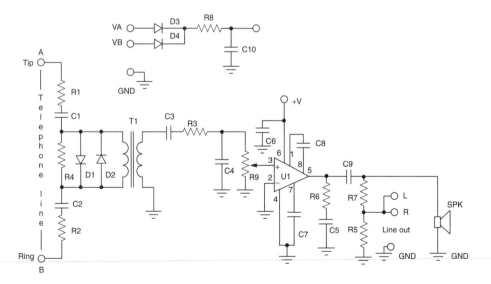

Figure 8-3 *Telephone amplifier-direct connect. Courtesy of Velleman Kits*

of the transformer is then fed to coupling capacitor C3 and then on to resistor R3. Components C3, R3, and C4 ahead of the potentiometer at R9 form an input filter. The input from the network is next fed to the potentiometer which is used to adjust the input signal to the IC amplifier chip, at U1. The output of the op-amp at pin is coupled to the output network formed by C5 and R6. The output coupling capacitor at C9 is fed to the resistor divider network and the speaker. The resistor network consists of resistors R5 and R7. This resistor network provides a line output. The line output is provided in the event that you wish to couple the telephone amplifier to a tape recorder or to a mixer circuit. Small radio station broadcasters may wish to feed the telephone amplifier into a mixer which is then coupled to the transmitter's audio input stage as a form of manual phone patch of sorts. The amplifier circuit is powered via diodes D3 and D4, which are fed from the telephone input connections through resistor R8 and C10. The telephone amplifier circuit was constructed on a small glass epoxy circuit board for reliable operation.

Circuit assembly

Locate a clean, well-lit worktable or workbench, a small pencil tip soldering iron, some 60/40 tin–lead solder and small can of Tip Tinner from your local RadioShack store. Tip Tinner is used to clean and prep the soldering tip. You will also want to locate some small tools such as a pair of small end cutters, a few small screwdrivers, and a small Exacto type knife. Heat up your soldering iron and place all the project parts in front of you, along with your printed circuit board. When building electronic circuit boards with integrated circuits it is a good idea to install IC sockets on the PC board. In the event of an IC failure it is then a simple matter to simply remove the defective IC and just install a new one. Integrated circuit sockets are inexpensive as well as a good insurance policy.

Place the your single-sided PC board in front of you, along with the schematic and any parts layout diagrams. Next, place all of the parts in front of you and create a check-off list to make sure you have all the components you will need for this project. It is usually a good idea to begin installing the lowest profile components such as resistors first. Resistors are color coded, and have three or four color bands, which begin at one end of the resistor body. See the resistor color code value chart shown in Table 8-1. The first resistor is the first number value, followed by the second number or code value. The third color band is the multiplier band, which is usually 20, 10 or 5%. If the resistor has no fourth color band then the resistor has a 20% value. If the fourth color band is silver then the resistor has a 10% value or tolerance, and if the fourth color band is gold then the resistor has a 5% tolerance value. Let us go ahead and begin installing the resistors. R1, R2, and R3 are all 10 K ohm or 10,000 ohm resistors. Since resistors have no polarity, you can install them in any direction in their proper locating holes (see the layout diagram). Next locate and install R4 and R5 the 1 K ohm resistors, followed by R6 and R7. Once all the resistors have been placed in their proper locations, you can solder them in place on the PC board. Take your small end-cutters and trim the excess component leads from the circuit board.

Now, let us begin installing the capacitors. Capacitors will have usually have their actual values printed on them. Often small capacitors will not have enough room for the entire value to be printed so they will have a three digit code (see Table 8-2). Let us begin with C1 and C2, these are 22 nF or 22 nanofarad capacitors. Go ahead and place them in their respective locations on the PC board. Next locate C3, an 18 nF capacitor and then locate capacitors C4 and C5 which are both 47 nF types. Finally locate C6, a 100 nF capacitor and then solder all of these capacitors to the circuit board. Now, we will identify and install the electrolytic capacitors. Capacitors C7 through C10 are all electrolytic types and will have a plus or minus marking at one end of the capacitor. The capacitors have polarity, and this must be observed in order for the circuit to operate correctly. When installing the electrolytic capacitors pay close attention to the marking and install them correctly, referring to both the schematic and parts layout diagrams. Once you have installed all of the electrolytic capacitors you can solder them in place on the circuit board. Remember to trim the excess leads, use your end-cutter to trim the leads flush to the edge of the circuit board.

The phone line amplifier circuit uses four silicon diodes located at D1, D2, D3, and D4. Diodes have polarity, so when installing them be sure to observe the correct polarity. Diodes will usually have a black or white band at one end of the diode body. The colored band specifies the cathode end of the diode. Observe the proper orientation using your schematic and parts layout diagrams.

Locate the potentiometer RV1 and install it on the circuit board and then solder it in place. Finally locate the isolation or matching transformer, this 600 to 600 ohm transformer will have both a primary and secondary winding. You will want to make sure that you use the outside windings and not the center tap windings, if your particular transformer has them. The matching transformer is installed between diode D1 and D2 and capacitor C3. Finally, use some wire to connect the speaker to the terminal block output.

Now is a good time to take a short rest and when we come back, we will inspect the circuit board for any coldsolder joints and any shorts. Pick up the circuit board with the foil side facing upwards toward you. Examine the circuit board carefully, looking for any solder joins that look dark or blobby. If you see a solder

joint that looks bad, use a desoldering tool and remove the solder and then resolder the joint all over again. Also inspect the foil side of the board for any short circuits which may be caused by stray or cut component leads,as they can often create a bridge across the circuit traces. Once you are satisfied that the circuit board looks good you can attach two wires from the phone line to the input of the phone line amplifier circuit at the locations marked A and B.

Your Direct Connection telephone amplifier is now ready to serve you. When the phone is on-the-hook the amplifier will be silent, but once the phone is picked up, you will be ready to hear the conversation and amplify it so others in the room can hear the phone conversation out loud. Your older relatives will be quite pleased to use this device, so now they can hear the phone conversation loud and clear. The direct phone amplifier is also good in a party or group conversation when a number of people wish to contribute to the conversation.

Chapter 9

Phone Recorder Switch Project

Parts list

Phone Record Switch

R1   220 ohm, $\frac{1}{4}$ W,
   5% resistor

R2   22 K ohm, $\frac{1}{4}$ W,
   5% resistor

R3, R4, R5   10 Megohm,
   $\frac{1}{4}$ W, 5% resistor

R6   1 Megohm trim-pot

C1, C2   100 nF,
   200 V mylar
   capacitors (104)

D1, D2, D3   1N4148
   silicon diode

D4, D5, D6   1N4148
   silicon diode

D7   10 V zener diode

D8   1N4004 silicon diode

Q1   BC548 transistor or
   equivalent

Q2   VN10 K N-channel
   DMOSFET

S1   DPDT switch – PCB –
   mount

J1 $\frac{1}{8}$" microphone plug
   – plug into recorder
   (specific to recorder)

J2   subminiature plug –
   REMOTE control

Misc   PC board, wire,
   hardware, enclosure

Model CK208 – Automatic
   Phone Record Switch

Ever want to tape that annoying or obscene phone call? Want to find out who is making those long distance calls from your phone? Or how about taping that call just to remember what was said? Many times we wish to remember exactly what was said during a phone conversation. Whether it is directions to grandma's house or repair instructions for some other manufacturer's electronic kits, the Phone Recorder Switch will become your "silent secretary," recording calls for future playback. The Phone Recorder Switch works in conjunction with a small tape recorder to record both sides of your phone conversation. When the phone is picked up from the switch-hook the Phone Recorder Switch activates your small tape recorder to begin recording the phone conversation (see Figure 9-1). Typical uses for the Phone Recorder Switch may include recording directions to a particular location, verifying calls made from or to your telephone, taping annoying or obscene calls, or to record the detail of a business call you do not want to forget.

This project serves well as an introduction to your telephone system and the tape recorder as well as the more basic operation of transistors, FETs as switches,

Figure 9-1 *Automatic telephone recording switch*

and diode operation. This circuit should work on virtually all phone systems. The "Hands Off" or automatic recorder switch project provides clean, clear audio quality, and ease of assembly of this build-it-yourself telephone recorder.

Note that your tape recorder must have both a microphone or MIC socket and a REMOTE jack on it so that this project can plug into the recorder and control it. Cheaper tape recorders tend to omit these external jacks and sockets, which are needed for this type of project. This circuit is designed to work for the newer 1.5-V and 3-V tape recorders as well as the usual 6-V or 12-V ones.

Circuit description

The Phone Recorder Switch circuit shown in Figure 9-2 is divided into two main parts and these can be easily seen in the diagram.

On the left are the connections to each telephone line and to the microphone or MIC socket of the tape recorder. The diode and capacitors ensure that no DC voltages pass through to the input of the MIC while the RC network clips large transients. On the right is the circuit which detects when the handset has been lifted and which then turns on the FET. The trim pot adjusts the voltage level of this circuit. The voltage of the normal telephone line is between 40 and 60 V (depending on country and telephone system).

When you pick up the handset of the telephone the voltage falls to between 6 and 12 V. It is this drop in voltage which is used to control the tape recorder through the REMOTE connector. When the line voltage

is high the base of the BC548 is pulled high so the transistor is turned "on." This pulls the gate of the FET down to less than 1 V, which then shuts off the FET. Note that N-channel enhancement FETs need drain bias positive and a positive gate to turn on. When the line voltage falls (that is, the hand-set is picked up) the BC548 must turn off; adjust the trim-pot if it does not. So the FET gate potential rises to the 10 volts set by the Zener diode. This turns the FET on to high efficiency conduction mode. Different recorders may have different polarities in their REMOTE sockets. To allow for this a PCB mounted switch has been added to the board which will reverse the polarity of the REMOTE just by switching it.

Circuit assembly

Before we begin building the Telephone Record Switch, let's first find a clean, well-lit worktable or workbench. Next we will gather a small 25–30 W pencil tipped soldering iron, a length of 60/40 tin–lead solder, and a small canister of Tip Tinner, available at your local RadioShack store; Tip Tinner is used to condition the soldering tip between solder joints. You drive the soldering tip into the compound and it cleans and prepares the soldering tip. Next grab a few small hand tools, and try and locate small flat blade and Phillips screwdrivers. A pair of needle-nose pliers, a pair of tweezers along with a magnifying glass, and a pair of end-cutters and we will begin constructing the project. Now locate your schematic diagram and parts layout diagram along with all of the components needed to build the project. Once all the components are in front

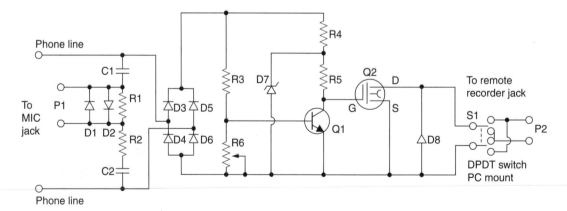

Figure 9-2 *Auto telephone switch for tape recording. Courtesy of DIY Electronics*

of you, you can check them off against the project parts list to make sure you are ready to start building the project. Now, locate Tables 9-1 and 9-2, and place them in front of you. Table 9-1 illustrates the resistor code chart and how to read resistors, while Table 9-2 illustrates the capacitor code chart which will aid you in constructing the project.

Refer now to the resistor color code chart shown in Table 9-1, which will help you identify the project resistors. Note that each resistor will have either three or four color bands, which begin at one end of the resistor's body. The first colored band represents the resistor's first digit, while the second color represents the second digit of the resistor's value. The third colored band represents the resistor's multiplier value, and the fourth color band represents the resistor's tolerance value. If there is no fourth colored band then the resistor has a 20% tolerance value. If the resistor's fourth band is silver then the resistor has a 10% tolerance value and if the fourth band is gold then the resistor's tolerance value will be 5%.

We are now ready to begin assembling the project, so let's get started. The prototype project was constructed on a single-sided printed circuit board or PCB. With your PC board in front of you, we can now begin populating the circuit board. We are going to place and solder the lowest height components first, the resistors

and the diodes. First locate our first resistor R1. Resistor R1 has a value of 220 ohms and is a 5% tolerance type. Now take a look for a resistor whose first color band is red, red, brown, and gold. From the chart you will notice that red is represented by the digit (2), and again the second band is red, which is represented by (2). Notice that the third color is brown and the multiplier is (10), so $(2)(2) \times 10 = 220$ ohms. Go ahead and place resistor R1 in its respective location on the PC board. Now solder the resistor to the circuit board and then use a pair of end-cutters to trim the excess component leads, flush to the edge of the circuit board. Now locate and install the remaining resistors on to the circuit board. Solder the resistors in place and remember to trim the excess lead lengths.

Note that capacitors generally come from two major groups or types. Non-polar types, such as the ones in this project, have no polarity and can be placed in either direction on the circuit board. The other type of capacitors are called polarized types; these capacitors have polarity and have to be inserted on the board, with respect to their polarity marking, in order for the circuit to operate correctly. On these types of capacitors, you will observe that the components actually have a plus or minus marking or a black or white band with a plus or minus marking on the body of the capacitors.

Table 9-1
Resistor color code chart

Color Band	1st Digit	2nd Digit	Multiplier	Tolerance
Black	0	0	1	
Brown	1	1	10	1%
Red	2	2	100	2%
Orange	3	3	1,000 (K)	3%
Yellow	4	4	10,000	4%
Green	5	5	100,000	
Blue	6	6	1,000,000 (M)	
Violet	7	7	10,000,000	
Gray	8	8	100,000,000	
White	9	9	1,000,000,000	
Gold			0.1	5%
Silver			0.01	10%
No color				20%

Next refer to Table 9-2, this chart helps to determine value of small capacitors.

Many capacitors are quite small in physical size. Therefore, manufacturers have devised a three-digit code which can be placed on the capacitor instead of the actual value. The code occupies a much smaller space on the capacitor and can usually fit nicely on the capacitor body, but without the chart the builder would have a difficult time determining the capacitor's value.

Now locate the capacitors from the component pile and try and locate capacitor C1, which should be marked with a (104); this capacitor has a 100 nF value. Let us go ahead install C1 on the circuit board and then solder it in place on the circuit board. Now take your end-cutters and cut the excess component leads flush to the edge of the circuit board. Once you have installed C1, you can move on to installing the remaining capacitors.

Table 9-2
Three-digit capacitor codes

pF	nF	μF	CODE	pF	nF	μF	CODE
1.0			**1R0**	3,900	3.9	.0039	**392**
1.2			**1R2**	4,700	4.7	.0047	**472**
1.5			**1R5**	5,600	5.6	.0056	**562**
1.8			**1R8**	6,800	6.8	.0068	**682**
2.2			**2R2**	8,200	8.2	.0082	**822**
2.7			**2R7**	10,000	10	.01	**103**
3.3			**3R3**	12,000	12	.012	**123**
3.9			**3R9**	15,000	15	.015	**153**
4.7			**4R7**	18,000	18	.018	**183**
5.6			**5R6**	22,000	22	.022	**223**
6.8			**6R8**	27,000	27	.027	**273**
8.2			**8R2**	33,000	33	.033	**333**
10			**100**	39,000	39	.039	**393**
12			**120**	47,000	47	.047	**473**
15			**150**	56,000	56	.056	**563**
18			**180**	68,000	68	.068	**683**
22			**220**	82,000	82	.082	**823**
27			**270**	100,000	100	.1	**104**
33			**330**	120,000	120	.12	**124**
39			**390**	150,000	150	.15	**154**
47			**470**	180,000	180	.18	**184**
56			**560**	220,000	220	.22	**224**
68			**680**	270,000	270	.27	**274**
82			**820**	330,000	330	.33	**334**
100		.0001	**101**	390,000	390	.39	**394**
120		.00012	**121**	470,000	470	.47	**474**
150		.00015	**151**	560,000	560	.56	**564**
180		.00018	**181**	680,000	680	.68	**684**
220		.00022	**221**	820,000	820	.82	**824**
270		.00027	**271**		1,000	1	**105**
330		.00033	**331**		1,500	1.5	**155**

Chapter Nine: Phone Recorder Switch Project

Table 9-2—cont'd

Three-digit capacitor codes

pF	nF	µF	CODE	pF	nF	µF	CODE
390		.00039	**391**		2,200	2.2	**225**
470		.00047	**471**		2,700	2.7	**275**
560		.00056	**561**		3,300	3.3	**335**
680		.00068	**681**		4,700	4.7	**475**
820		.00082	**821**			6.8	**685**
1,000	1.0	.001	**102**			10	**106**
1,200	1.2	.0012	**122**			22	**226**
1,500	1.5	.0015	**152**			33	**336**
1,800	1.8	.0018	**182**			47	**476**
2,200	2.2	.0022	**222**			68	**686**
2,700	2.7	.0027	**272**			100	**107**
3,300	3.3	.0033	**332**			157	**157**

Code = 2 significant digits of the capacitor value + number of zeros to follow

For values below 10 pF, use "R" in place of decimal e.g. 8.2 pF = 8R2

10 pF = 100

100 pF = 101

1,000 pF = 102

22,000 pF = 223

330,000 pF = 334

1 µF = 105

This project contains seven silicon diodes at D1 through D7. These diodes also have polarity which must be observed if the circuit is going to work correctly. You will notice that each diode will have a black or white colored band at one edge of the diode's body. The colored band represents the diode's cathode lead. Check your schematic and parts layout diagram when installing these diodes on the circuit board. Note that D7 is a 10 V Zener diode, while D1 through D6 are 1N4148 diodes, and D8 is a 1N4004 silicon diode. Place all the diodes on the circuit board and solder them in place on the PC board. Then cut the excess component leads with your end-cutters.

The Telephone Recorder Switch circuit also utilizes one transistor at Q1 and an N-Channel DMOSFET at Q2. Transistors generally have three leads, a Base lead, a Collector lead, and an Emitter lead. Refer to the schematic and you will notice that there will be two diagonal leads pointing to a vertical line. The vertical line is the transistor's Base lead and the two diagonal

lines represent the Collector and the Emitter. The diagonal lead with the arrow on it is the Emitter lead, and this should help you install the transistor onto the circuit board correctly. The FET also has three leads but they are labeled differently, as seen in the schematic. There is a Gate lead, a Drain lead, and a Source lead, see pin-out diagram. Be sure not to mix up the VN10K FET with the BC548; make sure you can identify these correctly before installation. Go ahead and install the transistor and the FET on the PC board and solder them in place, then follow up by cutting the excess component leads.

Next, go ahead and mount the DPDT PC mounted switch on the circuit board. Now, cut it into two lengths of 10″ audio cable; these are used to make the connections to the telephone line and to the tape recorder. Connect one end of each length of wire into the PCB at the places marked TO REMOTE, and MIC. Connect the following connectors to the other ends of the wires: to the free end of the TO REMOTE wire attach the 2.5-mm diameter plug. (This plug is the smaller of the two plugs. It plugs into the REMOTE socket of the tape recorder; to the free end of the MIC wire attach the 3.5-mm diameter plug. (This plug is the bigger of the two plugs supplied.) It plugs into the MIC socket of the tape recorder. Use the alligator clip cable for the TO EACH LINE pads. Cut the cable in half. Attach each half into the pads. Each clip goes onto one of the two telephone lines coming into your house. Note that this Phone Recorder Switch is attached in parallel with the phone line. If you find you have four lines going to your phone then the two lines you want will either be the outer two (most likely) or the inner two. You will have to experiment.

When the project is completed place the phone record switch and your tape recorder next to the phone. Plug the two sockets (MIC and REMOTE) into the recorder. Attach one alligator clip to each phone line (you may have to remove some plastic or covering from the phone line cables to get a good electrical connection.) Put in a cassette tape and push "play." Provided the project has been assembled together correctly, either of two things will happen: the tape in the recorder will start to play or it will not.

If it does not play then pick up the phone handset. If the tape now starts to play then the Kit is working. Put the handset down, depress the play and record buttons and the tape will now record when the handset is raised. If the tape plays then either of two things needs adjustment: Move the position of the trim-pot across its range of positions and see if this stops the playing. If it does, then lift the handset to see if the playing starts. It should. The kit is ready for use. If adjustment of the trim-pot does nothing then the REMOTE switch needs to be switched to the other position. Do this and repeat the steps as outlined above. The kit should now work.

Troubleshooting

If your Telephone Recorder Switch does not appear to function correctly, you will need to disconnect the circuit from the phone line and carefully inspect the circuit board. It is likely that you made a soldering or parts placement mistake during construction. Pick up the circuit board with its foil side facing upwards toward you. Carefully inspect the circuit board against the parts placement diagram and the schematic to make sure that the proper component was inserted in the correct PC hole. You may have placed the wrong resistor at a particular location. You may have inserted a diode backwards.

Once this preliminary inspection is over, you can inspect the circuit board for any possible cold solder joints or short circuits. Take a careful look at each of the PC solder joints, they should all look clean and shiny. If any of the solder joints look dull, dark, or blobby, then you unsolder the joint, remove the solder and then resolder the joint all over again. Next, examine the PC board for any short circuits. Short circuits can be caused by two circuit traces touching each other due to a stray component lead that stuck to the PC board from solder residue or from solder blobs bridging the solder traces. Once you have fully examined the circuit board, you can reconnect the circuit board back to the phone line and try out the phone transmitter once again.

U.S. residents should note that it is illegal to tape someone's telephone conversation without informing all parties that the conversation is being taped. So remember, when you pick up the phone and say hello, also inform the caller the conversation is being recorded. Remember that this project is not a toy. You may need to consult local regulations to provide you with the information necessary to use the Telephone Recorder Switch's capabilities in accordance with the law.

Chapter 10

Telephone Conferencer Unit

Parts list

Parts Bin

Telephone Conferencer Circuit

R1 10 K ohm, $\frac{1}{4}$ W, 5% resistor

R2 100 ohm, $\frac{1}{4}$ W, 5% resistor

C1, C2 10 µF, 50 V electrolytic capacitor

C3 10 µF, 15 V electrolytic capacitor

Q1 2N3904 NPN silicon transistor

S1, S2 DPST toggle switch

T1 telephone coupling transformer (Microtran T2100 or equiv.)

Misc PC board, wire, connectors, enclosure, etc.

Figure 10-1 *Conferencer circuit*

Today it is common for many people to have more than one phone line coming into the house. Usually the second line is called a "teen-age telephone" and is provided by the telephone company at a specially reduced rate. While the second telephone allows young people to talk on the telephone for hours at a time without tying up the first telephone, there is an additional benefit which can be derived by taking advantage of the fact that two independent telephone lines are available. This is the conference call, which has been available to commercial businesses for years. Of course, the telephone company would be glad to equip your telephone for conference call capability, but

they will want to charge you a monthly tariff whether you use it or not. If you build the Conferencer shown in Figure 10-1, you can add conference call capability to your telephone and you won't have to pay for its use. Ma Bell will not even know that you have it, since it has no effect on telephone performance, except to allow simultaneous communications between your two telephone lines and two others.

The Conferencer can be built at very low cost, making it a valuable addition to your telephone. It requires no external power source for its operation and is always ready for use whenever you are. The only requirement for making use of the Conferencer is to have two independent telephone lines available at a single location in your home or business.

Circuit details

In order to connect two independent telephone lines together without disturbing the DC conditions on either line, a transformer must be used. This is the heart of the circuit which allows audio information on one line to be impressed on another line with no DC connection between them. T1 is a telephone line coupling transformer, which has been designed for telephone line use. As shown in the schematic diagram of the

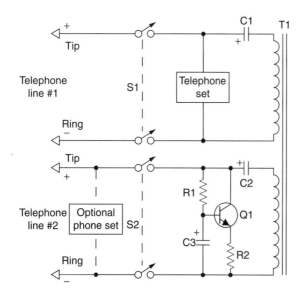

Figure 10-2 *Telephone conferencer. Courtesy of Thomson Corp*

Conferencer, Figure 10-2, each telephone line is connected to the windings of T1 through coupling capacitors.

Telephone line number #1 is connected across one winding of the transformer through S1 and electrolytic capacitor C1. The capacitor prevents any DC from flowing through the transformer winding, while coupling the audio frequency information appearing across the telephone line to the transformer winding. The DC isolation provided by C1 ensures that the normal DC conditions of the telephone line are not disturbed. In a similar manner, telephone line number two is connected to the second winding of the transformer. Each telephone line can be independently controlled through the action of S1 and S2, which have been provided to allow connection and disconnection of each telephone line as required.

An additional circuit has been included, in the second telephone line circuit. This is a DC holding circuit which provides the proper DC characteristic to hold telephone line number two in operation even though no telephone on that line is present. The circuit composed of R1, R2, C3, and Q1 acts like a resistance to DC, and as a high impedance to audio signals. The high impedance of the circuit is provided by C3, which prevents any audio signals from appearing at the base of Q1. Thus, any audio voltage appearing across the telephone line number #2 will not cause a corresponding current in Q1.

Note that only one telephone is required when using the Conferencer. If that telephone is equipped for Touch Tone dialing, it will be possible to dial telephone numbers to either line as required. If only rotary dialing is available and it is desirable to dial a number on line number #2, then a second telephone can be connected across that line for dialing purposes only. This is shown on the schematic diagram as an optional telephone set.

Once all the components are in front of you, you can check them off against the project parts list to make sure you are ready to start building the project. Now, locate Tables 10-1 and 10-2, and place them in front of you. Table 10-1 illustrates the resistor code chart and how to read resistors, while Table 10-2 illustrates the capacitor code chart, which will aide you in constructing the project.

Refer now to the resistor color code chart shown in Table 10-1, which will assist you in identifying the project resistors. Note that each resistor will have either three or four color bands, which begin at one end of the resistor's body. The first colored band represents the resistor's first digit, while the second color represents the second digit of the resistor's value. The third colored band represents the resistor's multiplier value, and the fourth color band represents the resistor's tolerance value. If there is no fourth colored band then the resistor has a 20% tolerance value. If the resistor's fourth band is silver then the resistor has a 10% tolerance value and if the fourth band is gold then the resistor's tolerance value will be 5%.

Finally we are ready to begin assembling the project, so let's get started. The prototype project was constructed on a single-sided printed circuit board or PCB. With your PC board in front of you, we can now begin populating the circuit board. We are going to place and solder the lowest height components first, the resistors and the diodes. First locate our first resistor R1. Resistor R1 has a value of 10,000 ohms and is a 5% tolerance type. Now take a look for a resistor whose first color band is brown, black, orange, and gold. From the chart you will notice that brown is represented by the digit (1), the second color band is black, which is represented by (0). Notice that the third color is orange and the multiplier is (1,000), so (1)(0) × 1,000 = 10,000 or 10 K ohms.

Go ahead and locate where resistor R1 goes on the PC and install it, next solder it in place on the PC board,

Table 10-1

Resistor color code chart

Color Band	1st Digit	2nd Digit	Multiplier	Tolerance
Black	0	0	1	
Brown	1	1	10	1%
Red	2	2	100	2%
Orange	3	3	1,000 (K)	3%
Yellow	4	4	10,000	4%
Green	5	5	100,000	
Blue	6	6	1,000,000 (M)	
Violet	7	7	10,000,000	
Gray	8	8	100,000,000	
White	9	9	1,000,000,000	
Gold			0.1	5%
Silver			0.01	10%
No color				20%

and then with a pair of end-cutters trim the excess component leads flush to the edge of the circuit board. Next locate the remaining resistors and install them in their respective locations on the PC board. Solder the resistors to the board, and remember to cut the excess lead lengths with your end-cutters.

Note that capacitors generally come from two major groups or types. Non-polar types, such as the ones in this project, which have no polarity, can be placed in either direction on the circuit board. The other type of capacitors are called polarized types, these capacitors have polarity and have to be inserted on the board with respect to their polarity marking in order for the circuit to operate correctly. On these types of capacitors, you will observe that the components actually have a plus or minus marking or a black or white band with a plus or minus marking on the body of the capacitors.

Next refer to Table 10-2. This chart helps to determine value of small capacitors. Many capacitors are quite small in physical size. Therefore, manufacturers have devised a three-digit code which can be placed on the capacitor instead of the actual value. The code occupies a much smaller space on the capacitor and can usually fit nicely on the capacitor body, but without the chart the builder would have a difficult time determining the capacitor's value.

Capacitors must be installed properly with respect to polarity in order for the circuit to work properly. Failure to install electrolytic capacitors correctly could cause damage to the capacitor as well as to other components in the circuit when power is first applied. Once you have installed these components, you can move on to installing the remaining capacitors.

Construction

The single-sided printed circuit board for the Conferencer circuit measures $2 \times 3''$. This size is based on using the transformer specified in the parts list. For this application you can use any audio coupling transformer which is rated at about 900 ohms both primary and secondary, over the frequency range of about 300 to 3,000 Hz. Be sure to allow extra room on the printed circuit board for the substitute transformer, if necessary.

Pay careful attention when inserting Q1 into the board so that the three leads are placed in the proper holes. The transistor's emitter lead connects to resistor R2, while the base lead is connected between the junction of R1 and C3. The transistor's collector lead is connected to the plus end of capacitor C2. Both capacitors C1 and C2, which are polarized components, must be installed correctly with respect to the polarity markings. If these capacitors are inadvertently placed in the board incorrectly, the telephone line DC conditions will be disturbed.

Table 10-2
Three-digit capacitor codes

pF	nF	μF	CODE	pF	nF	μF	CODE
1.0			**1R0**	3,900	3.9	.0039	**392**
1.2			**1R2**	4,700	4.7	.0047	**472**
1.5			**1R5**	5,600	5.6	.0056	**562**
1.8			**1R8**	6,800	6.8	.0068	**682**
2.2			**2R2**	8,200	8.2	.0082	**822**
2.7			**2R7**	10,000	10	.01	**103**
3.3			**3R3**	12,000	12	.012	**123**
3.9			**3R9**	15,000	15	.015	**153**
4.7			**4R7**	18,000	18	.018	**183**
5.6			**5R6**	22,000	22	.022	**223**
6.8			**6R8**	27,000	27	.027	**273**
8.2			**8R2**	33,000	33	.033	**333**
10			**100**	39,000	39	.039	**393**
12			**120**	47,000	47	.047	**473**
15			**150**	56,000	56	.056	**563**
18			**180**	68,000	68	.068	**683**
22			**220**	82,000	82	.082	**823**
27			**270**	100,000	100	.1	**104**
33			**330**	120,000	120	.12	**124**
39			**390**	150,000	150	.15	**154**
47			**470**	180,000	180	.18	**184**
56			**560**	220,000	220	.22	**224**
68			**680**	270,000	270	.27	**274**
82			**820**	330,000	330	.33	**334**
100		.0001	**101**	390,000	390	.39	**394**
120		.00012	**121**	470,000	470	.47	**474**
150		.00015	**151**	560,000	560	.56	**564**
180		.00018	**181**	680,000	680	.68	**684**
220		.00022	**221**	820,000	820	.82	**824**
270		.00027	**271**		1,000	1	**105**
330		.00033	**331**		1,500	1.5	**155**
390		.00039	**391**		2,200	2.2	**225**
470		.00047	**471**		2,700	2.7	**275**
560		.00056	**561**		3,300	3.3	**335**
680		.00068	**681**		4,700	4.7	**475**
820		.00082	**821**			6.8	**685**
1,000	1.0	.001	**102**			10	**106**
1,200	1.2	.0012	**122**			22	**226**
1,500	1.5	.0015	**152**			33	**336**
1,800	1.8	.0018	**182**			47	**476**

Table 10-2—cont'd

Three-digit capacitor codes

pF	nF	μF	CODE	pF	nF	μF	CODE
2,200	2.2	.0022	**222**			68	**686**
2,700	2.7	.0027	**272**			100	**107**
3,300	3.3	.0033	**332**			157	**157**

Code = 2 significant digits of the capacitor value + number of zeros to follow

For values below 10 pF, use "R" in place of decimal e.g. 8.2 pF = 8R2

10 pF = 100

100 pF = 101

1,000 pF = 102

22,000 pF = 223

330,000 pF = 334

1 μF = 105

There is no provision to mount S1 and S2 on the printed circuit board. The placement of these switches is left as an option for the builder. You may want to place the entire circuit within the confines of a telephone set if there is room, or you may want to house the Conferencer in a small cabinet to be located next to the telephone. When making connections to the telephone line, be sure to follow correct procedure and use the standard four prong telephone plugs or the new modular connectors, available from many electronics parts suppliers. These connectors allow easy connection to and disconnection from the telephone line, which is an important factor should the unit ever require servicing.

When making connections between the printed circuit and the telephone lines it will be necessary to observe the polarity of the telephone line. Use a DC voltmeter to determine the polarity of each line before wiring the telephone plugs. You must maintain correct polarity when connecting a Touch Tone telephone to the line and it is suggested that you follow the original wiring of the telephone set after it is disconnected to install the switch S1. Bear in mind that an optional telephone set connected across line number #2 is necessary only if you do not have Touch Tone dialing, and you wish to dial numbers on line number two.

Once the Conferencer circuit board has been completed, take a short break and then we will inspect the PC board for possible cold solder joints and short circuits. Pick up the PC board with the foil side facing upwards facing you. Carefully inspect the solder joints, they should look smooth, clean, and bright. If any of the solder joints look dull, dark, or blobby, then you should remove the solder with a solder sucker or a solder wick and then re-solder the joint all over again. Now we will inspect the PC board for possible short circuits. Short circuits can be caused from small solder blobs bridging the circuit traces or from cut component leads which can often stick to the circuit board once they have been cut. A sticky residue from solder will often trap component leads across the circuit traces. Look carefully for any bridging wires across circuit traces.

Once the inspection is complete, we can move on to installing the circuit board into an aluminum enclosure. Locate a suitable enclosure to house the circuit board within the enclosure. Arrange the circuit board and the transformer on the bottom of the chassis box. Now, locate four $1/4''$ standoffs, and mount the PC board between the standoffs and the bottom of the chassis box. If you selected a convention relay you will have to mount it on the chassis and wire it to the PC board. If you could not locate the small transformers specified, you will have to mount the transformers to the base of the chassis box and wire it to the circuit board. You could install two terminal strips to allow the two phone line connections to exit the rear of the chassis box. The two toggle switches can be mounted on the front of the chassis box.

Conferencer operation

You will be able to make and receive calls on either telephone line by closing the appropriate switch and using the telephone set connected across line number #1.

If you wish to initiate a conference call, use the following procedure: First, close S1 and make a call on line number #1. When you have established contact with party number one, ask him or her to hold while you contact party number two. Next, close S2, and dial the second number from the telephone set connected across telephone line number #1. When the second party answers, you will be able to carry on a three-way conversation. Finally, when using only rotary dial telephones, perform step two by dialing the second number from the optional telephone set connected across line No. #2 before S2 is closed. Once the number is dialed, you can close S2 and hang up telephone set number two and carry on the conference call in the normal manner.

Chapter 11

The Frustrator

Parts list

Parts Bin

The Frustrator

R1   22 ohms, $\frac{1}{4}$ W,
5% resistor

R2   150 K ohms, $\frac{1}{4}$ W,
5% resistor

R3   1 megohm, $\frac{1}{4}$ W,
5% resistor

R4   47 K ohms, $\frac{1}{4}$ W,
5% resistor

R5   2.2 K ohms, $\frac{1}{4}$ W,
5% resistor

R6   100 K ohms, $\frac{1}{4}$ W,
5% resistor

R7   2.7 megohm, $\frac{1}{4}$ W,
5% resistor

R8, R9   330 ohm, $\frac{1}{4}$ W,
5% resistor

R10   4.7 K ohms, $\frac{1}{4}$ W,
5% resistor

R11, R12   2.2 K ohms,
$\frac{1}{4}$ W, 5% resistor

R13   10 K ohms, $\frac{1}{4}$ W,
5% resistor

R14   47 K ohms, $\frac{1}{4}$ W,
5% resistor

R15, R17   100 K ohms, $\frac{1}{4}$ W,
5% resistor

R16   22 K ohms, $\frac{1}{4}$ w,
5% resistor

C1, C4, C7, C9   .1 µF,
200 V ceramic
capacitor

C2, C5   .01 µF, 10 V
ceramic capacitor

C3, C10   6.8 µF, 10 V
electrolytic capacitor

C6   10 µF, 10 V
electrolytic capacitor

C8   100 µF, 10 V
electrolytic capacitor

C11   1,000 µF, 10 V
electrolytic capacitor

D1, D2, D3, D4   1N2069
silicon diode

D5   1N5953A – 150 V
Zener diode

D6, D7, D8, D9   1N2069
silicon diode

D10, D11, D12,   1N2069
silicon diode

D13, D14   1N2069 silicon
diode

Q1   MPS A42 NPN
transistor

Q2   2N3904 NPN
transistor

Q3   2N3906 PNP
transistor

U1, U4   LM555CN timer IC

U2   CD4001B Quad 2-input
NOR gate IC

U3   CD4017B decade
Johnson counter IC

U5   LM386 audio IC

LED 1   red LED

SPKR   4-8 ohm speaker

```
T1  transformer
    115Pri/6-V s @
    300 mA

F1  ½-amp fast blow
    fuse

Misc  PC board, IC
    sockets, connectors,
    hardware, wire,
    enclosure, etc.
```

Have you been annoyed by unwanted telephone calls, especially at dinnertime, from people trying to sell you this or that? Is there ever a time when you just do not want to answer the phone, but worry that you might miss an important or urgent message? If so, the telephone Frustrator, shown in Figure 11-1, is for you.

It is called the Frustrator because that is exactly what it will do to anyone who dials your number and does not know the secret to get through to you. It will answer the call, on whatever ring you select, and the caller will hear nothing but silence. After about 30 seconds or so he or she will be automatically disconnected. Repeated attempts to get your number will result in the same thing, and your unwanted caller will go away very frustrated. It appears that your telephone or telephone line is defective.

However, if your caller knows the secret, all he or she needs to do is talk into the mouthpiece and identify him or herself. Even pressing any button (or a series of buttons in a predetermined code) on a Touch Tone phone will alert you that a friend or relative is calling. The Frustrator will connect an amplifier and speaker to the phone line and let you hear voice and Touch Tone signals. You do not even have to go near your phone. At this point, if you so desire, you have about 30 s to pick up the receiver and answer the call.

Otherwise the Frustrator will disconnect automatically. The Frustrator can be left connected to your phone line at all times, and will not interfere in any way with normal telephone operation. You always have the option of answering your telephone before or after the Frustrator does, and you can carry on a normal conversation. The Frustrator is automatically disconnected from your received call after 30 s, and resets itself to receive the next call each time you or it answers the phone.

How it works

There are four sections to the circuit, each with a specific task. Refer to the schematic in Figure 11-2.

The integrated circuit at U1 is an LM555 timer chip connected as a monostable (one-shot) multi-vibrator and detects the presence of the ringing signal (90 V rms, 20 Hz) on the telephone line. Each time the ringing signal appears, U1 is triggered and its output (pin 3) goes positive for a period of 3 s. In this way the 20 Hz high-voltage ringing signal with an on/off time of 2 s and 4 s is converted to a digital pulse train of 3 s on and 3 s off. This pulse train is used to clock decade Johnson counter U3, which counts from zero to whatever ring you have selected for the Frustrator to answer your telephone.

When the circuit is at rest, U3 is reset to a zero count by a positive pulse fed to pin 15 from either the last time the Frustrator answered the phone or the phone was used in the normal way. As a result pin 3 of U3, the decoded output for zero count, is positive (logic 1) and all other outputs are zero. Now, when the phone rings, each ring causes U3 to advance one count. Thus, the first ring causes pin 2 of U3 to go positive, the second

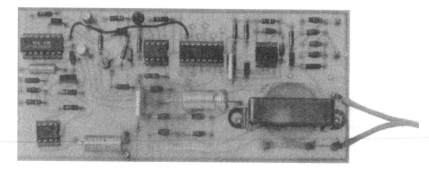

Figure 11-1 *Frustrator circuit. Courtesy of Cengage Learning*

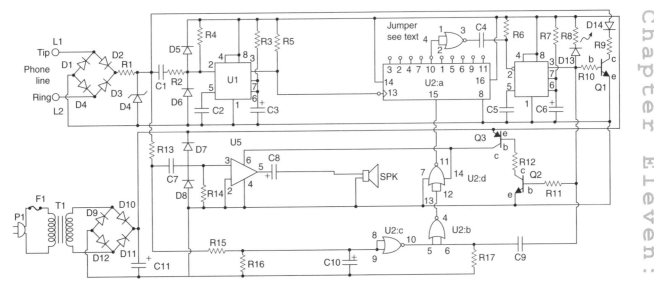

Figure 11-2 *The Frustrator. Courtesy of Cengage Learning*

ring causes pin 4 to go positive, and so on. When the pin you have selected goes positive the pulse is inverted by U2:a, differentiated, and fed to the trigger input of U4, another LM555 one-shot, multi-vibrator. U4 has a timing cycle of about 30 s. The positive output at pin 3 of this chip is used to forward bias Q1 and switch it on. The resulting current through R9 and D14, provided by the phone line, alerts the caller at the other end to hear a click, thus stopping the ring answering the phone. Thirty seconds later U4 pin 3 goes to zero, disconnecting the caller unless you have picked up the phone.

The positive output voltage of U4 is also used as a drive signal to switch on Q2 and Q3, thus applying B+ voltage to pin 6 of power amplifier U5. This amplifies the signals on the telephone line only during the "on" time of U4, and lets you hear what is going on even though you are not near your telephone. At all other times U5 is not powered, and the loudspeaker is silent.

Digital logic fed to U1:c or U2:b is used to reset U3 back to zero count after the Frustrator has done its job. U1:c detects when your telephone receiver is lifted off the hook, and U2:b detects when U4 has automatically set to receive the next call. An LED connected to the output of U4 alerts you to the status of the circuit at all times. When the Frustrator is on standby, waiting for a call, the LED will be lighted. During the 30-s call-answering sequence the LED will be extinguished.

Circuit assembly

Before we begin building the Frustrator, you will first need to find a clean, well-lit worktable or workbench. Next we will gather a small 25–30 W pencil tipped soldering iron, a length of 60/40 tin/lead solder, and a small canister of Tip Tinner, available at your local RadioShack store. Tip Tinner is used to condition the soldering tip between solder joints. You drive the soldering tip into the compound and it cleans and prepares the soldering tip. Next grab a few small hand tools, try and locate small flat blade and Phillips screwdrivers. A pair of needle-nose pliers, a pair of tweezers along with a magnifying glass, and a pair of end-cutters and we will begin constructing the project. Now locate your schematic diagram and parts layout diagram along with all of the components needed to build the project. Once all the components are in front of you can check them off again the project parts list to make sure you are ready to start building the project. The Frustrator can be constructed on a small printed circuit board or even a perf-board, since layout is not critical.

Locate Tables 11-1 and 11-2, and place them in front of you. Table 11-1 illustrates the resistor code chart and how to read resistors, while Table 11-2 illustrates the capacitor code chart, which will aid you in constructing the project.

Refer now to the resistor color code chart shown in Table 11-1, which will help you to identify the

Table 11-1
Resistor color code chart

Color Band	1st Digit	2nd Digit	Multiplier	Tolerance
Black	0	0	1	
Brown	1	1	10	1%
Red	2	2	100	2%
Orange	3	3	1,000 (K)	3%
Yellow	4	4	10,000	4%
Green	5	5	100,000	
Blue	6	6	1,000,000 (M)	
Violet	7	7	10,000,000	
Gray	8	8	100,000,000	
White	9	9	1,000,000,000	
Gold			0.1	5%
Silver			0.01	10%
No color				20%

Table 11-2
Three-digit capacitor codes

pF	nF	µF	Code	pF	nF	µF	Code
1.0			1R0	3,900	3.9	.0039	392
1.2			1R2	4,700	4.7	.0047	472
1.5			1R5	5,600	5.6	.0056	562
1.8			1R8	6,800	6.8	.0068	682
2.2			2R2	8,200	8.2	.0082	822
2.7			2R7	10,000	10	.01	103
3.3			3R3	12,000	12	.012	123
3.9			3R9	15,000	15	.015	153
4.7			4R7	18,000	18	.018	183
5.6			5R6	22,000	22	.022	223
6.8			6R8	27,000	27	.027	273
8.2			8R2	33,000	33	.033	333
10			100	39,000	39	.039	393
12			120	47,000	47	.047	473
15			150	56,000	56	.056	563
18			180	68,000	68	.068	683
22			220	82,000	82	.082	823
27			270	100,000	100	.1	104
33			330	120,000	120	.12	124
39			390	150,000	150	.15	154
47			470	180,000	180	.18	184
56			560	220,000	220	.22	224

Table 11-2—cont'd
Three-digit capacitor codes

pF	nF	μF	Code	pF	nF	μF	Code
68			**680**	270,000	270	.27	**274**
82			**820**	330,000	330	.33	**334**
100		.0001	**101**	390,000	390	.39	**394**
120		.00012	**121**	470,000	470	.47	**474**
150		.00015	**151**	560,000	560	.56	**564**
180		.00018	**181**	680,000	680	.68	**684**
220		.00022	**221**	820,000	820	.82	**824**
270		.00027	**271**		1,000	1	**105**
330		.00033	**331**		1,500	1.5	**155**
390		.00039	**391**		2,200	2.2	**225**
470		.00047	**471**		2,700	2.7	**275**
560		.00056	**561**		3,300	3.3	**335**
680		.00068	**681**		4,700	4.7	**475**
820		.00082	**821**			6.8	**685**
1,000	1.0	.001	**102**			10	**106**
1,200	1.2	.0012	**122**			22	**226**
1,500	1.5	.0015	**152**			33	**336**
1,800	1.8	.0018	**182**			47	**476**
2,200	2.2	.0022	**222**			68	**686**
2,700	2.7	.0027	**272**			100	**107**
3,300	3.3	.0033	**332**			157	**157**

Code = 2 significant digits of the capacitor value + number of zeros to follow

For values below 10 pF, use "R" in place of decimal e.g. 8.2 pF = 8R2

10 pF = 100
100 pF = 101
1,000 pF = 102
22,000 pF = 223
330,000 pF = 334
1 μF = 105

project resistors. Note that each resistor will have either three or four color bands, which begin at one end of the resistor's body. The first colored band represents the resistor's first digit, while the second color represents the second digit of the resistor's value. The third colored band represents the resistor's multiplier value, and the fourth color band represents the resistor's tolerance value. If there is no fourth colored band then the resistor has a 20% tolerance value. If the resistor's fourth band is silver then the resistor has a 10% tolerance value and if the fourth band is gold then the resistor's tolerance value will be 5%.

Finally we are ready to begin assembling the project, so let's get started. The prototype project was constructed on a single-sided printed circuit board or PCB. With your PC board in front of you, we can now begin populating the circuit board. We are going to place and solder the lowest height components first, the resistors and diodes. First locate our first resistor R1. Resistor R1 has a value of 22 ohms and is a ½-W 5% tolerance type. Now take a look for a resistor whose first color band is red, red, black, and gold. From the chart you will notice that red is represented by the digit (2), and again the second band is red, which is represented by (2). Notice that the third color is black and the multiplier is (0), so $(2)(2) \times 1 = 22$ ohms. Install resistor R1 in its respective location on the PC board. Now take a pair of end-cutters and trim the excess resistor leads. Cut the leads flush to the edge of the circuit board. Now, locate and install the remaining resistors in their correct positions on the PC board. Then use your end-cutters and remove the extra lead lengths.

Note that capacitors generally come from two major groups or types. Non-polar types, such as the ones in this project, which have no polarity, can be placed in either direction on the circuit board. The other type of capacitors are called polarized types, these capacitors have polarity and have to be inserted on the board with respect to their polarity marking in order for the circuit to operate correctly. On these types of capacitors, you will observe that the components actually have a plus or minus marking or a black or white band with a plus or minus marking on the body of the capacitors.

Next refer to Table 11-2, this chart helps to determine value of small capacitors. Many capacitors are quite small in physical size. Therefore manufacturers have devised a three-digit code which can be placed on the capacitor instead of the actual value. The code occupies a much smaller space on the capacitor and can usually fit nicely on the capacitor body, but without the chart the builder would have a difficult time determining the capacitor's value.

Now locate the capacitors from the component pile and try and locate capacitor C1, which should be marked with a (104), this capacitor has a .1 μF value. Let's go ahead and install C1 on the circuit board and then solder it in place on the circuit board. Now take your end-cutters and cut the excess capacitor leads flush to the edge of the circuit board. Once you have installed C1, you can move on to installing the remaining low profile components. Now locate and install the larger electrolytic capacitors. Use good quality electrolytic or tantalum capacitors for the timing circuits (C3 and C6) to assure accurate timing. Capacitor C10 should have low leakage since it is placed in a high impedance circuit. Remember the larger electrolytic capacitors have polarity which must be observed if the circuit is to work properly. Failure to install the electrolytic capacitors correctly can damage the component or other components when power is first applied to the circuit upon power-up. Install the capacitors in their correct

positions on the PC board, then solder them in place. Next trim the excess component leads flush to the edge of the circuit board using your end-cutters.

This project contains thirteen silicon diodes and a Zener diode at D4. These diodes also have polarity which must be observed if the circuit is going to work correctly. You will notice that each diode will have a black or white colored band at one edge of the diode's body. The colored band represents the diode's cathode lead. Check your schematic and parts layout diagrams when installing these diodes on the circuit board. Place all the diodes on the circuit board and solder them in place on the PC board. Then cut the excess component leads with your end-cutters.

The telephone Frustrator circuit utilizes three NPN transistors at Q1, Q2, and Q3. Transistors generally have three leads, a Base lead, a Collector lead and an Emitter lead. Refer to the schematic and you will notice that there will be two diagonal leads pointing to a vertical line. The vertical line is the transistor's base lead and the two diagonal lines represent the collector and the emitter. The diagonal lead with the arrow on it is the emitter lead, and this should help you install the transistor onto the circuit board correctly. Transistor Q1 must be a high-voltage transistor, rated at 300 V, to withstand the high-voltage ringing signal on the telephone line. Go ahead and install the transistors on the PC board and solder them in place, then follow-up by cutting the excess components leads with your end-cutter.

The Frustrator circuit utilizes five ICs. ICs must be handled carefully and installed properly in order for the circuit to work perfectly. Use anti-static techniques when handling the ICs to avoid damage from static electricity when moving them about and installing them. It is wise to install integrated circuit sockets as a cheap form of insurance against a possible circuit failure at some later point in time. Install the IC sockets on the PC board and solder them in place. Integrated circuits have to be installed correctly for the circuit to work, so you must pay strict attention to inserting them correctly into their sockets. Each integrated circuit has a locator either in the form of a small indented circle or a notch, or cutout at the top end of the IC package. Pin 1 of the IC will be just to the left of the locator.

The Frustrator circuit employs a single LED, and it must be oriented correctly if the circuit is to function properly. An LED has two leads, a cathode and an anode lead. The anode lead will usually be the longer of the two leads and the cathode lead will usually be the shorter lead, just under the flat side edge of the LED; this should help orient the LED on the PC board. Solder the LEDs to the PC board and then trim the excess leads.

Now is a good time to take a short break. After the break we will inspect the circuit board for any possible "cold" solder joints or "short" circuits. Take a careful look at each of the PC solder joints; they should all look clean and shiny. If any of the solder joints look dull, dark, or blobby, then you unsolder the joint, remove the solder and then resolder the joint all over again. Next, examine the PC board for any short circuits. Short circuits can be caused from two circuit traces touching each other due to a stray component lead that stuck to the PC board from solder residue or from solder blobs bridging the solder traces.

Table 15-3 is a chart which illustrates the wiring for the jumper in accordance with your selection. For example, if you want the Frustrator to answer the telephone after the fourth ring, you would connect the jumper between pin 10 of U3 and pins 1 and 2 of U2. Be sure to use only one jumper.

The finished circuit board can be placed in a small metal chassis box along with a loudspeaker. The circuit board can be centered in the bottom of the chassis box, and mounted atop four ¼″ standoffs. A power switch can be installed ahead of the fuse at F1. The fuse holder and line cord can exit the rear of the chassis. A two-circuit terminal strip can be mounted on the rear panel to permit hookup to the phone line. The on-off

switch if used along with the speaker can be mounted on the front panel of the chassis box.

The Frustrator unit does not have to be installed near the telephone; it can be placed anywhere convenient to an AC power receptacle and the telephone line. For a professional installation job you should terminate the pair of telephone line wires in a modular plug, and install the unit as you would any telephone accessory. Since the Frustrator has a bridge rectifier circuit at the telephone line input terminals, the polarity of the connection to the telephone line is not significant, just be sure you connect to the red and green wires of the existing telephone line.

Troubleshooting

If your Frustrator circuit does not appear to function correctly once powered up, you will need to disconnect the circuit from the phone line and carefully inspect the circuit board. It is likely that you made a soldering or parts placement mistake during construction. Pick up the circuit board with the foil side facing upwards toward you. Carefully inspect the circuit board against the parts placement diagram and the schematic to make sure that the proper component was inserted in the correct PC hole. You may have placed the wrong resistor at a particular location. You may have inserted a diode backwards.

Examine the circuit board closely for "cold" solder joints. Solder joints should appear clean, bright and smooth. If you see any solder joints that look dark, rough, or blobby, then use a solder sucker or wick to remove the solder. Now go ahead and resolder the joint. Finally, examine the PC board once again for possible short circuits. Short circuits are often caused by stray or remainder component leads which may have stuck to the PC board, often from excess solder flux which is very sticky. Look for stray leads which may bridge components traces on the PC board. Once your examination is complete you can move on.

Check out

Before you attempt to place the unit in operation it must be checked out to ascertain that it is working properly. This can be done very easily using an ordinary DC voltmeter or oscilloscope.

Table 11-3
Ring jumper list

# Rings	Integrated Circuit	Pin Number
1	U3	Pin #2
2	U3	Pin #4
3	U3	Pin #7
4	U3	Pin #10
5	U3	Pin #1
6	U3	Pin #5
7	U3	Pin #6
8	U3	Pin #9
9	U3	Pin #11

It is recommended that you check out the power supply before you place any of the ICs in their sockets. With no ICs installed, plug the line cord into a 115-V AC receptacle. Be careful to stay away from the wires feeding the transformer primary! Measure the voltage across C11. It should be about 7.8 V DC. If you do not read the correct voltage (within 10%) you will have to troubleshoot the problem before you can proceed to the next part of the check-out procedure. If the voltage is correct, leave the negative side of the voltmeter connected to the negative side of C11, and measure the B+ voltage at pin 8 of U1 and U4, pin 14 of U2, and pin 16 of U3. Each of these terminals must read B+ voltage, about 7.8 V. When you are satisfied that all voltages are correct, disconnect the unit from the power line and allow C11 to discharge.

One final check must be made before you place the Frustrator into operation; the timing of U1. To do this, place only U1 in its socket, observing proper orientation. Connect a DC voltmeter or scope between pin 3 of U1 and circuit common. Now, with power applied to the circuit take a short piece of wire and momentarily short pin 2 of U1 to circuit common being very careful not to touch any other part of the circuit. This will cause U1 to "one-shot," and can be observed by the voltmeter of scope indicating B+ voltage for about three seconds. You may trigger U1 several times with the jumper wire, each time noting the number of seconds that the voltage remains high. This period must be more than two seconds and less than four. If the timing of U1 is too short or too long, change the value of R3 to bring the timing to within the 2–4-s timing range. Smaller values of R3 will reduce timing; larger values will increase it. When you are satisfied that the timing of U1 is correct, disconnect the power cord and allow C11 to discharge. Then place the remaining ICs in their sockets, observing orientation. Insert the remaining integrated circuits in their respective sockets. If you have any difficulty in installing the transistors or ICs, you could ask a knowledgeable electronics enthusiast to help you.

The Frustrator may be placed in operation by connecting the line input wires to the telephone line and plugging in the AC power cord. If the LED does not light when power is first applied, U4 has been triggered by the sudden application of power and is in its timing cycle. Allow about 30 s for the LED to come on. Reset the circuit to standby mode by picking up the telephone receiver for about 5 s and then replacing it. You can check the operation of the circuit by dialing your number from another line (not an extension phone on the same line), and observing how many rings it takes for the Frustrator to answer your phone. You might have a friend dial your number and then signal you with a Touch Tone rendition of Jingle Bells: 999 999 9#789. Once you have established that the Frustrator is working properly, you are ready to thwart all those would-be telephone pitchmen. Be sure to tell your friends and relatives about your secret, or they will be frustrated too.

Chapter 12

Hold Circuits and Projects

Parts list

Parts Bin

Telephone Hold and Line-in-Use Indicator

R1   180 K ohm, $\frac{1}{4}$ W,
   5% resistor

R2   10 K ohm, $\frac{1}{4}$ W,
   5% resistor

R3   100 K ohm, $\frac{1}{4}$ W,
   5% resistor

R4, R5   1 K ohm, $\frac{1}{4}$ W,
   5% resistor

R6   47 ohm, $\frac{1}{4}$ W,
   5% resistor

R7   2.2 K ohm, $\frac{1}{4}$ W,
   5% resistor

R8   470 ohm, $\frac{1}{4}$ W,
   5% resistor (optional)

Q1, Q2   MPSA42 NPN
   transistor

SCR1   2N5064 −.8 amp
   200-PIV SCR

LED1, LED   LEDs

S1   momentary
   pushbutton N/O

PL1   modular 4-circuit
   telephone jack

Misc   PC board, wire,
   enclosure, hardware,
   etc.

Basic Music-on-Hold

R1   100 ohm, $\frac{1}{4}$ W,
   5% resistor

R2   330 ohm, $\frac{1}{2}$ W,
   5% resistor

C1   47 μF, 50 V
   electrolytic capacitor

C2   10 μF, 50 V
   electrolytic capacitor

T1   600 to 600 audio
   matching transformer

S1   SPST toggle switch

J1   mini $\frac{1}{8}$″ earphone jack

Misc   PC board, wire,
   enclosure, hardware,
   etc.

Universal Hold Circuit

R1, R4   56 K, $\frac{1}{8}$ W,
   5% resistor

R2, R5   10 K, $\frac{1}{8}$ W,
   5% resistor

R3   0.056 K, $\frac{1}{8}$ W,
   5% resistor

R6, R7   1 MEG, $\frac{1}{8}$ W,
   5% resistor

C1, C3   .1 MFD 50 V
   metalized film
   capacitor

C2   1,000 MFD 16 V
   electrolytic capacitor

C4   .1 MFD 250 V
   polyester capacitor

C5   1 MFD 50 V
   electrolytic capacitor

C6, C8   .01 MFD 50 V
   metalized film
   capacitor

C7 4.7 MFD 35 V
electrolytic
capacitor

D1 light emitting diode

D2 1N4744 15 V
1 watt Zener diode

D3, D4, D5 1N914
silicon diode

D6 1N4739 9.1 V
1 watt Zener diode

BR1 DF04 1 AMP
400 V DIP bridge
rectifier

IC1 H11C1 SCR
optoisolator

IC2 4N28 transistor
optoisolator

IC3 555 timer
DC Electronics

IC4 CD22204 DTMF
decoder DC Electronics
or Circuit Specialists

IC5 78L05 voltage
regulator

XTAL 3.579545 MHz
crystal

T1 600 to 600 ohm
transformer

T2 12 V DC 100 MA
wall transformer

Misc PC board, IC
sockets, wire
RJ-11phone plug,
chassis, etc.

Kit available:

Glolab Corp.

307 Pine Ridge Drive

Wappingers Falls,
NY 12590

Fax (845) 297-9772

www.glolab.com/order/
parts.html

Automatic Music-on-hold

R1, R2, R5, R6
100 K ohm, $\frac{1}{4}$ W,
5% resistor

R3, R4, R8, R14
10 K ohm, $\frac{1}{4}$ W,
5% resistor

R7 300 K ohm, $\frac{1}{4}$ W,
5% resistor

R9 39 K ohm, $\frac{1}{4}$ W,
5% resistor

R10 2 K ohm, $\frac{1}{4}$ W,
5% resistor

R11 1.2 K ohm, $\frac{1}{4}$ W,
5% resistor

R12 2.7 K ohm, $\frac{1}{4}$ W,
5% resistor

R13 6.6 K ohm, $\frac{1}{4}$ W,
5% resistor

R15 18 K ohm, $\frac{1}{4}$ W,
5% resistor

R16 10 ohm, $\frac{1}{4}$ W,
5% resistor

R17 680 ohm, $\frac{1}{4}$ W,
5% resistor

VR1 10 K PC trimmer
potentiometer

C1, C2, C30 .02 μF,
250 V, ceramic
capacitor

C3, C4, C33, C34, C37
.1 μF, 50 V ceramic
capacitor

C5, C6, C12, C13, C29
10 μF, 35 V
electrolytic capacitor

C7 47 μF, 35 V
electrolytic capacitor

C9, C14, C35 100 μF,
35 V electrolytic
capacitor

C10 100 nF, 35 V,
capacitor

C11 47 nF, 35 V
  capacitor

C15 680 pF, 35 V
  ceramic capacitor

C16 10 nF, 35 V
  polystyrene capacitor

C17 3.9 nF, 35 V
  capacitor

C18 3.3 nF, 35 V
  capacitor

C19 220 pF, 35 V
  capacitor

C20 68 pF, 35 V
  capacitor

C21 82 pF, 35 V
  capacitor

C22 470 pF, 35 V
  capacitor

C23, C24, C28 100 nF,
  35 V capacitor

C25 22 nF, 35 V
  capacitor

C26 180 pF, 35 V
  capacitor

C27 330 pF, 35 V
  capacitor

U1 Teltone 8870-01
  (C.P. Claire)

U2 MC14082 – dual
  4-input AND gate

U3 LM324N – Op-Amp
  (National)

U4 TDA7088T – FM
  receiver chip
  (Phillips-NXP)

U5 LM366 audio
  amplifier IC

Q1 MPSA13 dual
  transistor

D1, D2, D3, D4 1N4003
  silicon diode

D5, D6, D9, D12 1N4003
  silicon diode

D8 Red LED

D10 BB909 - VHF
  varactor diode

D11 3 V Zener diode

L1 78 nH metal can coil

L2 70 nH metal can coil

XTL 3.58 MHz color
  burst crystal

S1, S2 normally open
  pushbutton switches

RY1, RY2 12 V DPDT
  mini relays

T1 600 to 600 mini
  isolation transformer

Misc PC board,
  IC sockets, wire,
  enclosure, etc.

Automatic Music-on-hold Power Supply

BR1 silicon diode
  bridge −100 V, −1-amp

R18, R19 470 ohm,
  $\frac{1}{4}$ w, 5%

C31 450 μF, 50 V
  electrolytic capacitor

C33, C37 .1 μF, 50 V
  ceramic capacitor

C35 100 μF, 35 V
  electrolytic capacitor

C32, C36 10 μF, 50 V
  electrolytic capacitor

U6 LM78L09 +9 V
  regulator

U7 LM7805 +5 V
  regulator

U8 LM7812 +12 V
  regulator

F1 1 Amp fuse fast blow

T2 110 V AC to 18 V AC
  transformer .500 mA.

Misc PC board, wire,
  standoffs, etc.

Project 1: Telephone hold and Line-in-Use

In this project, two simple circuits combine in order to take the hassle out of sharing phone lines! Now it is convenient to inform others when we want to use the phone or to see when it is in use. The simple circuit revolves around a pair of MPSA42 transistors, a silicon controlled rectifier and a pair of light emitting diodes. The schematic for the Telephone Hold and Line-in-Use monitor is shown in Figure 12-1.

The Line-in-Use section of the circuit revolves around the Q1 and Q2, which were chosen for their collector to emitter breakdown voltage of 300 V, while the telephone hold section is built around the 2N5064 SCR, which has a blocking voltage of 200 V. Each section draws about 5 mA when active. The phone line has three states; on-hook, off-hook, and on-hook ringing. When on-hook, the line voltage is typically 40 to 60 V; when off-hook the line voltage ranges from 5 to 6 V, and when ringing 70 to 140 V. The reasons for the great variation are the length of lines from the phone central office, and the resistance of the long lines or wires. Typically the positive (+) wire is known as RING and is red, while the (−) minus wire or common is called TIP.

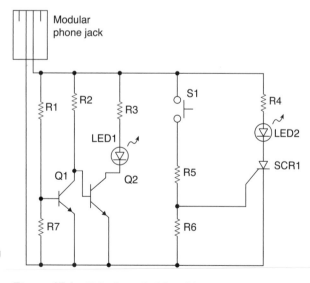

Figure 12-1 *Telephone hold and Line-in-Use monitor. Courtesy of Poptronix*

Most new phones and modems work even if the polarity is reversed. For this circuit to work properly polarity must be observed.

Line-in-Use circuit

The Line-in-Use section of the circuit consists of Q1 and Q2, LED1 and R1 through R4. When the phone is on-hook, R1 and R2, which form a simple voltage divider, supply 600 mV to the Base of Q1. That causes Q1 to turn on, pulling its collector to near ground potential. The low voltage at the collector of Q1 is applied to the Base of Q2, keeping Q2 in the cut-off state, Since no current flows through Q2, LED1 does not light. But when the phone is taken off-the-hook the normal telephone line voltage drops, thereby reducing the bias voltage applied to Q1. The reduction in bias current causes Q1 to turn off. With Q1 turned off, its collector voltage rises to a level sufficient to bias transistor Q2 into conduction. With Q2 now conducting, current flows through Q2 and LED1, causing the LED to light, indicating that the line is in use.

Line hold circuit

The line hold portion of the project consists of S1, R5 through R7, LED2, and the SCR. Two of the resistors, R5 and R6, form a voltage divider that is connected in series with switch S1.

When S1 is pressed and held with the phone off-hook, current flows through the voltage divider to the gate of SCR1. That provides sufficient current to cause SCR1 to fire, but not enough current through it to cause it to latch. When the phone is placed back on-hook, the line voltage increases, causing a current of sufficient magnitude to flow through R7 to latch SCR1 and light LED1. Even after releasing S1, the LED remains on, signaling that the line is on hold. When any phone on the line is picked up, the line voltage drops again to around 5 V, so that the SCR is starved for current and turns off, releasing the hold on the line.

Resistor R8 is optional; that resistor is included in the circuit only if a flashing unit is used for LED2 in the line hold portion of the circuit to attract more attention. If you do use a flashing LED in the line hold circuit, change the value of R7 to 1,000 ohms and add

R8 across the LED, as indicated. Resistor R8 helps to keep the latching current flowing through the SCR during the blinking LED's off cycle, so as not to disrupt the operation of the line hold portion of the circuit. It is ill advised to use a blinking LED for LED1, as it can generate annoying clicking sounds.

Circuit assembly

Before we begin building the Telephone Hold and Line-in-Use Monitor, you will first need to find a clean, well-lit worktable or workbench. Next we will gather a small 25–30-W pencil tipped soldering iron, a length of 60/40 tin–lead solder, and a small canister of Tip Tinner, available at your local RadioShack store. Tip Tinner is used to condition the soldering tip between solder joints. You drive the soldering tip into the compound and it cleans and prepares the soldering tip. Next grab a few small hand tools, try and locate small flat blade and Phillips screwdrivers. A pair of needle-nose pliers, a pair of tweezers along with a magnifying glass, and a pair of end-cutters and we will begin constructing the project.

The author's prototype of the project was assembled on a small printed circuit board, measuring $2 \times 2 \frac{1}{2}''$. If you have created your own PC board, now you can begin populating the circuit board with components.

Now locate your schematic diagram and parts layout diagram along with all of the components needed to build the project. Once all the components are in front of you, you can check them off against the project parts list to make sure you are ready to start building the project.

Refer now to the resistor color code chart shown in Table 12-1, which will help you identify the resistors for the project.

Note that each resistor will have either three or four color bands, which begin at one end of the resistor's body. The first colored band represents the resistor's first digit, while the second color represents the second digit of the resistor's value. The third colored band represents the resistor's multiplier value, and the fourth color band represents the resistor's tolerance value. If there is no fourth colored band then the resistor has a 20% tolerance value. If the resistor's fourth band is silver then the resistor has a 10% tolerance value and if the fourth band is gold then the resistor's tolerance value will be 5%.

Finally, we are ready to begin assembling the project. So let's get started. The prototype project was constructed on a single-sided printed circuit board or PCB. With your PC board in front of you, we can now begin populating the circuit board. We are going to place and solder the lowest height components first,

Table 12-1
Resistor color code chart

Color Band	1st Digit	2nd Digit	Multiplier	Tolerance
Black	0	0	1	
Brown	1	1	10	1%
Red	2	2	100	2%
Orange	3	3	1,000 (K)	3%
Yellow	4	4	10,000	4%
Green	5	5	100,000	
Blue	6	6	1,000,000 (M)	
Violet	7	7	10,000,000	
Gray	8	8	100,000,000	
White	9	9	1,000,000,000	
Gold			0.1	5%
Silver			0.01	10%
No color				20%

the resistors and diodes. First, locate our first resistor R1. Resistor R1 has a value of 180k ohms and is a 5% tolerance type. Now take a look for a resistor whose first color band is red, red, brown, and gold. From the chart you will notice that brown is represented by the digit (1), and the second band is gray, which is represented by (8). Notice that the third or multiplier band is yellow or (10,000), so $(1)(8) \times 10,000 = 180,000$ ohms. Go ahead and place R1 on the PC board in its respective location and solder it in place on the board. Take a pair of end-cutters and trim the excess resistor leads, cutting the leads flush to the edge of the circuit board. Now install the remaining resistors and solder them in place on the circuit board. Remember to trim the excess component leads with your end-cutters.

It is advisable to install the lowest profile components first, so let's go ahead and install the remaining resistors onto the circuit board. Go ahead and solder the resistors in their respective locations on the PC board, then trim the excess component leads using a small pair of end-cutters. Cut the excess leads flush to the edge of the circuit board. This project has no capacitors, so we will not have to worry about identifying them.

The Telephone Hold and Line-in-Use circuit contains two NPN silicon transistors Q1 and Q2. Transistors generally have three leads, a Base lead, a Collector lead, and an Emitter lead. Refer to the schematic and you will notice that there will be two diagonal leads pointing to a vertical line. The vertical line is the transistor's base lead and the two diagonal lines represent the collector and the emitter. The diagonal lead with the arrow on it is the emitter lead, and this should help you install the transistor on to the circuit board correctly. Once the two identical transistors have been identified you can go ahead and install them on the PC board. Remember to trim the excess component leads flush to the edge of the circuit board.

The Hold and Line-in-Use circuit also utilizes a sensitive gate SCR shown at SCR1. SCR devices also have three leads, but unlike a transistor, which has a base, collector, and emitter, an SCR has an anode, cathode, and a gate lead. The SCR for this project looks much like a transistor, but make sure you can properly identify it before installing it on the PC board. The symbol for an SCR is a diode with an angled lead connected to the cathode end. Once you have identified and determined the leads, you can go ahead and install the SCR.

The LEDs for this project can be either the same or different colors, as desired. The Line-Hold LED could be a red blinking LED, while the in-Use LED could be a green one.

As mentioned, the Hold and Line-in-Use indicator circuit could be mounted inside your existing telephone base if this is large enough, or it could be mounted in a small plastic enclosure. You could elect to mount the project in a small plastic box with a telephone jack at both ends, so that you could plug your telephone into the project, or you could simply connect a 4-wire cord with a 4-conductor modular plug and plug it into a telephone wall jack or extension jack if desired. You could build a Hold and Line-in-Use indicator circuit and place one at each phone for convenience.

Installation and use

As mentioned earlier, the polarity of your telephone lines may be not be properly indicated by the coloring of the wiring. Measure the actual polarity of the telephone line with a multi-meter, so that you will be sure you are connecting the positive (+) circuit-board trace to the RING input or PL1 and the negative trace to the TIP terminal. The board can be installed inside your telephone housing or mounted inside a small plastic box. Using the Line-in-Use circuit requires no instruction and using the Hold circuit is simple; with the phone off-hook, press and hold S1, and then place the phone on-hook and release S1. The LED will stay lit until any phone is taken off the hook again.

Project 2: Basic Music-on-Hold

What is more annoying than calling someone and getting placed on-hold? Being placed on-hold is a bit more tolerable when you can listen to some music while you are waiting. The Basic Music-on-Hold circuit requires no external power source and uses the voltage across the phone line to power the circuit. You do not have to spend a lot of money to construct your own Music-on-Hold project. Now you can build a simple, low cost manual on-hold circuit in just a few minutes by constructing the Music-on-Hold circuit depicted

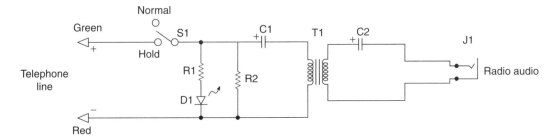

Figure 12-2 *Basic manual Music-on-Hold circuit*

in Figure 12-2. With a small handful of components and about a half hour of your time you can build this simple Music-on-Hold circuit.

The Music-on-Hold circuit consists of two resistors, a capacitor, an LED, a matching transformer and an SPST toggle switch. You simply connect the Music-on-Hold circuit to the phone line and connect the output of the matching transformer to a radio earphone jack on an FM radio and you are ready to go!

Circuit description

The input to the Basic Music-on-Hold circuit is fed to the telephone line via the red and green telephone wires. The red telephone line is connected directly to SPST switch S1. You can connect the circuit directly to a telephone jack via RJ11 plug, or wire it directly into a telephone wall jack. The opposite side of S1 is coupled to 100 ohm resistor in series with a red LED connected across the telephone line. This forms the on-hold indicator, while resistor R1 forms the actual "hold" circuit. Capacitor C1 is used to couple the audio from the phone line to the matching transformer at T1. The 600 to 600 ohm Transformer T1 is used to isolate the phone line from your radio sound source. The secondary output from the audio matching transformer is fed to capacitor C2. The opposite end of C2 is then fed directly to a mini ⅛″ audio chassis jack. The Basic Music on-hold circuit is simple but quite efficient and will perform flawlessly.

Circuit assembly

Before we get started building the Basic Music-on-Hold circuit, you will need to locate a clean, well-lit worktable or workbench. Next we will gather a small

25–30-W pencil tipped soldering iron, a length of 60/40 tin–lead solder, and a small canister of Tip Tinner, available at your local RadioShack store, Tip Tinner is used to condition the soldering tip between solder joints. You drive the soldering tip into the compound and it cleans and prepares the soldering tip. Next grab a few small hand tools; try and locate small flat blade and Phillips screwdrivers. A pair of needle-nose pliers, a pair of tweezers along with a magnifying glass, and a pair of end-cutters and we will begin constructing the project. Now locate your schematic diagram along with all of the components needed to build the project. Once all the components are in front of you, you can check them off against the project parts list to make sure you are ready to start building the project. Now, locate Tables 12-1 and 12-2, and place them in front of you. Table 12-1, illustrates the resistor code chart and how to read resistors, while Table 12-2 illustrates the capacitor code chart which will aide you in constructing the project.

Refer now to the resistor color code chart in Table 12-1, which will help you identify the resistors for this project. Note that each resistor will have either three or four color bands, which begin at one end of the resistor's body. The first colored band represents the resistor's first digit, while the second color represents the second digit of the resistor's value. The third colored band represents the resistor's multiplier value, and the fourth color band represents the resistor's tolerance value. If there is no fourth colored band then the resistor has a 20% tolerance value. If the resistor's fourth band is silver then the resistor has a 10% tolerance value and if the fourth band is gold then the resistor's tolerance value will be 5%.

Finally, we are ready to begin assembling the project. Let's finally get started. The Basic Music-on-Hold project was constructed on a single-sided printed circuit

Table 12-2
Three-digit capacitor codes

pF	nF	µF	Code	pF	nF	µF	Code
1.0			**1R0**	3,900	3.9	.0039	**392**
1.2			**1R2**	4,700	4.7	.0047	**472**
1.5			**1R5**	5,600	5.6	.0056	**562**
1.8			**1R8**	6,800	6.8	.0068	**682**
2.2			**2R2**	8,200	8.2	.0082	**822**
2.7			**2R7**	10,000	10	.01	**103**
3.3			**3R3**	12,000	12	.012	**123**
3.9			**3R9**	15,000	15	.015	**153**
4.7			**4R7**	18,000	18	.018	**183**
5.6			**5R6**	22,000	22	.022	**223**
6.8			**6R8**	27,000	27	.027	**273**
8.2			**8R2**	33,000	33	.033	**333**
10			**100**	39,000	39	.039	**393**
12			**120**	47,000	47	.047	**473**
15			**150**	56,000	56	.056	**563**
18			**180**	68,000	68	.068	**683**
22			**220**	82,000	82	.082	**823**
27			**270**	100,000	100	.1	**104**
33			**330**	120,000	120	.12	**124**
39			**390**	150,000	150	.15	**154**
47			**470**	180,000	180	.18	**184**
56			**560**	220,000	220	.22	**224**
68			**680**	270,000	270	.27	**274**
82			**820**	330,000	330	.33	**334**
100		.0001	**101**	390,000	390	.39	**394**
120		.00012	**121**	470,000	470	.47	**474**
150		.00015	**151**	560,000	560	.56	**564**
180		.00018	**181**	680,000	680	.68	**684**
220		.00022	**221**	820,000	820	.82	**824**
270		.00027	**271**		1,000	1	**105**
330		.00033	**331**		1,500	1.5	**155**
390		.00039	**391**		2,200	2.2	**225**
470		.00047	**471**		2,700	2.7	**275**
560		.00056	**561**		3,300	3.3	**335**
680		.00068	**681**		4,700	4.7	**475**
820		.00082	**821**			6.8	**685**
1,000	1.0	.001	**102**			10	**106**
1,200	1.2	.0012	**122**			22	**226**
1,500	1.5	.0015	**152**			33	**336**
1,800	1.8	.0018	**182**			47	**476**

Table 12-2—cont'd

Three-digit capacitor codes

pF	nF	μF	Code	pF	nF	μF	Code
2,200	2.2	.0022	**222**			68	**686**
2,700	2.7	.0027	**272**			100	**107**
3,300	3.3	.0033	**332**			157	**157**

Code = 2 significant digits of the capacitor value + number of zeros to follow

For values below 10 pF, use "R" in place of decimal e.g. 8.2 pF = 8R2

10 pF = 100

100 pF = 101

1,000 pF = 102

22,000 pF = 223

330,000 pF = 334

1 μF = 105

board, but you could elect to build the circuit on a perf-board. With your PC board in front of you, you can now begin populating the circuit board. We are going to place and solder the lowest height components first, the resistors and diodes. First locate our first resistor R1. Resistor R1 has a value of 330 ohms and is a 5% tolerance type. Now take a look for a resistor whose first color band is orange, orange, brown, and gold. From the chart you will notice that red is represented by the digit (3), and again the second band is red, which is represented by (3). Notice that the third color is brown and the multiplier is (10), so $(3)(3) \times 10 = 330$ ohms. Go ahead and install R1 on the PC board and solder it in place. Now, take a pair of end-cutters and trim the excess resistor leads flush to the edge of the PC board. Now install the remaining resistors and solder them in their respective locations. Remember to trim the excess component leads.

Note that capacitors generally come from two major groups or types. Non-polar types, such as the ones in this project, which have no polarity, can be placed in either direction on the circuit board. The other type of capacitors are called polarized types, these capacitors have polarity and have to be inserted on the board with respect to their polarity marking in order for the circuit to operate correctly. On these types of capacitors, you will observe that the components actually have a plus or minus marking or a black or white band with a plus or minus marking on the body of the capacitors.

Next, refer to Table 12-2. This chart helps to determine the value of small capacitors. Many capacitors are quite small in physical size, therefore manufacturers have devised a three-digit code which can be placed on the capacitor instead of the actual value. The code occupies a much smaller space on the capacitor and can usually fit nicely on the capacitor body, but without the chart the builder would have a difficult time determining the capacitor's value.

Now locate the capacitors from the component pile and try and locate capacitor C1, which should be marked 47 μF. Note that both C1 and C2 are electrolytic capacitors, so they will have polarity which must be observed when mounting these components on the PC board. Go ahead and place C1 and C2 on the circuit board and solder them in place. Now take your end-cutters and cut the excess component leads flush to the edge of the circuit board. Once you have installed the resistors and capacitors, you can move on to installing the remaining components.

Now install the LED, observing that the cathode is connected to the bottom end of resistor R2. You may want to leave at least a half inch of lead length on the LED, so that it can be placed flush to the enclosure and the LED can shine outside the mounting enclosure. You could also elect to mount the LED directly on the chassis or enclosure.

Now locate transformer T1. Since the transformer is a 600 to 600 ohm transformer both the primary and secondary are the same, so you can designate either side as primary or secondary. Most small matching transformers have two small mounting tabs. Drill two small holes for the mounting tabs, then place the transformer on the PC board, and bend the tabs under the board to secure the transformer to the PC board. You can solder the tabs to the ground plane of the PC board, if there is a common ground plane.

The on-off switch at S1 can be either a PC mount switch which can be mounted directly on the PC board, or you could elect to mount the switch off the circuit board directly on the enclosure. The ⅛ inch mini audio jack can be mounted directly on the chassis box if desired.

Once the circuit board is finished, you will need to decide how you will connect the circuit to the phone line. You could elect to attach a 1–2′ length of 2-conductor phone wire from the Music-on-Hold circuit to the telephone line, or you could place a 2-conductor jack on the Music-on-Hold chassis and then wire a plug to a phone jack at the wall.

Next, you will have to construct or buy a ⅛″ plug to ⅛″ plug audio coax cable. You can build you own cable if desired by purchasing two ⅛″ mini plugs and a 1–3′ length of mini audio coax cable. One end of the cable will plug into the jack on the Music-on-Hold circuit enclosure and the other end of the cable will plug into the earphone jack of an FM radio tuned to your favorite station.

Circuit operation

In order to utilize the Basic Music-on-Hold circuit, you will have to connect the circuit to the phone line as shown and to the earphone jack of an FM radio tuned to your favorite station. Then you simply have to establish your telephone call, and when you place the phone call "on-hold" you simply have to switch the toggle switch from the "NORMAL" position to the "HOLD" position. When you return to the phone call you will need to switch S1 back to the normal position. If the phone is not switched back to the normal position, you will not be able to hang-up the phone line or receive new phone calls.

Your Basic Music-on-Hold project is now ready to serve you; it will perform flawlessly for many years to come and you will pleased with your new low-cost Music-on-Hold feature with a minimum cash outlay.

Project 3: Universal Hold Circuit

If you have a touch tone telephone service, you can now put a call on hold from any phone in the house, even from cordless phones and phones without a hold button, by plugging this device into any telephone jack, anywhere in the house. Just one of these Universal Hold Circuits adds the call hold feature to every phone that has a keypad with a # (pound) key. To put a call on hold, you press the # key and hang the phone up. The # key function remains activated for five seconds after the key is pressed and released so you can hang up phones that have a keypad in the handset.

The Universal Hold Circuit detects the dual tone, multi frequency (DTMF) tone that is generated when the # key is pressed, and then activates a circuit that partially loads the telephone line so that the central office thinks a phone is still off-hook even after it is hung up. When any phone is again picked up, the hold function is canceled. The Universal Hold project is shown in Figure 12-3.

The Universal Hold schematic diagram is illustrated in Figure12-4. The telephone line is connected to the hold components through bridge rectifier BR1 so the input is not polarity sensitive and the input lines may be connected either way. BR1 always connects the positive side of the line to Zener diode D2 and the negative side of the line to U2. If L1 is positive, current flows into the circuit through bridge diode section A and back out through diode section D into L2. If L2 is

Figure 12-3 *Universal Hold Circuit. Courtesy of GloLab*

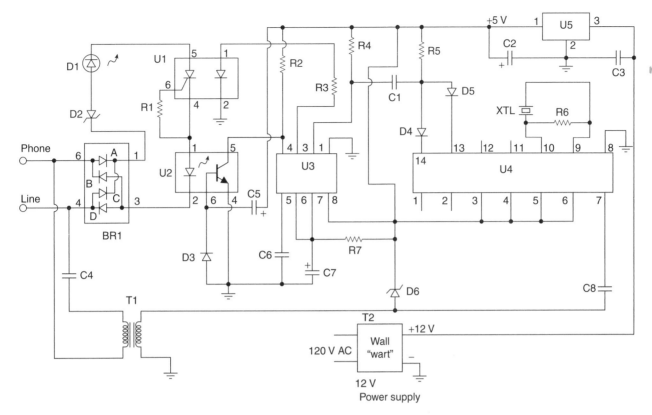

Figure 12-4 *Universal Hold Circuit (Courtesy GloLab Corp)*

positive, current flows in through diode section C and out through diode section B into L1. The telephone line also connects to tone decoder U4 through C4 and T1. Opto-isolator IC1 contains a silicon controlled rectifier (SCR) having a characteristic that makes it continue to conduct current, or latch on, once turned on by current flowing through an optically coupled LED within the IC. The SCR conduction continues even after the LED current is removed, providing that enough anode current is available to sustain it. When the SCR anode voltage is removed and conduction stops, it will not resume when voltage is re-applied until again triggered by LED current. This characteristic makes the SCR opto-isolator ideal for a line hold application. When all phones are hung up (on-hook) the voltage across the telephone line is about 48 V. When a phone is picked up (off-hook) it places a load on the line and a current of about 20 mA flows through the phone which causes the line voltage to drop to about 3 to 8 V, depending on the telephone unit. Current also flows through circuits in the central office, indicating that the phone has been picked up.

A call is put on hold when all phones are hung up and the SCR in U1 is triggered on by circuits driving its

LED (we'll tell you more about this later). When the SCR conducts, current flows through BR1, D2, D1, U1, U2 and back through BR1, placing a load on the line that keeps the central office circuits active. A 15-V drop across D2, a 2-V drop across D1, a 1.5-V drop across the LED in U2, and about a 1.5-V drop across the diodes in BR results in the normal 48-V open circuit line voltage across L1; L2 is clamped to about 20 V, which allows line current to flow but is a higher voltage than when a telephone loads the line. Now when a phone is picked up and the line voltage drops to about 8 volts or less, there is no longer enough voltage to produce conduction through D2, D1, U1, U2, and BR1 and the hold function is canceled. LED D1 is provided to indicate that a call is on hold.

The telephone line also connects through C4 and transformer T1 to capacitor C8 which couples the DTMF signal to the input of U4, a Harris CD22204 dual tone, multi-frequency tone decoder integrated circuit used to detect the # key tone (see block diagram shown in Figure 12-5).

The 3.57 MHz crystal XTAL generates a clock required by the decoder. Zener diode D6 clamps its

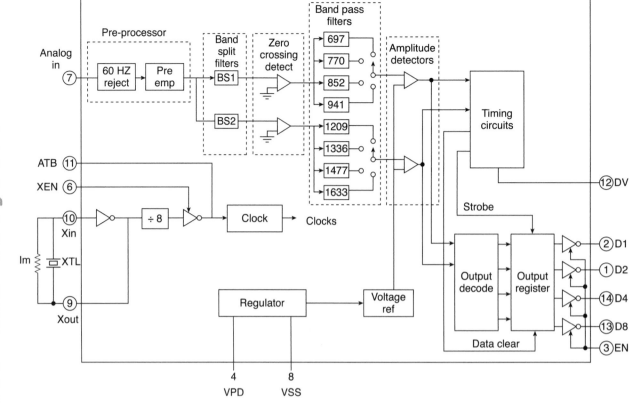

Figure 12-5 *CD22204 touch-tone decoder. Courtesy of Intersil Corp*

input to prevent damage from transients. The decoder has four outputs that produce a hexadecimal code corresponding to the tones it receives when a key is pressed as shown in Table 12-3.

When these outputs are further decoded, the "1" or "0" status of all four outputs usually must be considered to get a unique hexadecimal number for each DTMF tone. Notice, however, that a "1" occurs simultaneously in both the pins 13 and 14 outputs only when the # key is pressed. This makes using the outputs easier, since only two hexadecimal bits must be considered to decode the # key. The binary decoding is done by a logical AND circuit consisting of two diodes, D4 and D5, whose cathode terminals must both be at an up level or logical "1" for their anodes, which are connected together, to be pulled up by R5. This up level discharges C1 through R4 into the positive supply.

When the # key is released the DTMF signal stops, pins 13 and 14 go down again and a down level pulse is produced by time constant C1, R4 at the pin 2 trigger input of timer U3. Since the IC3 trigger input must

Table 12-3
DTMF decoder output codes

DIGIT	D4	D3	D2	D1
1	0	0	0	1
2	0	0	1	0
3	0	0	1	1
4	0	1	0	0
5	0	1	0	1
6	0	1	1	0
7	0	1	1	1
8	1	0	0	0
9	1	0	0	1
0	1	0	1	1
#	1	1	0	0
A	1	1	0	1
B	1	1	1	0
C	1	1	1	1
D	0	0	0	0

receive a negative transition to start a timing cycle, the cycle does not start until the # key is released. When the timing cycle starts, U3 output pin 3 goes up for 5 s and drives current through R3 into the LED within U1 which places its SCR in a potentially conducting state where it needs only to have sufficient voltage applied to its anode terminal to make it heavily conduct and latch on, a condition that occurs when the phone is hung up. The 5 s output from U3 keeps U1 LED current flowing long enough to hang up the phone and have the required anode voltage applied to maintain conduction. The time duration is controlled by R7 and C7 and may be altered by changing the value of R7. U2 is an opto-isolator with a transistor output. After the # key is pressed and the phone is hung up, telephone line current flowing through the LED in U2 turns its transistor on, which resets timer IC3 and terminates its timing cycle. This reset function removes current from the LED in IC1 and allows a hold to be canceled immediately after initiating it without having to wait for the 5-s cycle to complete. The transistor within U2 also performs a power-on reset by having its base coupled through C5 to the positive supply. When power is turned off, C5 discharges through D3.

Power is supplied by a 12-V DC wall transformer and by U5, a 78L05 5-V regulator, and is filtered by C2 and C3. The power supply, DTMF decoder, and timing circuits are isolated from the telephone line by the opto-isolators and transformer T1.

Circuit assembly

Before we begin constructing the Universal Hold Circuit, you will need to locate a clean, well-lit worktable or workbench. Next, we will gather a small 25–30-W pencil tipped soldering iron, a length of 60/40 tin–lead solder, and a small canister of Tip Tinner, available at your local RadioShack store; Tip Tinner is used to condition the soldering tip between solder joints. You drive the soldering tip into the compound and it cleans and prepares the soldering tip. Next grab a few small hand tools, and try and locate small flat blade and Phillips screwdrivers. A pair of needle-nose pliers, a pair of tweezers along with a magnifying glass, and a pair of end-cutters and we will begin constructing the project. Now locate your schematic diagram and parts layout diagram along with all of the components needed to build the project.

Once all the components are in front of you, you can check them off against the project parts list to make sure you are ready to start building the project. Now, locate Tables 12-1 and 12-2, and place them in front of you.

Table 12-1 illustrates the resistor code chart and how to read resistors, while Table 12-2 illustrates the capacitor code chart, which will aid you in constructing the project.

Finally, we are ready to begin assembling the project. So let's get started. The prototype was constructed on a single-sided printed circuit board or PCB. With your PC board in front of you, we can now begin populating the circuit board. Now refer to the resistor color code chart shown in Table 12-1. You will note that each resistor will have either three or four color bands, which begin at one end of the resistor's body. The first colored band represents the resistor's first digit, while the second color represents the second digit of the resistor's value. The third colored band represents the resistor's multiplier value, and the fourth color band represents the resistor's tolerance value. If there is no fourth colored band then the resistor has a 20% tolerance value. If the resistor's fourth band is silver then the resistor has a 10% tolerance value and if the fourth band is gold then the resistor's tolerance value will be 5%.

We are going to place and solder the lowest height components first, the resistors and diodes. First locate our first resistor R1. Resistor R1 has a value of 10k ohms and is a 5% tolerance type. Now take a look for a resistor whose first color band is brown, black, orange, and gold. From the chart you will notice that brown is represented by the digit (1), the second color band is black, which is represented by (0). Notice that the third color is orange and the multiplier is (1,000), so $(1)(0) \times 1,000 = 10,000$ or 10 K ohms.

Go ahead and locate where resistor R1 goes on the PC and install it, next solder it in place on the PC board, and then with a pair of end-cutters trim the excess component leads flush to the edge of the circuit board. Next locate the remaining resistors and install them in their respective locations on the main controller PC board. Solder the resistors to the board, and remember to cut the extra lead with your end-cutters.

Next refer to Table 12-2; this chart helps to determine the value of small capacitors. Many capacitors are quite small in physical size. Therefore manufacturers have devised a three-digit code which can be placed on the capacitor instead of the actual value. The code occupies a much smaller space on the capacitor and can usually fit nicely on the capacitor body, but without the chart the builder would have a difficult time determining the capacitor's value.

Note that capacitors generally come from two major groups or types. Non-polar types, such as the ones in this project, have no polarity and can be placed in either direction on the circuit board. The other type of capacitors are called polarized types, these capacitors have polarity and have to be inserted on the board with respect to their polarity marking in order for the circuit to operate correctly. On these types of capacitors, you will observe that the components actually have a plus or minus marking or a black or white band with a plus or minus marking on the body of the capacitors. Capacitors must be installed properly with respect to polarity in order for the circuit to work properly. Failure to install electrolytic capacitors correctly could cause damage to the capacitor as well as to other components in the circuit when power is first applied.

Now look for the lowest profile capacitors from the component stack and locate capacitor C1, which is labeled (100) or 100 pF. Note that there are a number of small capacitors in the circuit that may not be labeled with their actual value, so you will have to check Table 12-2 to learn how to use the three-digit capacitor code. Go ahead and place C1 on the circuit board and solder it in place. Now take your end-cutters and cut the excess component leads flush to the edge of the circuit board. Next, locate the remaining low profile capacitors and install them on the printed circuit board, then solder them in place. Remember to trim the extra component leads. There are a number of electrolytic capacitors in the circuit and placing them with respect to their proper polarity is essential for proper operation of the circuit. Once you have located the larger electrolytic capacitors install them in their respective locations and solder them in place. Finally, trim the excess component leads.

The Universal Hold Circuit contains a number of silicon diodes and a Zener diode. Diodes, as you will remember, always have some type of polarity markings, which must be observed if the circuit is going to work correctly. You will notice that each diode will have a black or white colored band at one edge of its body. The colored band represents the diode's cathode lead. Check your schematic and parts layout diagrams when installing these diodes on the circuit board. Place all the diodes on the circuit board in their respective locations and solder them in place on the PC board. The diode bridge on the right side of the schematic can be installed now. After soldering the diodes and the diode bridge in place in their respective locations, remember to cut the excess component leads with your end-cutters flush to the edge of the circuit board.

The Universal Hold Circuit contains five integrated circuits (ICs). The two main integrated circuits are the 14 pin Touch-Tone decoder at U4 and the 555 Timer chip at U3. The Universal Hold Circuit also utilizes two opto-isolator packages for interfacing. You will notice that both opto-isolators are different; the 4N28 uses a transistor, while the H11C1 uses an SCR. Both packages will be smaller 6-pin DIP sized packages and they should be handled like integrated circuits. The LM78L05 IC is a 5-V regulator at U5.

ICs are often static sensitive, so they must be handled with care. Use a grounded anti-static wriststrap and stay seated in one location when handling the ICs. Take out a cheap insurance policy by installing IC sockets for each of the ICs. In the event of a possible circuit failure, it is much easier to simply unplug a defective IC, than trying to unsolder 14 or more pins from a PC board without damaging the board. Integrated circuits have to be installed correctly if the circuit is going to work properly. IC packages will have some sort of markings which will help you orient them on the PC board. An IC will have either a small indented circle, a cutout, or notch at the top end of the IC package. Pin 1 of the IC will be just to the left of the notch, or cutout. Refer to the manufacturer's pin-out diagram, as well as the schematic when installing these parts. Accidental reversal of pin insertion can cause the integrated circuit to fail. If you doubt your ability to correctly orient the ICs, seek the help of a knowledgeable electronics enthusiast.

Next, identify the quartz crystal, which connects between pins 9 and 10 of U4. Place the 3.579545 MHz crystal (XTAL) in its proper location and solder it into place.

Next, locate transformer T1 and place it on the circuit board. Check the manufacturer's specification sheet for the actual wiring of your particular transformer. Many 600 to 600 ohm transformers will have three windings for both the primary winding and the secondary winding. You will want to make sure that you use the correct wires for both the primary and secondary connections. Install the transformer then solder the leads to the PC board, remembering to trim the excess leads.

Once the main Universal Hold Circuit board has been completed, take a short break and then we will inspect the PC board for possible "cold" solder joints and "short" circuits. Pick up the PC board with the foil side facing upwards facing you. Carefully inspect the solder joints, which should look smooth, clean, and bright. If any of the solder joints look dull, dark, or blobby, then you should remove the solder with a solder sucker or a solder wick and then resolder the joint all over again. Now we will inspect the PC board for possible short circuits. Short circuits can be caused from small solder blobs bridging the circuit traces or from cut component leads which can often stick to the circuit board once they have been cut. A sticky residue from solder will often trap component leads across the circuit traces. Look carefully for any bridging wires across circuit traces.

Once this take is complete, we can move on to installing the circuit board into a plastic box, as shown in Figure 12-6.

Locate a suitable enclosure within which to house the circuit board. Arrange the circuit board and the transformer on the bottom of the chassis box. Now, locate four ¼″ standoffs, and mount the PC board between the standoffs and the bottom of the chassis box. If you selected a convention relay you will have to mount it on the chassis and wire it to the PC board. If you selected a non-PC board transformer then you will have to mount the transformer to the base of the chassis box and wire it to the circuit board. Locate a 9-V

Figure 12-6 *Universal Hold Circuit with case. Courtesy of GloLab*

battery holder and mount it to the bottom of the plastic box and solder the battery clip wire to the circuit board.

Testing

Connect the hold circuit to the telephone line and plug the wall transformer into an AC outlet. Pick a phone up, press the # key and hang it up within 5 s. D1 should light. Pick a phone up and D1 should go out. Press the # key, wait about 8 s and hang the phone up. D1 should not light.

Operation

The 5 s timing cycle that starts after the # key is released gives ample time to hang up a phone with a keypad on the handset. If the keypad is on the telephone base the # key can be pressed while the phone is being hung up. If the # key is pressed for reasons other than to place a call on hold, such as to signal the end of a number entry after using a fax back service, the timing cycle will end after 5 s and will probably not be active when the phone is hung up; to be sure that the line is not on hold after using the # key just pick the phone up and hang up again. Have fun and enjoy your new Universal Hold Circuit project.

Project 4: Automatic Music-on-Hold

Impress your callers by adding an FM Music-on-Hold feature to your telephone. If you thought a Music-on-Hold feature for your telephone was only for high-budget professions, think again. You will learn how you can add FM Music-on-Hold to any analog telephone line with a Touch Tone telephone. It is ideal for homes, offices, or small businesses. Some of the features of this design include: LED status indication audio volume control, built-in antenna, only one operating adjustment, and a mute function to eliminate "hiss" in between stations. You can build this Music-on-Hold project in three or so hours and for less than about $70.

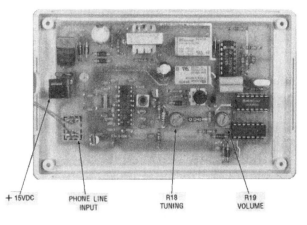

Figure 12-7 *Music on hold circuit. Courtesy of Poptronix*

Construction, test, and alignment are made easy due to the use of specialized ICs, namely a single FM receiver chip, U4, and a DTMF decoder U1. There are no special coils to wind, and no tricky circuit adjustments are required. All you need is a DMM to test and align the circuit; see Figure 12-7.

On-hold circuit

Let's take a look at how the Music-on-Hold unit works. A block diagram of the unit is shown in Figure 12-8, and the schematic in Figure 12-9.

The FM on-hold device connects to an analog telephone line via an RJ11 modular jack. The automatic Music-on-Hold circuit is powered by a separate multi-voltage output power supply, which is plugged into a standard 120-V AC outlet.

When a key on any Touch Tone telephone is depressed, the signal is passed through U3-d, an LM324N balanced amplifier The purpose of this amplifier is two fold; it acts as a balance to unbalanced matching network, and its gain is set to 0.1 to act as a line-voltage attenuator. Capacitors C1 and C2 block the phone line's 48 V DC from entering the amplifier. The ringing voltage is limited by R1 and R2. The ratio of R3 to R1 sets the gain of U3:d to 0.1. Resistor R4 biases U3:d between supply voltage and ground. This allows the circuit to operate from the single +6.0 V DC power supply line. The output of the balance amplifier passes through coupling capacitor C3 and is then decoded by U1, a Teltone 8870-01 dual multi-frequency (DTMF) decoder IC.

The output of U1 is a 4-bit word whose codes are listed in Table 12-3.

It is connected to U2:b, a 4,082 dual quad-input AND gate, so that the output of that IC (pin 13) is normally low and goes high only when the "*" key is pressed. Therefore, when the "*" key is decoded by U1, pins 1, 2, and 13 are high while pin 14 is low. To switch the output of U2-b high, four logic-high inputs present. The high inputs are provided by U1 pins 1, 2, and l3 and U2:a pin 1.

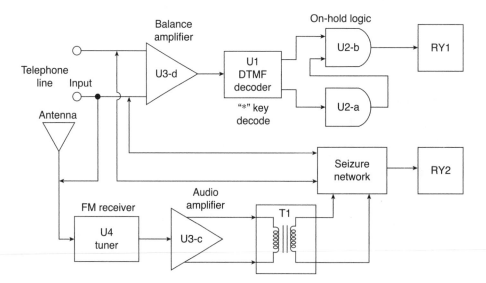

Figure 12-8 *Music-on-Hold block diagram. Courtesy of Poptronix*

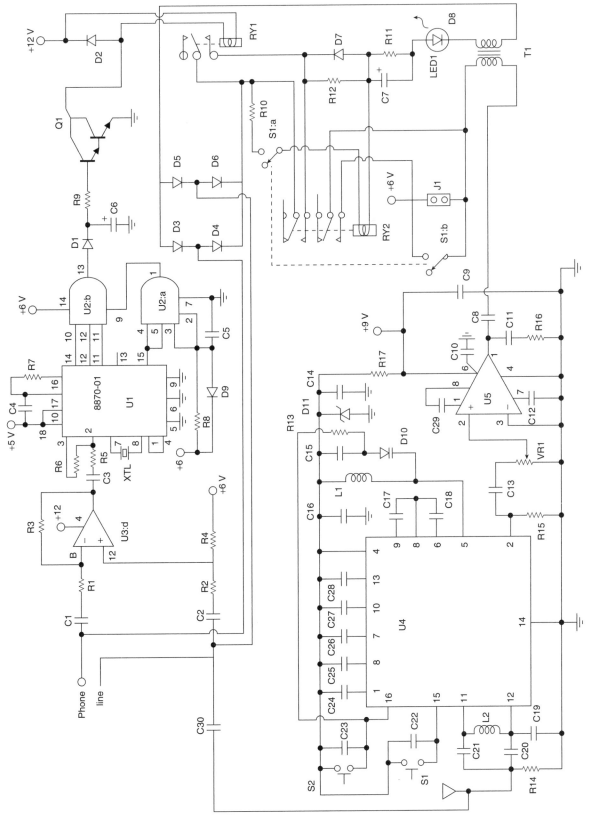

Figure 12-9   *Automatic Music-on-Hold project. Courtesy of Poptronix*

In order for U2:a's output to go high, it must also have four logic inputs. Two of those are provided by R7, D10, and C27. Those components ensure that the internal power supply is operating. That will prevent the unit from seizing the phone line if power is lost or removed while it is connected to the phone line. The remaining two inputs are provided by a logic high from U1 pin 12, which is the data valid output pin, assuring proper operation of U1 by providing internal checks. When those checks are valid, DV will output a logic high. That prevents false triggering due to voice or other tones that occur during normal telephone usage. When the "*" key is depressed, U2:b goes high, which in turn charges C4 and turns on switching transistor Q1. That activates relay RY1. Diode D1 prevents DC voltage from bleeding back into U2:b pin 13. The time-base oscillator for U1 is formed from a 3.58-MHz crystal XTL and R5.

The normally open contacts of RY1 close, and D7, R9, RY2, R10, C5. LED1, transformer (T1) and the four diodes from the polarity bridge (D3-D6) are connected across the telephone line and effectively "seize" it. That combination of components is referred to as the seizure network. The unit is now in a STANDBY mode. If jumper J1 is in the IN position and a station is tuned in to the FM tuner, that station will be heard on the telephone line. If it is in the OUT position, the station will not be heard until the phone is hung up.

Relay RY1 will stay activated for approximately four seconds. That delay is determined by the RC network of R6/C4. Diode D2 prevents relay-coil induction-induced "spikes" from appearing on the +12-V DC power supply line. If the telephone is hung up within the four second time-out period, additional loop current will flow through the seizure network and activate RY2. That causes the normally open contacts of RY2 to close. The project is now in the "on-hold" mode, LED1 will be brightly lit, and the selected radio station will be heard in the telephone line regardless of the position of jumper J1. After the four second time-out period, RY1 will deactivate. The loop current flowing through RY2 keeps the seizure network across the telephone line and the unit remains on-hold.

To return to the call, the telephone can be picked up. The loop current flowing through the seizure network is reduced because of the double termination. RY2 deactivates, and the seizure network is disconnected.

Kick-back capacitor C5 ensures the loop current is reduced below the drop-out current for RY2. That reduction in current turns off LED1, disconnects the music, and reconnects the caller.

If the telephone is not hung up within the 4 s time-out period, RY1 will deactivate and the project will be taken out of the "STANDBY" mode and placed in the "NORMAL" mode. LED1 will not be lit, and the caller will be disconnected if the telephone is hung up.

Latching push-button switch S1 is used to tune in the desired station. When it is in the IN position, the seizure network is placed across the telephone line and the output of the tuner is also connected (regardless of the status of J1). That allows you to hear the output of the FM tuner and adjust the station tuning and volume. (A feature of the receiver is the elimination of inter-station "hiss," therefore no audio will be present until a station is tuned in.

FM receiver circuit

At the heart of the FM receiver circuit is U4, a TDA7088T FM receiver chip. This IC has a frequency-locked loop system with intermediate frequency (IF) 70 KHz. The IF can be chosen by active RC filters. The receiver is tuned via pushbutton switches, which select the desired radio stations. A block diagram of the FM receiver chip is shown in Figure 12-10.

The FM receiver contains a mixer section, where the antenna enters the circuit and the input filters reside. The VCO is fed directly to the mixer section as shown. The tuning section of the circuit is centered around pins 1, 15, and 16. The IF limiter is coupled to the all-pass filter and demodulator circuit, which feeds the loop filter. The loop filter is then coupled into the search tuning circuit. The receiver chip pin-outs are illustrated in Table 12-4.

The antenna "piggy-backs" or utilizes one telephone line and the RJ11 cable. The RF signal travels through that path and is coupled via DC blocking capacitor C30 to RF input bandpass filter. This broadband low-Q filter consists of C20, C21, and L2. Its primary purpose is to pass RF energy in the 8.0- to 108.0-MHz range while attenuating RF energy from above and below that frequency range. The bandpass filter serves to suppress potential interfering energy from outside the commercial FM broadcast band.

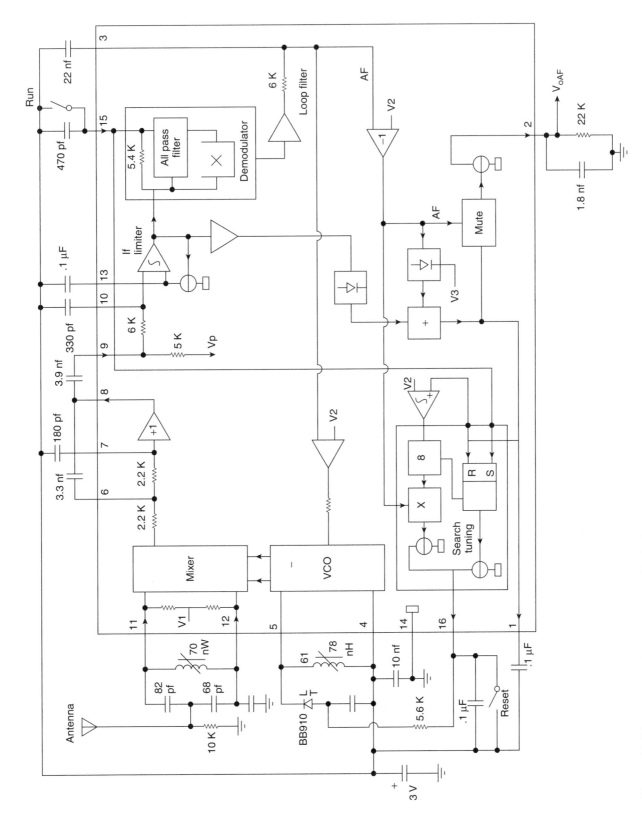

Figure 12-10 *TDA-7088T FM receiver block diagram*

Table 12-4

TDA7088T FM receiver chip pin-outs

Symbol	Pin	Description
MUTE	1	mute output
V_{oAF}	2	audio frequency output signal
LOOP	3	AF loop filter
V_p	4	+3 V supply voltage
OSC	5	oscillator resonant circuit
IFFB	6	IF feedback
C_{LP1}	7	low pass capacitor of 1-dB amplifier
V_{oIF}	8	IF output to external coupling capacitor (high-pass)
V_{ilf}	9	IF input to limiter amplifier
C_{LP2}	10	low-pass capacitor of IF limiter
V_{iRF}	11	radio frequency input
V_{iRF}	12	radio frequency input
C_{LIM}	13	limiter offset voltage capacitor
GND	14	ground (0 V)
C_{AP}	15	all-pass filter capacitor/input for search tuning
TUNE	16	electrical tuning/AFC output

The FM receiver is tuned using S1, the manual tune search pushbutton switch, and the reset pushbutton switch at S2. The AFC or automatic frequency control of the local oscillator is accomplished using the Varicap diode at D10.

The audio signal from the receiver chip passes through C13 and VR1 to the inverting input of audio amplifier U5. Feedback resistor C29 controls the gain of the amplifier from 0 to 20. Transformer T1 matches the amplifier's output impedance to the telephone line impedance.

The multi-output power supply for the automatic Music-on-Hold circuit is shown in Figure 12-11.

The power supply consists of a transformer which steps down the 110 V AC to 18 V AC; the output from the transformer is sent to a rectifier and filter capacitors at C31 and C32. The output from the filter is next sent to diode D12 and then on to U8, a 12-V regulator. The 12 volts from this first regulator are then used to feed the remaining two regulators which provide 9 V, through U6 and 5 volts through U7. A 6-V output is provided from a voltage divider off the 12-V regulator.

Construction

The entire FM hold circuit is mounted on one double-sided PC board. The use of a single-sided board will work as long as the jumper wires are added to the top where necessary. We recommend that a PC board be used because of the VHF range involved in this project. The importance of proper soldering cannot be emphasized enough for VHF circuits. It is recommended that the flux residue be removed from the completed PC board using a mild on-CFC cleaner that is not harmful to plastics. Always read the manufacturer's label.

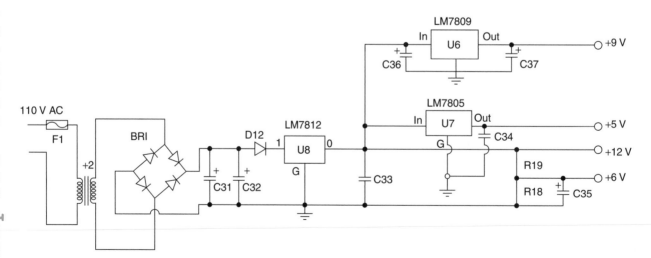

Figure 12-11 *Automatic Music-on-Hold power supply*

Static sensitive devices

Observe electrostatic discharge precautions when handling individual semiconductors as well as the completed circuit board. Component leads should be pre-formed before inserting them in the board. Before you begin construction, there are few things to keep in mind.

Before we begin constructing the Music-on-Hold circuit, you will need to locate a clean, well-lit worktable or workbench. Next, we will gather a small 25–30-W pencil tipped soldering iron, a length of 60/40 tin–lead solder, and a small canister of Tip Tinner, available at your local RadioShack store; Tip Tinner is used to condition the soldering tip between solder joints. You drive the soldering tip into the compound and it cleans and prepares the soldering tip. Next grab a few small hand tools, try and locate small flat blade and Phillips screwdrivers. A pair of needle-nose pliers, a pair of tweezers along with a magnifying glass, and a pair of end-cutters and we will begin constructing the project. Now locate your schematic diagram and parts layout diagram along with all of the components needed to build the project.

Once all the components are in front of you, you can check them off against the project parts list to make sure you are ready to start building the project. Now, locate Tables 12-1 and 12-2, and place them in front of you. Table 12-1, illustrates the resistor code chart and how to read resistors, while Table 12-2 illustrates the capacitor code chart which will aide you in constructing the project.

Finally, we are ready to begin assembling the project, so let's get started. The prototype was constructed on a single-sided printed circuit board or PCB. With your PC board in front of you, we can now begin populating the circuit board. Now refer to the resistor color code chart shown in Table 12-1. You will note that each resistor will have either three or four color bands, which begin at one end of the resistor's body. The first colored band represents the resistor's first digit, while the second color represents the second digit of the resistor's value. The third colored band represents the resistor's multiplier value, and the fourth color band represents the resistor's tolerance value. If there is no fourth colored band then the resistor has a 20% tolerance

value. If the resistor's fourth band is silver then the resistor has a 10% tolerance value and if the fourth band is gold then the resistor's tolerance value will be 5%. We are going to place and solder the lowest height components first, the resistors and diodes. First locate our first resistor R1. Resistor R1 has a value of 10 K ohms and is a 5% tolerance type. Now take a look for a resistor whose first color band is brown, black, orange, and gold. From the chart you will notice that brown is represented by the digit (1), and the second color band is black, which is represented by (0). Notice that the third color is orange and the multiplier is (1,000), so $(1)(0) \times 1,000 = 10,000$ or 10 K ohms.

Go ahead and locate where resistor R1 goes on the PC and install it; next solder it in place on the PC board, and then with a pair of end-cutters trim the excess component leads flush to the edge of the circuit board. Next, locate the remaining resistors and install them in their respective locations on the main controller PC board. Solder the resistors to the board, and remember to cut the extra lead with your end-cutters.

Next refer to Table 12-2. This chart helps to determine the value of small capacitors. Many capacitors are quite small in physical size, so manufacturers have devised a three-digit code which can be placed on the capacitor instead of the actual value. The code occupies a much smaller space on the capacitor and can usually fit nicely on the capacitor body, but without the chart the builder would have a difficult time determining the capacitor's value.

Note that capacitors generally come from two major groups or types. Non-polar types, such as the ones in this project, have no polarity and can be placed in either direction on the circuit board. The other type of capacitors are called polarized types, these capacitors have polarity and have to be inserted on the board with respect to their polarity marking, in order for the circuit to operate correctly. On these types of capacitors, you will observe that the components actually have a plus or minus marking or a black or white band with a plus or minus marking on the body of the capacitors. Capacitors must be installed properly with respect to polarity in order for the circuit to work properly. Failure to install electrolytic capacitors correctly could cause damage to the capacitor as well as to other components in the circuit when power is first applied.

Now look for the lowest profile capacitors from the component stack and locate capacitor C1, it is labeled (100) or 100 pF. Note that there are a number of small capacitors in the circuit that may not be labeled with their actual value, so you will have to check Table 12-2 to learn how to use the three-digit capacitor code. Go ahead and place C1 on the circuit board and solder it in place. Now take your end-cutters and cut the excess component leads flush to the edge of the circuit board. Next, locate the remaining low profile capacitors and install them on the printed circuit board, then solder them in place. Remember to trim the extra component leads. There are a number of electrolytic capacitors in the circuit and placing them in the circuit with respect to their proper polarity is essential for proper operation of the circuit. Once you have located the larger electrolytic capacitors install them in their respective locations and solder them in place. Finally, trim the excess component leads. When installing polarized capacitors be sure to orient them so their values can be easily read. This will help if troubleshooting is needed later on.

The Music-on-Hold circuit uses a number of diodes, including a Zener diode and a Varicap diode. Diodes are generally two lead devices which have polarity. Usually you will find a white or black band at one side of the diode package. The colored band denotes the diode's cathode lead. When installing the diodes make sure to observe the correct polarity; failure to do so will result in the circuit not working correctly. Note that diodes D1 through D9 are convention silicon diodes, while D10 is a Varicap diode and D11 is a Zener diode. Locate and install LED1, which is D8 and located at one end of the transformer and the junction of R11 and C7. Solder the diodes in their respective locations and then remember to trim the excess lead with your end-cutters.

The circuit uses a single dual transistor at Q1. Transistors are three lead devices, which have a Collector, a Base and Emitter lead. The base lead is generally in the input lead which is drawn perpendicular to a vertical line in its symbol. The emitter lead will always have an arrow shown on it. In this project the base lead is connected to R9, while the emitter lead is connected and ground and finally the collector lead is connected to the anode of diode D2. Go ahead and install transistor Q1 and solder it in place, then trim the excess leads.

You should use IC sockets for all ICs. Integrated circuit sockets are a low cost form of insurance against circuit failure at possible later date. Integrated circuits usually have some form of "key" at the top edge of the IC package. This key will either be a small indented circle at the top left of the IC package or a cut-out at the top edge of the package. Usually pin 1 of the IC is just to the left of the cut-out or indented circle. Insert the ICs carefully into the IC sockets using a grounded anti-static wristband. When installing the transformer T1, bend the tabs flush against the PC board. The audio matching transformer has a "P" indicating the primary side. The primary mounts towards the outside of the board. C6 should be mounted vertically with the body in the hold closest to D4. The shields of coils L1 and L2 should be soldered to the ground on the circuit board. The FM radio chip should be soldered directly to the PC board. This is a surface mounted component so take care in handling this small component. Be sure to use a small pencil tip soldering and be sure to quickly solder the connections to the PC board. The LED should be mounted about a ½″ away from the PC board. Mount D10 flush against the PC board to avoid stray capacitance. The crystal has no polarity so it can be mounted in either direction. The relays are both the same type and are interchangeable.

The following pre-test steps should be done after all components have been installed. Check that all components are mounted in their proper location. Verify polarized components are properly oriented and that all pads and connections have been properly soldered and de-fluxed. Once those steps have been completed, you can begin bench testing.

Testing and alignment

The only instrument needed to test the unit is a DMM. Connect the power pack or a +15- to +28-V DC power source to the input. Connect AC power to the power pack. Do not connect the unit to the phone line at this time.

Once the phone line is connected, dial your own number to eliminate the signal tone and off-hook warning tone. Turn the receiver on by depressing pushbutton switch S1. Set the tuning potentiometer to the extreme counter-clockwise position (low end of the band). Note that due to the mute function, there is silence until a station is received.

Turn the volume control potentiometer (VR1) ½ to ¾ clock-wise. Adjust the slug in L2 until the station operating at the lowest dial setting in your area is received with the loudest audio output. Use care when adjusting the slug as it is quite delicate and easily broken. Next, set the tuning potentiometer to the extreme clockwise position (top end of the band). Tune back down towards the bottom end of the band (counter-clockwise) until the station operating at the highest frequency is received.

Tune through the entire range to verify all stations available to your area are being received. The receiver section was designed with a mute function built-in to allow only the strongest stations to be received. That makes tuning easier and suppresses images or "ghost" stations that appear in the wrong part of the tuning dial. Release the pushbutton and hang up the phone. You can check for proper operation by having a friend call and be placed on hold by depressing the star "∗" key.

Installation and use

A special feature of this project allows you to select when the music is present in the handset. Some telephone services (call waiting, call forwarding, voice mail) require the use of the "∗" key. With J1 in the

OUT position (circuit open), music will not be heard in the handset when the "∗" key is depressed. It will, however be heard by the caller when the phone is hung up. With J1 in the IN position (circuit closed), music will be heard every time the "∗" key is depressed.

If you would like to connect an external antenna or RF source, such as cable, to the tuner, you connect it to the junction of C20 and C30. It may be advantageous to disconnect the phone-line antenna by breaking the connection at C30.

It is easy to use the FM on-hold unit. To place a caller on hold, press the star "∗" key on any Touch Tone telephone. That places the unit in a standby mode and the LED lights dimly. The telephone must be hung up within four seconds for the caller to be placed on hold. When that is done, the LED lights brightly. If it is not hung up within four seconds, the unit resets itself and the LED goes out. The caller will be disconnected if the phone is hung up.

After a caller has been placed on hold, all you have to do is pick up the telephone to return to the conversation (any telephone connected to the line, Touch Tone, or rotary). When the handset is picked up, the brightly lit LED will extinguish, the music will go off, and you will be connected to the caller.

Chapter 13

Telephone Ringer Projects

Parts list

Silent Cell Phone Ringer

R1   100 K ohm, $\frac{1}{4}$ W,
   5% resistor

R2   3.9 K ohm, $\frac{1}{4}$ W,
   5% resistor

R3   1 megohm, $\frac{1}{4}$ W,
   5% resistor

C1, C2   100 nF, 63 V
   polyester capacitor

C3   220 µF, 25 V
   electrolytic capacitor

D1   ultra-bright LED

D2   BAT42 schottky-
   barrier diode or
   1N5819 diode

Q1   BC547 NPN transistor

U1   TS555CN or 7555 CMOS
   timer IC

L1   sensor coil –
   130–150T of .2 mm
   enameled wire
   (see text)

B1   1.5 volt AAA battery

Misc   PC board, wire,
   IC socket, enclosure,
   battery holder/
   clip, etc.

Remote Telephone Ringer

R1   1 megohm, $\frac{1}{2}$ W,
   5% resistor

R2   2.2 K ohm, $\frac{1}{4}$ W,
   5% resistor

C1   470 nF, 250 volt
   mylar capacitor

C2   10 µF, 15 V
   electrolytic capacitor

D1, D2   20 V Zener
   diodes

D3, D4   1N4007 silicon
   diodes

D5, D6   1N4007 silicon
   diodes

D7   super bright red LED

D8   12 V Zener diode

BZ   12 V electronic
   buzzer

J1   RJ-11 PC board type
   telephone socket

Misc   PC board, wire,
   IC socket, hardware,
   connectors, enclosure,
   etc

Kit is available from
Velleman, Inc

Kit - K8087

Telephone Ring Detector

R1   1 megohm, $\frac{1}{2}$ W,
   5% resistor

R2   2.2 K ohm, $\frac{1}{4}$ W,
   5% resistor

R3   15 K ohm, $\frac{1}{4}$ W,
   5% resistor

R4   100 K ohm, $\frac{1}{4}$ W,
   5% resistor

R5   1 K ohm, $\frac{1}{4}$ W,
   5% resistor

C1 470 nF, 250 volt
mylar capacitor

C2 100 µF, 25 V
electrolytic capacitor

C3 47 µF, 25 V
electrolytic capacitor

D1, D2 20 volt Zener
diodes

D3, D4 1N4007 silicon
diodes

D5, D6, D8 1N4007
silicon diodes

D7 super bright red LED

D9 1N4148 silicon diode

Q1 BC557 PNP transistor

U1 4N27 opto-coupler IC

J1 RJ-11 PC board type
telephone socket

J2 coaxial switched
power jack

RY1 VR15M121C – SPDT –
12 V DC - 360 ohm -
PC board type relay

Misc PC board, wire,
IC socket, hardware,
enclosure, etc

Ext 12 V - 100 mA—
"wall-wart" power
supply

Kit is available from
Velleman, Inc

Kit - K8086

**Wireless Telephone
Ringer**

R1 68 K ohms, $\frac{1}{4}$ W,
5% resistor

R2 150 ohms, $\frac{1}{4}$ W,
5% resistor

R3, R5 47 K ohms, $\frac{1}{4}$ W,
5% resistor

R4, R6 1 K ohms, $\frac{1}{4}$ W,
5% resistor

R7 10 K ohms, $\frac{1}{4}$ W,
5% resistor

R8 4.7 K ohms, $\frac{1}{4}$ W,
5% resistor

R9 100 K ohms, $\frac{1}{4}$ W,
5% resistor

R10, R11 4.7 K ohms,
1 W, 5% resistor

R12 5-10 K ohms, 1 W,
5% resistor (see text)

R13 1 megohm, $\frac{1}{4}$ W,
5% resistor

C1 6,800 pF, 35 V
ceramic or mylar
capacitor

C2, C10, C12 .1 µF,
35v ceramic capacitor

C3, C11 .001 µF, 35 V
ceramic capacitor

C4 22 pF, 35 V NPO
ceramic capacitor

C5 47 pF, 35 V NPO
ceramic capacitor

C6 2.2 µF, 250 V mylar
capacitor

C7 33 µF, 35 V
electrolytic capacitor

C8 .47 µF, 35 V
ceramic capacitor

C9 10 pF, 35 V NPO
ceramic capacitor

D1, D2 1N4148 silicon
diode

Q1 2N2646 uni-junction
transistor

Q2, Q3 2N5179 NPN RF
silicon transistor

Q4 2N6659 N-Channel
Field-Effect
transistor

Q5 2N2907 PNP silicon
transistor

U1 HCPL-3700
Opto-Coupler IC

B1 9-V transistor radio
battery

L1 inductor hand-wound
(see text)

S1 SPST switch

Misc PC board, whip
antenna, enclosure,
battery clip,
hardware, etc.

Telephone Ring Generator

R1 150 K ohm $\frac{1}{4}$ W,
5% resistor

R2 100 ohm $\frac{1}{4}$ W,
5% resistor

R3, R13, R14 470 ohm
$\frac{1}{2}$ W, 5% resistor

R4 75 ohm $\frac{1}{4}$ W,
5% resistor

R5, R7, R8 10 K ohm
$\frac{1}{4}$ W, 5% resistor

R6 300 K ohm $\frac{1}{4}$ W,
5% resistor

R9 50 K ohm trim
potentiometer

R10 22 K ohm $\frac{1}{4}$ W,
5% resistor

R11, R12, R15, R16
1 K ohm $\frac{1}{2}$ W,
5% resistor

C1 .01 µF, 35 V
ceramic disk capacitor

C2, C3 1,000 µF,
35 V electrolytic
capacitor

C4 22 µF, 35 V
electrolytic capacitor

C5 4 µF, 35 V
electrolytic capacitor

C6 1 µF, 35 V
electrolytic capacitor

D1, D2, D3, D4 1N4004
silicon diode

Q1 2N3053 transistor
NPN

Q2 TIP47 or NTE287
transistor NPN

Q3, Q6, Q7, Q8 NTE287
transistor NPN

Q4, Q5 NTE288
transistor PNP

U1 74HC14 IC

L1 10 mH coil

S1 momentary pushbutton
switch

Misc PC board, wire,
connectors, chassis
box, etc.

Project 1: Silent Cellular Phone Ringer

The Silent Cellular Phone Ringer is a novel yet simple circuit designed to work with your personal cell phone. Suppose you are in an important meeting, at the doctor's office, and it is not permitted to have your cell ring in public. This circuit will allow you to monitor a cell phone quietly or remotely when silence is needed.

The heart of the Silent Cell Phone Ringer circuit is the integrated circuit (IC) at U1, which is shown in Figure 13-1.

The circuit is designed to detect when an incoming cellular phone is arriving but will alert you silently via an LED. This circuit must be placed within 6″ or so from your cell phone, so its sensor coil at L1 can detect the field emitted by the cell phone receiver during an incoming call.

The signal detected by the sensor coil is amplified by transistor Q1 and drives the mono-stable input if U1. The integrated circuit's output voltage is doubled by C2 and D2 in order to drive the high efficiency ultra-bright LED at a suitable peak-voltage. The Silent Cell Phone Ringer circuit is powered by a single 1.5-V battery. The circuit draws less than 200 µA, so no power on-off

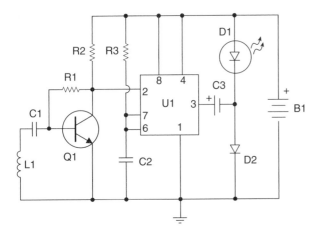

Figure 13-1 *Silent Cellular Phone Ringer (Courtesy Redcircuits.com)*

switch was used. Sensitivity of the circuit depends on the sensor coil. The sensor Coil L1 consists of 130 to 150 turns of .2 mm enameled wire on a 5 cm diameter form. The coil is an air core type, so you will have to remove the form once the coil is constructed. Once the form is removed, wrap the coil in some electrical tape. You could also elect to use a standard commercial 10 mH miniature inductor but sensitivity will be lower. The integrated circuit at U1 must be a CMOS IC timer such as a TS555CN or a 7555 which operates 1.5 V. A BAT46 Schottky-barrier diode should be used for diode D2, although you could substitute a 1N5819 diode.

Circuit assembly

Before we begin building the Silent Cellular Phone Ringer, you will need to locate a clean, well-lit worktable or workbench. Next we will gather a small 25–30-W pencil tipped soldering iron, a length of 60/40 tin–lead solder, and a small canister of Tip Tinner available at your local RadioShack store; Tip Tinner is used to condition the soldering tip between solder joints. You drive the soldering tip into the compound and it cleans and prepares the soldering tip. Next grab a few small hand tools; try and locate small flat blade and Phillips screwdrivers. A pair of needle-nose pliers, a pair of tweezers along with a magnifying glass, and a pair of end-cutters and we will begin constructing the project. Now locate your schematic diagram and parts layout diagram along with all of the components needed to build the project. Once all the components are in front of you, you can check them off against the project parts list to make sure you are ready to start building the project. Now, locate Tables 13-1 and 13-2, and place them in front of you.

Table 13-1 illustrates the resistor code chart and how to read resistors, while Table 13-2 illustrates the capacitor code chart which will aide you in constructing the project.

Refer now to the resistor color code chart shown in Table 13-1, which will assist you in identifying the

Table 13-1
Resistor color code chart

Color Band	1st Digit	2nd Digit	Multiplier	Tolerance
Black	0	0	1	
Brown	1	1	10	1%
Red	2	2	100	2%
Orange	3	3	1,000 (K)	3%
Yellow	4	4	10,000	4%
Green	5	5	100,000	
Blue	6	6	1,000,000 (M)	
Violet	7	7	10,000,000	
Gray	8	8	100,000,000	
White	9	9	1,000,000,000	
Gold			0.1	5%
Silver			0.01	10%
No color				20%

Table 13-2
Three-digit capacitor codes

pF	nF	µF	Code	pF	nF	µF	Code
1.0			**1R0**	3,900	3.9	.0039	**392**
1.2			**1R2**	4,700	4.7	.0047	**472**
1.5			**1R5**	5,600	5.6	.0056	**562**
1.8			**1R8**	6,800	6.8	.0068	**682**
2.2			**2R2**	8,200	8.2	.0082	**822**
2.7			**2R7**	10,000	10	.01	**103**
3.3			**3R3**	12,000	12	.012	**123**
3.9			**3R9**	15,000	15	.015	**153**
4.7			**4R7**	18,000	18	.018	**183**
5.6			**5R6**	22,000	22	.022	**223**
6.8			**6R8**	27,000	27	.027	**273**
8.2			**8R2**	33,000	33	.033	**333**
10			**100**	39,000	39	.039	**393**
12			**120**	47,000	47	.047	**473**
15			**150**	56,000	56	.056	**563**
18			**180**	68,000	68	.068	**683**
22			**220**	82,000	82	.082	**823**
27			**270**	100,000	100	.1	**104**
33			**330**	120,000	120	.12	**124**
39			**390**	150,000	150	.15	**154**
47			**470**	180,000	180	.18	**184**
56			**560**	220,000	220	.22	**224**
68			**680**	270,000	270	.27	**274**
82			**820**	330,000	330	.33	**334**
100		.0001	**101**	390,000	390	.39	**394**
120		.00012	**121**	470,000	470	.47	**474**
150		.00015	**151**	560,000	560	.56	**564**
180		.00018	**181**	680,000	680	.68	**684**
220		.00022	**221**	820,000	820	.82	**824**
270		.00027	**271**		1,000	1	**105**
330		.00033	**331**		1,500	1.5	**155**
390		.00039	**391**		2,200	2.2	**225**
470		.00047	**471**		2,700	2.7	**275**
560		.00056	**561**		3,300	3.3	**335**
680		.00068	**681**		4,700	4.7	**475**
820		.00082	**821**			6.8	**685**
1,000	1.0	.001	**102**			10	**106**
1,200	1.2	.0012	**122**			22	**226**
1,500	1.5	.0015	**152**			33	**336**
1,800	1.8	.0018	**182**			47	**476**

Table 13-2—cont'd

Three-digit capacitor codes

pF	nF	µF	CODE	pF	nF	µF	CODE
2,200	2.2	.0022	222			68	686
2,700	2.7	.0027	272			100	107
3,300	3.3	.0033	332			157	157

Code = 2 significant digits of the capacitor value + number of zeros to follow

For values below 10 pF, use "R" in place of decimal e.g. 8.2 pF = 8R2

10 pF = 100

100 pF = 101

1,000 pF = 102

22,000 pF = 223

330,000 pF = 334

1 µF = 105

resistor values. Note that each resistor will have either three or four color bands, which begin at one end of the resistor's body. The first colored band represents the resistor's first digit, while the second color represents the second digit of the resistor's value. The third colored band represents the resistor's multiplier value, and the fourth color band represents the resistor's tolerance value. If there is no fourth colored band then the resistor has a 20% tolerance value. If the resistor's fourth band is silver then the resistor has a 10% tolerance value and if the fourth band is gold then the resistor's tolerance value will be 5%.

So finally we are ready to begin assembling the project, so let's get started. The prototype project was constructed on a single-sided printed circuit board or PCB; with your PC board in front of you, we can now begin populating the circuit board. We are going to place and solder the lowest height components first, the resistors and diodes. First locate our first resistor R1. Resistor R1 has a value of 100 K ohms and is a 5% tolerance type. Now take a look for a resistor whose first color band is brown, black, yellow, and gold. From the chart you will notice that brown is represented by the digit (1), and the second band is black, which is represented by (0). Notice that the third color is brown and the multiplier is (10,000), so $(1)(0) \times 10,000 = 100,000$ ohms.

Go ahead and place R1 on the circuit board and solder it in place. Next take your end-cutters and trim the excess component leads from the resistor. Locate the remaining resistors and solder them in place on the

board and remember to trim the excess leads from the board.

Note that capacitors generally come from two major groups or types. Non-polar types, such as the ones in this project, have no polarity and can be placed in either direction on the circuit board. The other type of capacitors are called polarized types, these capacitors have polarity and have to be inserted on the board with respect to their polarity marking in order for the circuit to operate correctly. On these types of capacitors, you will observe that the components actually have a plus or minus marking or a black or white band with a plus or minus marking on the body of the capacitors.

Next, refer to Table 13-2; this chart helps to determine the value of small capacitors. Many capacitors are quite small in physical size, so manufacturers have devised a three-digit code which can be placed on the capacitor instead of the actual value. The code occupies a much smaller space on the capacitor and can usually fit nicely on the capacitor body, but without the chart the builder would have a difficult time determining the capacitor's value.

Now locate the capacitors from the component pile and try and locate capacitor C1, which should be marked with a (104), this capacitor has a 100 nF value. Now that you can read the resistor and capacitor charts, let's go ahead and place R1 and C1 on the circuit board and solder them in place. Now take your end-cutters and cut the excess component leads flush to the edge of the circuit board. Once you have installed the first two components, you can move on to installing the remaining resistors and capacitors.

This project contains Schottky-barrier silicon diode at D2. These diodes also have polarity, which must be observed if the circuit is going to work correctly. You will notice that each diode will have a black or white colored band at one edge of the diode's body. The colored band represents the diode's cathode lead. Check your schematic and parts layout diagrams when installing the diode on the circuit board, then solder it in place, remembering to cut the excess component leads.

The Silent Cell Phone Ringer circuit also utilizes one transistor at Q1, a PNP transistor. These generally have three leads, a Base lead, a Collector lead and an Emitter lead. Refer to the schematic and you will notice that there will be two diagonal leads pointing to a vertical line. The vertical line is the transistor's base lead and the two diagonal lines represent the collector and the emitter. The diagonal lead with the arrow on it is the emitter lead, and this should help you install the transistor onto the circuit board correctly. Go ahead and install the transistor on the PC board and solder it in place, then follow-up by cutting off the excess component leads.

The Silent Cellular Phone Ringer circuit contains a single IC at U1. Note that the IC is a 555 timer, but it is a low power version of the 555. The IC timer is a CMOS version of the timer; use a TS555CN or 7555 which runs on a single 1.5-V AAA cell. ICs have to be installed in a particular way in order to function properly. You will note that at the top end of the package, you will see a small cutout of notch. Pin 1 will be just to the left of this notch. When installing the IC pay close attention to orienting the chip correctly and refer to the schematic and parts layout diagrams. It would be very prudent to install an IC socket for the timer IC, in the unlikely event of a circuit failure. It is much easier to simply unplug an IC rather than trying to unsolder a number of pins from the circuit board in the event of a circuit failure.

When installing the indicator LED, D1, be sure to pay particular attention to the orientation. The symbol for an LED is a diode, it therefore has an anode and cathode. Usually the anode lead is longer and this will help orient the LED when mounting it on the PC board. Depending upon how you mount the circuit board in

your enclosure, you may want to leave ½" of lead length on the diodes.

Finally, it is time to install the coil, L1. Remember coil L1 is composed of 130–150 turns of enameled wire. The enameled wire has a coating on its entire length, so you will have to scrape about a ⅛" of coating off each end of the wire, so you can solder the wire to the circuit board. It is usually recommended to mount the coil at one end of the circuit board, so it can face outwards. Now locate your 1.5-V AAA single cell battery holder and mount it in your chosen enclosure and wire the holder to the circuit board.

Now that the circuit board has been completed, you can take a short well deserved break and when we return we will inspect the circuit board for any "cold" solder joints or "short" circuits.

Pick up the circuit board with the foil side facing upwards toward you. First we will attempt to locate any cold solder joints on the circuit board. Take a look at all the solder joints. The solder joints should all look clean, bright, and shiny. If any of the solder joints look dark, dull, or blobby then you will have to remove the solder from the joint with a solder sucker or wick, and re-solder the joint all over again.

Next we will examine the PC board for any possible short circuits. Short circuits can be caused by a solder blob bridging across the circuit traces or it could be caused by stray component leads. Often solder can leave a sticky residue, and cut component leads can attach themselves to this residue and cause a short between circuit traces. Look carefully for any solder blobs or stray leads bridging the circuit traces. When everything looks good, you can move on to mounting the circuit into your favorite enclosure.

Operation

Once the circuit has been inspected, you can mount it in a plastic enclosure, and you must use a plastic enclosure since metal would block the radio waves between the phone and your ringer circuit. Once the circuit is mounted, insert an AAA battery into the battery holder to apply power to the circuit. Place the Silent Cellular Phone Ringer close to your cell phone and have someone call your cell phone. Once the cell phone rings the LED on your Silent Cellular Phone Ringer should begin blinking. By letting the phone ring for a while you can experiment with the distances between the cell phone and your Silent Cellular Phone Ringer circuit.

In the event that your circuit does not function, you will have to remove power and once again inspect the circuit board for any errors. The most common errors for circuit failure include improper orientation of electrolytic capacitors, diodes, and semiconductors such as transistors.

Another cause for failure is the incorrect orientation on any integrated circuits. Once you have carefully re-inspected the circuit board, you can apply power and re-test the circuit all over again.

Have fun using your new Silent Cellular Phone Ringer circuit. The Silent Cell Phone Ringer is ideal for those important business meetings or for when the baby is sleeping.

Figure 13-2 *Remote telephone ring indicator. Courtesy of Velleman Kits*

Project 2: Remote telephone ringer – w/buzzer and LED

The Remote Telephone Ringer project is illustrated in Figure 13-2. This handy circuit can be placed in a remote location such as a garage or patio to alert you of an incoming telephone call. It could also be used to alert you of an incoming call with the buzzer off when someone is sleeping. The remote ringer project is great for noisy environments, for the hearing impaired, as additional ringer, or to replace an existing ringer. The Remote Telephone Ringer is simply connected in parallel with the phone line and requires no external power supply, and consumes a modest 10 mA from the phone line. The circuit simply plugs into a standard RJ-11 telephone jack at any location.

The clever ringer/flasher circuit shown in Figure 13-3 plugs into any RJ-11 phone jack and will alert you when the phone is ringing with a bright LED and electronic buzzer.

The circuit begins at the RJ-11 phone jack at the left of the diagram. Pin 4 of the phone jack is first coupled

to a 1 megohm resistor in parallel with a 470 nF capacitor. The ring signal is then passed on through two 20-V Zener diodes; back-to-back in series with a 2.2 K ohm resistor. Resistor R2 is next coupled to a diode bridge consisting of diodes D3 through D6. The diode bridge converts the AC ringing signal to a DC output to drive the LED on and off with the ringing pulses. The output of the LED is then passed on to a 12-V Zener diode which limits the voltage to the buzzer at BZ. A filter capacitor is placed across the buzzer leads to help smooth the voltage signal to the DC powered buzzer. The plus lead of the buzzer is connected to the plus side of the filter capacitor, while the minus lead of the buzzer goes to ground.

Circuit assembly

Before we begin building the Remote Telephone Ringer, let's first find a clean, well-lit worktable or workbench. Next we will gather a small 25–30-W pencil tipped soldering iron, a length of 60/40 tin–lead solder, and a small canister of Tip Tinner, available at your local RadioShack store; Tip Tinner is used to condition the

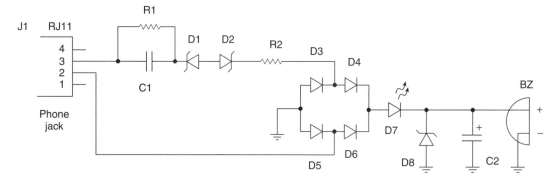

Figure 13-3 *Remote telephone ringer with LED and buzzer. Courtesy of Velleman Kits*

soldering tip between solder joints. You drive the soldering tip into the compound and it cleans and prepares the soldering tip. Next grab a few small hand tools, try to locate small flat blade and Phillips screwdrivers. A pair of needle-nose pliers, a pair of tweezers along with a magnifying glass, and a pair of end-cutters and we will begin constructing the project. Now locate your schematic diagram and parts layout diagram along with all of the components needed to build the project. Once all the components are in front of you, you can check them off against the project parts list to make sure you are ready to start building the project. Now, locate Tables 13-1 and 13-2, and place them in front of you.

Table 13-1 illustrates the resistor code chart and how to read resistors, while Table 13-2 illustrates the capacitor code chart which will aid you in constructing the project.

Refer now to the resistor color code chart shown in Table 13-1, which will help you to identify the resistor values for the project. Note that each resistor will have either three or four color bands, which begin at one end of the resistor's body. The first colored band represents the resistor's first digit, while the second color represents the second digit of the resistor's value. The third colored band represents the resistor's multiplier value, and the fourth color band represents the resistor's tolerance value. If there is no fourth colored band then the resistor has a 20% tolerance value. If the resistor's fourth band is silver then the resistor has a 10% tolerance value and if the fourth band is gold then the resistor's tolerance value will be 5%.

Finally, we are ready to begin assembling the project, so now let's get started. The prototype project was constructed on a single-sided printed circuit board or PCB. With your PC board in front of you, we can now begin populating the circuit board. We are going to place and solder the lowest height components first, the resistors and diodes. First locate our first resistor R1. Resistor R1 has a value of 1 megohm and is a 5% tolerance ½-W type. Now take a look for a resistor whose first color band is brown, black, green, and gold. From the chart you will notice that brown is represented by the digit (1), the second band is black, which is represented by (0). Notice that the third color is green and the multiplier is (100,000), so (1)(0) × 100,000 = 1,000,00 or 1 megohm.

Go ahead and install R1 on the circuit board and solder it in place in its respective position. Locate your end-cutters and trim the excess component leads. Cut the excess leads flush to the edge of the circuit board. Now locate the remaining resistors and place them on the circuit board and solder them in place, remembering to trim the excess component leads with your end-cutters.

Note that capacitors generally come from two major groups or types. Non-polar types, such as the ones in this project, have no polarity and can be placed in either direction on the circuit board. The other type of capacitors are called polarized types; these capacitors have polarity and have to be inserted on the board with respect to their polarity marking in order for the circuit to operate correctly. On these types of capacitors, you will observe that the components actually have a plus or minus marking or a black or white band with a plus or minus marking on the body of the capacitors.

Next refer to Table 13-2; this chart helps to determine the value of small capacitors. Many capacitors are quite small in physical size, so manufacturers have devised a three-digit code which can be placed on the capacitor instead of the actual value. The code occupies a much smaller space on the capacitor and can usually fit nicely on the capacitor body, but without the chart the builder would have a difficult time determining the capacitor's value.

Now locate the capacitors from the component pile and try and locate capacitor C1, which should be marked with a three-digit code marked (474) or 470 nF. Capacitor C1 will be rated at 250 V, and may be somewhat large. Next locate C2, a 10 µF electrolytic capacitor. Now, let's go ahead and place C1 and C2 on the circuit board and solder them in place. Now take your end-cutters and cut the excess component leads flush to the edge of the circuit board. Once you have installed the first two components, you can move on to installing the remaining components. Remember to trim the excess component leads after the solder joints are completed.

The Remote Telephone Ringer project utilizes two 20-V and one 12-V Zener diode. The project also contains four silicon diodes at D3 through D6. Diodes have polarity which must be observed if the circuit is going to work correctly. You will notice that each diode will have a black or white colored band at one edge of the diode's body. The colored band represents the diode's cathode lead. Check your schematic and parts layout diagrams when installing these diodes on the circuit board. Place all the diodes on the circuit board and solder them in place on the PC board. Then cut the excess component leads with your end-cutters flush to the edge of the PC board.

The Remote Telephone Ringer circuit employs a single super bright LED at D7. Note that the symbol for an LED is similar to a diode which has two leads, a cathode lead and an anode lead. The cathode lead will be the shorter of the two leads and schematically is the flat line in the symbol.

Finally, you can go ahead and mount the PC type Piezo buzzer on the circuit board. Pay close attention to the pinouts on the Piezo buzzer, one lead will be the plus or red lead, while the other lead is the ground or minus lead. In order for the circuit to work properly, the plus buzzer lead must be connected to the cathodes of D7 and D8 and to the plus side of capacitor C2.

Since the circuit is telephone line powered, you will not have to worry about powering the circuit.

Once all the components have been mounted and soldered on to the circuit board, you can take a short break and when we return, we will inspect the circuit board for any possible "cold" solder joints and "short" circuits. Pick up the circuit board with the foil side facing upwards toward you. Take a close look at the solder joints. They should all look clean, bright, and shiny. If any of the solder joints look dull, dark, or blobby then you should remove the solder from the joint with a solder sucker or wick and resolder the joint all over again.

Now, let's move to inspect the circuit board for any possible short circuits. Short circuits can be caused from solder blobs bridging across the circuit traces or from stray component leads bridging across the circuit traces. Solder residue often traps cut component leads to the foil side of the PC board; unchecked, this could cause damage to the circuit if not removed before power is applied to the circuit for the first time. Now we are ready to install the circuit board into the enclosure of your choice. Mount the circuit board at the bottom of your chassis box, on ¼" standoffs. Mount the circuit board at one edge of the enclosure, so that the telephone jack can protrude from the case. If you chose to use a non-PC board type buzzer, then you will have to mount the buzzer on the front panel of the enclosure. Mount the super bright LED, on the front panel of the case, so that it can protrude from the case

Operation

Once your circuit is mounted in your favorite enclosure and the circuit attached to the phone line, you are ready to test the circuit. Remember, the phone line has a polarity and it will affect how the circuit operates. Connect the Remote Phone Ringer to the phone line in your chosen area and then have someone call your telephone number and listen for the Remote Ringer circuit to come alive. In the event that the circuit does not work immediately, you will have to first check the polarity of the phone line connection or possibly recheck the circuit for any errors. The most common reason for circuit failure is improper orientation of the electrolytic capacitors, diodes, and semiconductors such as transistors and of course ICs. Recheck your circuit carefully and once inspected and errors fixed, you install the circuit into its enclosure and retest the circuit. Now you can go ahead and install the Remote Telephone Ringer in your shop, barn or basement.

Project 3: Telephone ring detector – w/LED and relay output

The Telephone Ring Detector project is illustrated in Figure 13-4; this handy circuit can be placed anywhere to alert you of an incoming telephone call with both LED and relay control which can be used to drive a siren, load noise maker, or strobe light.

The Telephone Ring Detector project is great for noisy environments, for the hearing impaired, as an additional ringer, or to replace an existing ringer. The Telephone Ring detector is connected in parallel with the phone line at any location that uses a standard RJ-11 telephone jack. Power to the ring detector circuit is provided by a 12-V external power supply.

The clever Ring Detector circuit shown in Figure 13-5 plugs into any RJ-11 phone jack and will alert you when the phone is ringing with a bright LED and relay control which can be used to control a siren, strobe lamp, or loud bell.

The circuit begins at the RJ-11 phone jack at the left of the diagram. Pin 3 of the phone jack is first coupled to a 1 megohm resistor in parallel with a 470 nF

Figure 13-4 *Remote telephone ring relay controller. Courtesy of Velleman Kits*

capacitor. The ring signal is then passed on through 22-V Zener diodes; back-to-back in series with a 2.2 K ohm resistor. Resistor R2 is next coupled to a diode bridge consisting of diodes D3 through D6. The diode bridge converts the AC ringing signal to a DC output to drive the LED on and off with the ringing pulses. The output of the bright LED is then passed on to an LED driver in an opto-coupler IC. The LED inside the opto-coupler activates the NPN transistor when the phone signal is present. The more powerful the transistor inside the opto-coupler drive, the more powerful the switching transistor at Q1, which in turn operates the relay at RLY1. Power for Q1 and the relay are provided by a 12-V DC, 100 mA "wall-wart" power supply cube. If jumper J1 is installed the relay will be operated on a continuous basis, if the jumper is left open the relay closes when the phone line rings. The relay used has both normally open and normally closed contacts which can be wired in series with an external power supply used to drive a siren, bell, or strobe lamp if desired.

Circuit assembly

Before we begin building the Telephone Ring Detector circuit, let us first find a clean, well-lit worktable or workbench. Next we will gather a small 25–30-W pencil tipped soldering iron, a length of 60/40 tin–lead solder, and a small canister of Tip Tinner, available at your local RadioShack store; Tip Tinner is used to condition the soldering tip between solder joints. You drive the soldering tip into the compound and it cleans and prepares the soldering tip. Next grab a few small hand tools; try and locate small flat blade and Phillips screwdrivers. A pair of needle-nose pliers, a pair of tweezers along with a magnifying glass, and a pair of end-cutters and we will begin constructing the project. Now locate your schematic diagram and parts layout diagram along with all of the components needed to build the project. Once all the components are in front of you, you can check them off against the project parts list to make sure you are ready to start building the project. Now, locate Tables 13-1 and 13-2, and place them in front of you.

Table 13-1 illustrates the resistor code chart and how to read resistors, while Table 13-2 illustrates the capacitor code chart which will aid you in constructing the project.

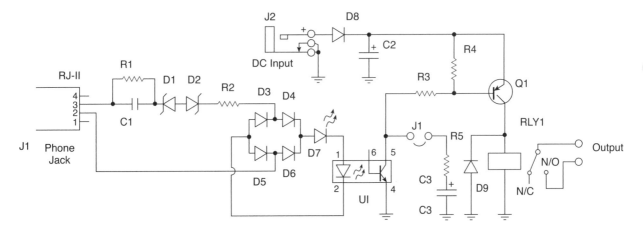

Figure 13-5 *Telephone ring detector with LED and relay control. Courtesy of Velleman Kits*

Refer now to the resistor color code chart shown in Table 13-1, which will help you to identify the resistor values for the project. Note that each resistor will have either three or four color bands, which begin at one end of the resistor's body. The first colored band represents the resistor's first digit, while the second color represents the second digit of the resistor's value. The third colored band represents the resistor's multiplier value, and the fourth color band represents the resistor's tolerance value. If there is no fourth colored band then the resistor has a 20% tolerance value. If the resistor's fourth band is silver then the resistor has a 10% tolerance value and if the fourth band is gold then the resistor's tolerance value will be 5%.

Finally, we are ready to begin assembling the project, so now let's get started. The prototype project was constructed on a single-sided printed circuit board or PCB. With your PC board in front of you, we can now begin populating the circuit board. We are going to place and solder the lowest height components first, the resistors and diodes. First locate our first resistor R1. Resistor R1 has a value of 1 megohm and is a 5% tolerance ½-W type. Now take a look for a resistor whose first color band is brown, black, green, and gold. From the chart you will notice that brown is represented by the digit (1), the second band is black, which is represented by (0). Notice that the third color is green and the multiplier is (100,000), so (1)(0) × 100,000 = 1,000,00 or 1 megohm.

Go ahead and install R1 on the circuit board and solder it in place in its respective position. Locate your

end-cutters and trim the excess component leads. Cut the excess leads flush to the edge of the circuit board. Now locate the remaining resistors and place them on the circuit board and solder them in place, remembering to trim the excess component leads with your end-cutters.

Note that capacitors generally come from two major groups or types. Non-polar types, such as the ones in this project, have no polarity and can be placed in either direction on the circuit board. The other type of capacitors are called polarized types, these capacitors have polarity and have to be inserted on the board with respect to their polarity marking in order for the circuit to operate correctly. On these types of capacitors, you will observe that the components actually have a plus or minus marking or a black or white band with a plus or minus marking on the body of the capacitors.

Next refer to Table 13-2; this chart helps to determine the value of small capacitors. Many capacitors are quite small in physical size, so manufacturers have devised a three-digit code which can be placed on the capacitor instead of the actual value. The code occupies a much smaller space on the capacitor and can usually fit nicely on the capacitor body, but without the chart the builder would have a difficult time determining the capacitor's value.

Now locate the capacitors from the component pile and try and locate capacitor C1, which should be marked with a three-digit code marked (474) or 470 nF. Capacitor C1 will be rated at 250 V, and may be somewhat large. Next locate C2 and C3. Now, let us go ahead and place C1, C2, and C3 on the circuit board and solder them in place. Now take your end-cutters and cut the excess component leads flush to the edge of the circuit board. Once you have installed the first two components, you can move on to installing the remaining components. Remember to trim the excess component leads after the solder joints are completed.

The Telephone Ring Detector project utilizes two 20-V and one 12-V Zener diode. The project also contains six silicon diodes at D3 through D6, as well as D8 and D9. Diodes have polarity, which must be observed if the circuit is going to work correctly. You will notice that each diode will have a black or white colored band at one edge of the diode's body. The colored band represents the diode's cathode lead. Check your schematic and parts layout diagrams when installing these diodes on the circuit board. Place all the diodes on the circuit board and solder them in place on the PC board. Then cut the excess component leads with your end-cutters flush to the edge of the PC board.

The Telephone Ring Detector utilizes single conventional transistor. Transistors are generally three lead devices, which must be installed correctly if the circuit is to work properly. Careful handling and proper orientation in the circuit is critical. A transistor will usually have a Base lead, a Collector lead, and an Emitter lead. The base lead is generally a vertical line inside a transistor circle symbol. The collector lead is usually a slanted line facing towards the base lead and the emitter lead is also a slanted line facing the base lead but it will have an arrow pointing towards or away from the base lead. Now, identify the two transistors, refer to the manufacturer's specification sheets as well as the schematic and parts layout diagrams when installing the two transistors. If you have trouble installing the transistors, get the help of an electronics enthusiast with some experience. Install the transistors in their respective locations and then solder.

The ring detector circuit contains an ICs opto-coupler, at U1. Integrated circuits are often static sensitive, so they must be handled with care. Use a grounded anti-static wriststrap and stay seated in one location when handling the ICs. Take out a cheap insurance policy by installing IC sockets for each of the ICs. In the event of a possible circuit failure, it is much easier to simply unplug a defective IC, that trying to unsolder 14 or 16 or more pins from a

PC board without damaging the board. ICs have to be installed correctly if the circuit is going to work properly. IC packages will have some sort of markings which will help you orient them on the PC board. An IC will have either a small indented circle, a cut-out or notch at the top end of the IC package. Pin 1 of the IC will be just to the left of the notch, or cutout. Refer to the manufacturer's pin-out diagram, as well as the schematic when installing these parts. If you doubt your ability to correctly orient the ICs, then seek the help of a knowledgeable electronics enthusiast to help you.

The Telephone Ring Detector circuit employs a single super bright LED at D7. Note that the symbol for an LED is similar to a diode which has two leads, a cathode lead and an anode lead. The cathode lead will be the shorter of the two leads and schematically is the flat line in the symbol.

Finally, you can go ahead and mount the PC type relay on the circuit board. Pay close attention to the pinouts when installing the relay, do not mix-up the coil and the contact pins.

Once all the components have been mounted and soldered on to the circuit board, you can take a short break and when we return, we will inspect the circuit board for any possible "cold" solder joints and "short" circuits. Pick up the circuit board with the foil side facing upwards toward you. Take a close look at the solder joints. They should all look clean, bright, and shiny. If any of the solder joints look dull, dark, or blobby then you should remove the solder from the joint with a solder sucker or wick and resolder the joint all over again.

Now, let's move to inspect the circuit board for any possible short circuits. Short circuits can be caused from solder blobs bridging across the circuit traces or from stray component leads bridging across the circuit traces. Solder residue often traps cut component leads to the foil side of the PC board; unchecked, this could cause damage to the circuit if not removed before power is applied to the circuit for the first time. Now we are ready to install the circuit board into the enclosure of your choice. Mount the circuit board in the bottom of the chassis box using ¼" plastic standoffs. Mount the circuit board so the RJ-11 phone jack is closest to one end of the box, so that you can cut a hole for the jack to protrude. Next install the coaxial power jack; if you chose to use a chassis mounted jack, then you will have to line up the jack with the hole drilled on the side or rear of the chassis box. You can install an RCA jack or screw terminals for the relay output pins on the rear of the chassis box. You can install the LED on the front panel of the chassis box if desired.

Operation

Once your circuit is mounted in your favorite enclosure and attached to the phone line, you are ready to test the circuit. Remember, the phone line has a polarity and this will affect how the circuit operates. Connect the Telephone Ring Detector to the phone line in the area of your choice and apply power to the ring detector circuit. Next, have someone call your telephone number, then listen and watch for the telephone ring circuit to come alive. In the event that the circuit does not work immediately, you will have to first check the polarity of the phone line connection or possibly recheck the circuit for any errors. The most common reason for circuit failure is improper orientation of the electrolytic capacitors, diodes, and semiconductors such as transistors and of course integrated circuits. Recheck you circuit carefully and once inspected and errors fixed, you can install the circuit into its enclosure and retest the circuit. Now you can go ahead and install the Telephone Ring Detector in your shop, barn or basement, or anywhere you like, for long reliable operation.

Project 4: Wireless telephone ringer

This poor man's cordless telephone works with an FM radio to remotely signal when your wired telephone rings. A cordless telephone is a useful consumer electronic device. Though it lets you originate calls, its major benefit is in alerting you to incoming calls when you are not near enough to a wired tone to hear it ring. If the alert feature is all you really need, the Wireless Telephone Ringer presented will do the job inexpensively, though you will have to get to a wired phone to answer the call (see Figure 13-6).

This Wireless Ringer is easy to build because the "receiver" section already exists. It is an ordinary portable FM radio that can be picked up for $10 or less.

Figure 13-6 *Wireless telephone ringer. Courtesy of CQ Communication*

Only one radio is needed in any given installation, but because this is an RF transmitter project, any number of radios can be used if more than one person wants to listen for an incoming call signal from different remote locations. Transmitter power from the Ringer is low enough so that the project will not interfere with your neighbors' FM reception.

Totally automatic in operation, this project does not even require a power switch, though you can include one to permit you to turn it off when it will not be used for weeks or months at a time.

Circuit description

The Wireless Telephone Ringer schematic is shown in Figure 13-7. The cordless telephone transmitter consists of a telephone ring signal detector U2, which is followed by a two stage transmitter consisting of Q2 and Q3 that operates on the 88 to 108 MHz FM broadcast band. The transmitter's carrier frequency is modulated by uni-junction transistor oscillator Q1, whose operating frequency is set at about 2 KHz.

Power for the transmitter and audio oscillator is provided by 9-V battery B1. This battery is automatically switched through to the project's circuits

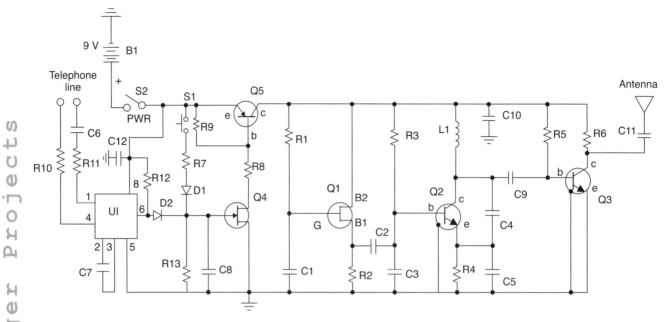

Figure 13-7 *Wireless telephone ringer. Courtesy of CQ Communications*

to deliver current only in the presence of the ring signal that announces an incoming call. Current demand from the battery is very modest. Because of the automatic power-up feature, useful battery life should therefore be extremely long.

Transistor Q2 and its associated components make up a traditional grounded base Colpitts oscillator whose frequency of oscillation is determined by the values of L1, C4, and C5. To meet the criteria to sustain oscillation, it is necessary that the base and collector of the transistor be connected to opposite ends of an LC tank circuit, with the transistor's emitter connected somewhere between these two points. In the circuit, this is accomplished by placing the base of Q2 at RF ground potential by means of C3 and the emitter about one-third the way up from ground, which is accomplished through the voltage divider action of C4 and C5.

A traditional Colpitts oscillator has two capacitors connected in series with each other to form the voltage divider needed in an oscillator circuit of this type. Contrast this to the Hartley oscillator, which is similar in design to the Colpitts oscillator, except that a tapped inductor is used to provide voltage division. Transistor Q2 is forward biased by R3 to sustain oscillation when power is applied to the circuit.

Uni-junction transistor Q1 is connected into the circuit to forma free running relaxation oscillator. When 9-V DC power is applied to the circuit, CI charges up at a rate determined by the RC time constant of R1 and C1. When the potential across C1 reaches about 50% of the supply voltage, Q1 suddenly conducts and dumps most of the charge on C1 into R2.

Now partially discharged, C1once again begins to charge, as it did initially, until Q1 breaks into conduction and dumps the charge into R2. This charge/discharge action repeats at the rate of about 2,000 times per second for as long as power is applied to the circuit.

When Q1 conducts, a voltage spike appears across R2 as a result of the discharge action of C1. This voltage is capacitively coupled, via C2, to the base of Q2 and causes a very slight variation in this transistor's operating current at the 2 KHz audio-frequency rate at which Q1 is operating. As a result, the frequency of operation of Q2 is modulated so that its output signal, amplified by Q3 and radiated by the antenna, can be detected by the discriminator circuit in any FM radio or receiver that is within range and is tuned to the project's carrier frequency. When this signal is detected, the FM radio will produce a pulsed 2 KHz tone to inform you that your telephone is ringing.

IC U1 is a special opto-coupler device, which is utilized as a telephone ring signal detector to respond to the pulses on the telephone line to which it is connected. Contained inside this chip are a bridge rectifier, 5-V regulator, and transient suppression circuits that prevent damage to the IC and false operation in the event of a random voltage spike on the telephone line.

During standby, U1 presents a high impedance to the telephone line and does not affect either outgoing or incoming calls. When the 90-V, 20-Hz ring signal appears across the line to announce an incoming call, the energy of the incoming signal triggers the opto-coupler and +5 V DC appears at the output of U1 at pin 6. This regulated voltage is then fed to the gate of enhancement mode N-channel field effect transistor Q4.

Transistors Q4 and Q5 and their associated circuitry make up an automatic electronic switch that permits current from B1 to flow to the rest of the project's circuitry only in the presence of the telephone ring signal. At all other times, the switch "opens" and disconnects the battery from the circuit.

Battery voltage is fed to the emitter of Q5, which cannot conduct current until its base is forward-biased. During standby, Q4's gate-to-source voltage is zero because U1 is idle and thus is delivering no voltage to the circuit. The drain-to-source resistance of Q4 is essentially infinite under this condition, so no base or collector current flows through Q5. An output or load resistor was placed between the output pin 6 and pin 8. This resistor should have a value between 5 K and 10 K ohms for best results.

When Q4 is switched on by the voltage that appears at its gate in response to the ring signal, this transistor's drain-to-source resistance becomes almost zero, allowing base current to flow out of Q5 and causing this transistor to saturate. This places almost the full battery voltage in the collector of Q5 and allows the transmitter to operate.

Since the time constant of C8 and R11 is relatively short, Q4 operates only during each two second burst of ring-signal energy produced by telephone company equipment. This produces a sound in the portable FM radio that is similar in cadence to the normal telephone ring signal.

To verify that the circuit is operating normally, open TEST pushbutton switch S1 has been included

in the transmitter. When closed, this switch applies forward-bias to Q4 without the need for the ring signal to be present. Pressing S1 verifies the operating condition of the transmitter by allowing you to listen for the 2-KHz tone (this time continuous as long as S1 is closed) in the FM radio being used as the system's receiver.

Construction

Since this project operates in the VHF radio range, it is important that you adhere to RF construction techniques to obtain proper operation. Printed circuit board wiring, therefore, is mandatory. The PC board must be copper clad on both sides. One side has no copper removed from it during the etching process (except for small areas, which will be discussed presently), while the other is etched to remove all copper from it except as needed for interconnecting conductors. The solid copper side of the board serves as both an RF ground plane and circuit, common to which all grounded component leads are to be soldered. You can fabricate your own printed circuit board.

If you fabricate your own board, be sure to first coat one entire side with etch resist or mask it with tape to prevent the etchant from eating away any copper. After etching the board to bring out the required copper trace pattern, remove the etch resist (or tape) from both sides of it. Then drill all component lead holes, as indicated by the solder pad centers. Though you do not etch the top (ground plane) side of the board, you must still clear copper from around all holes through which component leads must pass and are *not* to connect to the ground plane.

Use about a 1/16″ drill bit to remove about a 1/8″ diameter circle of copper around each indicated hole. After placing the bit in the hole, gently rotate it by hand to clear way the copper. Do *not* use a power or pin drill for this operation; the danger of accidentally drilling clear through and ruining the board is too great. Instead, use a very sharp, preferably new, bit. The sharper the bit, the faster and easier will be the manual drilling task.

Bear in mind that the holes into which component leads that are to be grounded must not be cleared of copper. Otherwise, it will be difficult, if not impossible, to solder these leads to the copper on the top of the board. When the board is ready to be populated, start by plugging the leads of the resistors into the indicated holes and soldering them into place. (Note: All components mount on the ground plane side of the board.) Next, install and solder into place the capacitors and diodes, taking care to properly orient the latter. Then go ahead and install the transistors in their respective locations. Remember that just one orientation-sensitive component incorrectly installed on the board will render the circuit inoperative and may even result in damage to itself and/or nearby components.

Before we begin building the Wireless Telephone Ringer, you will need to locate a clean, well-lit worktable or workbench. Next we will gather a small 25–30-W pencil tipped soldering iron, a length of 60/40 tin–lead solder, and a small canister of Tip Tinner, available at your local RadioShack store; Tip Tinner is used to condition the soldering tip between solder joints. You drive the soldering tip into the compound and it cleans and prepares the soldering tip. Next grab a few small hand tools, try to locate small flat blade and Phillips screwdrivers. A pair of needle-nose pliers, a pair of tweezers along with a magnifying glass, and a pair of end-cutters and we will begin constructing the project. Now locate your schematic diagram and parts layout diagram along with all of the components needed to build the project. Once all the components are in front of you, you can check them off against the project parts list to make sure you are ready to start building the project. Now, locate Tables 13-1 and 13-2, and place them in front of you.

Table 13-1 illustrates the resistor code chart and how to read resistors, while Table 13-2 illustrates the capacitor code chart which will aid you in constructing the project.

Refer now to the resistor color code chart shown in Table 13-1, which will assist you in identifying the resistor values. Note that each resistor will have either three or four color bands, which begin at one end of the resistor's body. The first colored band represents the resistor's first digit, while the second color represents the second digit of the resistor's value. The third colored band represents the resistor's multiplier value, and the fourth color band represents the resistor's tolerance value. If there is no fourth colored band then the resistor

has a 20% tolerance value. If the resistor's fourth band is silver then the resistor has a 10% tolerance value and if the fourth band is gold then the resistor's tolerance value will be 5%.

So finally we are ready to begin assembling the project; let's get started. The prototype project was constructed on a single-sided printed circuit board or PCB. With your PC board in front of you, we can now begin populating the circuit board. We are going to place and solder the lowest height components first, the resistors and diodes. First locate our first resistor R1. Resistor R1 has a value of 100 K ohms and is a 5% tolerance type. Now take a look for a resistor whose first color band is brown, black, yellow, and gold. From the chart you will notice that brown is represented by the digit (1), and the second band is black, which is represented by (0). Notice that the third color is brown and the multiplier is (10,000), so $(1)(0) \times 10,000 = 100,000$ ohms.

Go ahead and place R1 on the circuit board and solder it in place. Next take your end-cutters and trim the excess component leads from the resistor. Locate the remaining resistors and solder them in place on the board and remember to trim the excess leads from the board.

Note that capacitors generally come from two major groups or types. Non-polar types, such as the ones in this project, have no polarity and can be placed in either direction on the circuit board. The other type of capacitors are called polarized types; these capacitors have polarity and have to be inserted on the board with respect to their polarity marking in order for the circuit to operate correctly. On these types of capacitors, you will observe that the components actually have a plus or minus marking or a black or white band with a plus or minus marking on the body of the capacitors.

Next refer to Table 13-2, this chart helps to determine value of small capacitors. Many capacitors are quite small in physical size, so manufacturers have devised a three-digit code which can be placed on the capacitor instead of the actual value. The code occupies a much smaller space on the capacitor and can usually fit nicely on the capacitor body, but without the chart the builder would have a difficult time determining the capacitor's value.

Now locate the capacitors from the component pile and try and locate capacitor C1, which should be marked with a (104); this capacitor has a 100 nF value.

Now that you can read the resistor and capacitor charts, go ahead and place R1 and C1 on the circuit board and solder them in place. Now take your end-cutters and cut the excess component leads flush to the edge of the circuit board. Once you have installed the first two components, you can move on to installing the remaining resistors and capacitors.

It is important that you use the specified components for C4, C5, C6, Q2, and Q3. The transistors are rated for high-frequency amplifier applications, as in this project, and the capacitors are temperature-stable to maintain stable operating frequency.

Note that Q2 and Q3 have four leads: one each for emitter, base, collector and case. The case lead must be soldered to the ground plane on the board to provide an RF shield for the transistor in both cases. All transistor grounded leads and grounded pads of the other components are soldered to a common ground.

Use a socket for IC U1. It will not be necessary to solder grounded pins 3 and 7 to the top of the board ground plane. These two pins will automatically be grounded through the copper paths on the bottom of the board after soldering the socket into place. Be sure, however, that all holes for the socket, except those for pins 3 and 7, are cleared of copper on the top surface. Though you must hand-wind inductor L1, the procedure for doing so is very simple. Start with a length of solid bare hookup wire or enameled magnet wire. If the latter, use fairly heavy wire so that it readily holds its shape without having to use a coil form, and scrape away the enamel coating a distance of ¼″ at both ends of the wire. This wire must be exactly 4½″ long to ensure that the circuit will tune to the center of the FM broadcast band. Wrap the wire around an ordinary lead pencil so that you end up with 3½″ closely spaced turns with two tails that are parallel to each other and of equal length on opposite sides of the pencil. Slide the coil off the pencil and gently spread the turns evenly so that the inductor's two leads drop into the L1 holes in the board.

Adjust the height of the coil above the ground plane surface of the board so that ⅟₁₆″ of lead length protrudes from the center of each pad on the bottom of the board. Solder both leads into place. If you use hookup wire to wind the coil, inspect the installation to make

sure that no turns short to each other, the ground plane on the board, or any nearby components.

When you have completed wiring the board, examine the assembly; try carefully for inadvertent "short" circuits between closely spaced pads inductors. Pay particular attention to the top of the board to verify all grounded leads are properly soldered to the ground plane and that all leads that are not supposed to be grounded are fully insulated from the ground plane. If in doubt about the latter, use an ohmmeter set to its lowest range or an audible continuity tester to verify that you have the required insulation between component lead and ground plane in each case.

Check all joints for good soldering. If any connection appears to be suspicious, re-flow the solder on it. If you locate a solder bridge, re-flow the solder and use solder wick or a solder sucker to remove it. Most problems in projects like this can be attributed to poor soldering.

Plug the IC into the socket, making sure you properly orient it. Also, make certain that no pins overhang the socket or fold under between the socket and IC as you push it home. For optimum circuit stability, it is best to house the circuit assembly inside an all-metal enclosure that has a movable top to permit internal access for fine tuning the tank circuit. Use an enclosure that is at least 2″ high so that you can mount the circuit board assembly horizontally and spaced about ¼ or ½″ above the bottom panel. Use metal spacers to mount the assembly so that the copper border around the solder side is electrically connected to the metal of the enclosure, to ensure that the metal box is also at RF ground potential.

A telescoping-type transistor radio replacement antenna that extends to 12 to 18″ is suitable for use with this project. Bear in mind, however, that you should use as short an antenna as needed for consistent operation at the desired range over which the project is to be used.

Use plastic hardware to mount the antenna to the enclosure so that it does not short to any part of the metal box. Also, locate the antenna close to the connecting pad on the printed circuit board where the connection between the two is as short as possible. Capacitor C11 will be used to make the actual connection between antenna and the collector of Q3.

Power switch S2 should be included if you anticipate long periods of time (weeks or more) during which the project will not be used. This switch and TEST pushbutton switch S1 should be located on the top or side of the enclosure where it will not physically or electrically interfere with the circuit board assembly, and as far as possible from the antenna so that when operating either there will be minimum hand capacitance effect coupled to the antenna circuit. Use a snap-type battery connector to make connection to the 9-V battery used for B1. Be sure to wire this connector so that the battery's negative terminal (connector's black insulated lead) connects to the ground plane on the circuit board. Then plug the red insulated connector wire into the hole at B1, and solder into place. Secure the battery inside the enclosure with either a standard 9-V battery clip or a length of double-sided foam tape.

Connection to the telephone line must be made via a telephone cord that has a modular connector at the end that attaches to the telephone line. Such cords are readily available from any outlet that stocks telephone accessories. Use of a modular connector is an FCC requirement, aside from the fact that it makes it easy to plug in and remove the project from any phone connector block.

If the only cord you can obtain has connectors at both ends, cut off one connector. Then remove about 1″ of outer plastic jacket and strip an inch of insulation from the ends of the exposed red and green insulated conductors. (If there are other conductors besides the red and green lines, clip them off flush with the plastic jacket.) This project's circuit is AC coupled to the telephone line. Therefore, it is necessary to observe any particular polarity when wiring the telephone cord into the circuit via the indicated holes in the circuit board. Route the cord into the enclosure through a rubber grommet-lined hole and tie a knot in it about 3″ from the prepared end inside the box, to serve as a strain relief, before lugging the wires into the board's holes and soldering them into place.

Operation

Use an ordinary FM broadcast-band radio (portable or otherwise) to check out circuit operation. For initial tests, do not connect the project to the telephone line.

Also, U1 need not be installed in its socket at this time. During testing, the circuit board assembly and antenna should be mounted in and on the project's enclosure to avoid any shift in frequency after circuit tuning. With a 9-V battery plugged into the snap connector, operate the transmitter either by having an assistant hold down the TEST button continuously or by temporarily connecting a wire jumper across the terminals of S1.

Place the FM radio 10 or 20′ away from the transmitter to obviate overloading the radio with excessive signal strength. Now very slowly tune across the FM band until you hear the transmitted 2-KHz audio tone. If you are unable to detect any signal from the transmitter, its carrier may be tuned outside the FM radio's band. In this event, change the transmitter's operating frequency by spreading or compressing the turns of L1 to raise or lower, respectively, the frequency until you are able to find the tone within the FM band. Be careful to avoid shorting the turns to each other or the circuit board's ground plane! When you do hear the 2-KHz tone, tune the FM radio to a dead spot on the dial and place it about 20′ away from the transmitter. Now use a plastic tuning tool to adjust the spread of the turns of L1 until you once again hear the 2-KHz tone. Work very carefully because only a very slight adjustment of the coil's turns should yield the proper results. Remember, to raise the frequency, spread the turns; and to lower it, compress them. Check that the signal is still present in the radio when you place the cover on the project's enclosure. If necessary, tweak the coil slightly to keep reception of the tone signal at the dead spot on the FM dial. If you are unable to hear the 2-KHz audio tone in the FM radio at any setting of the dial and with any spreading or compression of the coil's turns, you will have to troubleshoot the circuit. Start by checking the electronic switch, using a DC voltmeter to check the reading at the collector of Q5 to verify that the Q4/Q5 switching circuit is operating. Use a multi-meter with at least a 20,000-ohms input sensitivity for this test. You should obtain a reading between 5 and 9 V when the TEST button is closed, assuming the battery is reasonably fresh.

Carefully check the circuit board assembly to make sure you installed all components in their respective locations, that the components are of the correct value or number in each case, and that the diodes and transistors are properly oriented. Also make sure that none of the turns of L1 are touching each other. Review the winding instructions for the coil to ascertain that you have wound it correctly.

Use the DC voltmeter to measure the voltage across R4, which should yield a reading of 4 V when Q2 is drawing current. To check the Q1 circuit, you need an oscilloscope. Use the scope to observe the sawtooth waveform across C1; presence of this waveform indicates that the 2-KHz oscillator is operating as it should. Once you are satisfied that the circuit is performing properly with the TEST pushbutton switch closed, unsnap the connector from the battery and install U1 in its socket. Observe correct orientation. This done, snap the connector back onto the battery and plug the project's phone cord into any convenient telephone line outlet. Now have a friend call your number. When your phone rings, you will hear the project's 2-KHz ring signal in your FM radio, verifying that the project is operating properly. Once you have tuned the project to dead spot on the FM dial, it is best to change the length of the antenna. If you do change its length, a very slight change in operating frequency may result, which will require slight adjustment of the radio's tuning. To use the project, connect it to the telephone line and place your FM radio some distance away from it. To verify that the system is working, press the TEST button. That is all there is to it. You can now carry the radio with you and have confidence that you will not miss a call as long as you are within range of the project's transmitter.

Project 5: Telephone ring generator

A Telephone Ring Generator is extremely useful for testing telephones and telephone circuits, as well as for creating your own telephone intercom systems. You can easily create your own very useful ring generator by building this transistorized project.

The Telephone Ring Generator is highlighted in the schematic diagram in Figure 13-8.

The ringing generator circuit generates the necessary high voltage from a simple switching mode power supply which employs a CMOS Schmitt Trigger square

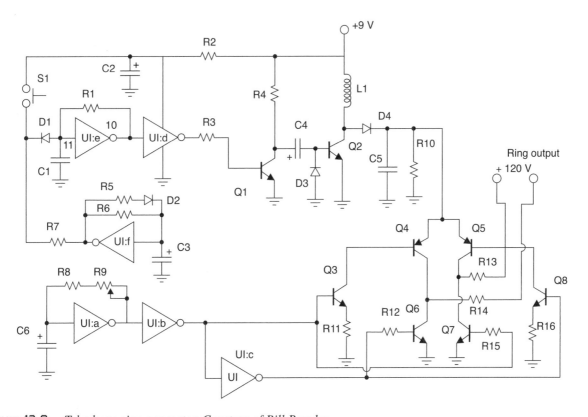

Figure 13-8 *Telephone ring generator. Courtesy of Bill Bowden*

wave oscillator, 10 mH inductor, high voltage switching transistor (TIP47 or other high voltage, 1 amp transistor, and a driver transistor (2N3053). The inductor should have a low DC resistance of 1.5 ohms or less. The switching supply must have a load connected to prevent the voltage from rising too high, so a 22 K resistor is used across the output which limits the voltage output to about 120 V DC with the phone ringer disconnected and about 90 V DC connected.

The output voltage can be adjusted by changing the value of the 150 K resistor between pins 10 and 11 which will alter the oscillator frequency (frequency is around 800 Hz as shown). The supply is gated on and off by a second Schmitt Trigger oscillator (pins 12 and 13) so that the phone rings for about 2 s and then the circuit idles for about 1 min between rings. These times can be adjusted with the 10 K and 300 K resistors connected to pin 12. The push button shown is used to manually ring the phone.

The 25 Hz ringing frequency is generated by another Schmitt Trigger oscillator (pins 1and 2) which controls the H bridge transistor output circuit. The six transistors in the output stage, i.e., the four NPN and the two PNP types, should be high voltage types rated at 200 V collector to emitter or more. The ringer will only draw around 10 mA, so the output transistors can have a low current rating but must have a high voltage rating. I used TIP47s and small signal PNPs of unknown numbers that I had on hand, but other types such as NTE287 (NPN) and NTE288 (PNP) should work. Both have a 300 V C-E rating and cost less than a dollar.

The two 470 ohm resistors connected to the output serve to limit the current in case the output is shorted. They should limit the surge to around 120 mA, which should be low enough to prevent damage. The circuit draws around 250 mA when the ring signal is present so if you want to operate it from batteries, six 'D' type alkaline cells are recommended. It probably will not work with a small 9-V battery.

Circuit assembly

Before we begin building the Telephone Ring Generator, let's first locate a clean, well-lit worktable or workbench. Next we will gather a small 25–30-W pencil tipped soldering iron, a length of 60/40 tin–lead solder, and a

small canister of Tip Tinner, available at your local RadioShack store; Tip Tinner is used to condition the soldering tip between solder joints. You drive the soldering tip into the compound and it cleans and prepares the soldering tip. Next grab a few small hand tools, try to locate small flat blade and Phillips screwdrivers. A pair of needle-nose pliers, a pair of tweezers along with a magnifying glass, and a pair of end-cutters and we will begin constructing the project. Now locate your schematic diagram and parts layout diagram along with all of the components needed to build the project. Once all the components are in front of you, you can check them off against the project parts list to make sure you are ready to start building the project. Now, locate Tables 13-1 and 13-2, and place them in front of you.

Table 13-1 illustrates the resistor code chart and how to read resistors, while Table13-2, illustrates the capacitor code chart, which will aid you in constructing the project.

Refer now to the resistor color code chart shown in Table 13-1, which will help you to identify the resistor values for the project. Note that each resistor will have either three or four color bands, which begin at one end of the resistor's body. The first colored band represents the resistor's first digit, while the second color represents the second digit of the resistor's value. The third colored band represents the resistor's multiplier value, and the fourth color band represents the resistor's tolerance value. If there is no fourth colored band then the resistor has a 20% tolerance value. If the resistor's fourth band is silver then the resistor has a 10% tolerance value and if the fourth band is gold then the resistor's tolerance value will be 5%.

So let's get started; we are ready to begin assembling the project. The prototype project was constructed on a single-sided printed circuit board or PCB. With your PC board in front of you we can now begin populating the circuit board. We are going to place and solder the lowest height components first, the resistors and diodes. First locate our first resistor R1. Resistor R1 has a value of 150 K ohms and is a 5% tolerance type. Now take a look for a resistor whose first color band is brown, green, yellow, and gold. From the chart you will notice that brown is represented by the digit (1), the second band is green, which is represented by (5). Notice that the third color is yellow and the multiplier is (10,000), so $(1)(5) \times 10,000 = 150,000$ ohms.

Go ahead and place R1 on the circuit board and then solder it in its respective location on the circuit board. Now locate your end-cutters and trim the excess resistor leads. Cut the excess leads flush to the edge of the circuit board. Now install the remaining resistors and solder them in place, remembering to trim the excess component leads.

Note that capacitors generally come from two major groups or types. Non-polar types, such as the ones in this project, have no polarity and can be placed in either direction on the circuit board. The other type of capacitors are called polarized types, these capacitors have polarity and have to be inserted on the board with respect to their polarity marking in order for the circuit to operate correctly. On these types of capacitors, you will observe that the components actually have a plus or minus marking or a black or white band with a plus or minus marking on the body of the capacitors.

Next refer to Table 13-2; this chart helps to determine the value of small capacitors. Many capacitors are quite small in physical size, so manufacturers have devised a three-digit code which can be placed on the capacitor instead of the actual value. The code occupies a much smaller space on the capacitor and can usually fit nicely on the capacitor body, but without the chart the builder would have a difficult time determining the capacitor's value.

Now locate the capacitors from the component pile and try and locate capacitor C1, which should be marked with a (103), this capacitor has a .01 µF value. Now that you can read the resistor and capacitor charts, let go ahead and place R1 and C1 on the circuit board and solder them in place. Now take your end-cutters and cut the excess component leads flush to the edge of the circuit board. Once you have installed the first two components, you can move on to installing the remaining resistors and capacitors. There are a number of electrolytic type capacitors in this circuit, so pay particular attention to the polarity when installing them on the circuit board. Remember to trim the excess component leads using your end-cutters, cutting the leads flush to the edge of the circuit board.

This project contains four silicon diodes at D1 through D4. These diodes also have polarity which must be observed if the circuit is going to work correctly. You will notice that each diode will have a black or white colored band at one edge of the diode's body. The colored band represents the diode's cathode lead. Check

your schematic and parts layout diagrams when installing these diodes on the circuit board. Note that all the diodes are 1N4004 silicon diodes. Place all the diodes on the circuit board and solder them in place on the PC board. Then remember to cut the excess component leads with your end-cutters.

The Telephone Ring Generator circuit also utilizes eight transistors. Transistors generally have three leads, a Base lead, a Collector lead and an Emitter lead. Refer to the schematic and you will notice that there will be two diagonal leads pointing to a vertical line. The vertical line is the transistor's base lead and the two diagonal lines represent the collector and the emitter. The diagonal lead with the arrow on it is the emitter lead, and this should help you install the transistor onto the circuit board correctly. Go ahead and install the transistors on the PC board and solder them in place, then follow-up by cutting the excess component leads, flush to the edge of the circuit board.

The Telephone Ring Generator circuit also employs a single inductor at L1, it is a common 10 mH commercial inductor readily available from Mouser or Digi-Key Electronics.

The Telephone Ring Generator utilizes a single IC at U1. The IC is a 74HC14, which contains six inverter circuits. It is recommended that you use an IC socket for the IC, as a cheap insurance policy against possible later circuit failure. When installing the integrated circuit, you will have to pay close attention to the mounting of the device to ensure proper operation of the circuit. ICs will always have either a small indented circle, or a small cutout at the top end of the IC package. Pin 1 will always be just to the left of the cutout or circle. Refer to the schematic and parts layout diagrams when installing the IC to ensure proper orientation.

Once the circuit board has been completed, you can take a short well deserved break and when we return we will inspect the circuit board for any possible "cold" solder joints or "short" circuits. Pick up the circuit board with the foil side facing upwards toward you. Examine the board carefully. Look at all the solder joints. The solder joints should all look clean, shiny, and bright, if any solder joints look dull, dark, or blobby then you should use a solder sucker or wick to remove the solder and then re-solder the joint all over again. Once you have inspected the circuit for cold solder joints, we can move onto inspecting the board for possible short circuits. Short circuits can be caused from solder blobs bridging across the circuit traces or from cut components bridging across circuit traces. Often solder residue can trap or cut component leads which can bridge across circuit traces causing a short circuit. You want to try to avoid short circuits before power is applied for the first time in order not to damage the circuit.

Operation

Once the circuit has been built and inspected, you can mount it in a suitable metal chassis box. You will want to mount the circuit on standoffs above the metal bottom of the chassis box to avoid shorting to ground or metal of the box. You will likely want to install a power switch in series with the power supply to the power input of the circuit through L1. You could elect to power the circuit with a suitable power supply or from "D" or "C" cell batteries if desired. The momentary switch at S1 can be mounted at the side or top of the chassis box.

You will need to mount a 2-position terminal block for the ringer's output wires. The terminal block could be mounted at the front or rear of the chassis box for easy access.

Now connect up your power supply or batteries and power up the ringer circuit. Connect the output of the Telephone Ringer circuit to a telephone bell or AC voltmeter for test and once power is applied you can press switch S1 to read 120 V AC on your AC meter or if a telephone bell is attached to the output of the ringer circuit, you should hear the bell ring.

In the event of a circuit failure and no voltage is present at the output of the ringer circuit, you will have to remove power from the circuit and inspect it for any errors. The most common cause for circuit errors is incorrect placement of critical components such as electrolytic capacitors, diodes and transistors. Generally the builder has made a mistake in installing these components through making a polarity error. Also check the orientation of the integrated circuit in its socket. This is another common cause for error.

Once the circuit has been finally inspected, you can reconnect the power supply or batteries and retest the Telephone Ringer Generator for proper operation. Hopefully this time around the circuit will perform flawlessly. Good luck!

Chapter 14

Phone Line Status Circuits and Projects

Parts list

Phone In-Use Indicator

R1 3.3 K ohm, $\frac{1}{2}$ W,
   5% resistor

R2 33 K ohms, $\frac{1}{2}$ W,
   5% resistor

R3 56 K ohms, $\frac{1}{2}$ W,
   5% resistor

R4 22 K ohms, $\frac{1}{2}$ W,
   5% resistor

R5 4.7 K ohms, $\frac{1}{2}$ W,
   5% resistor

D1, D2, D3, D4 1N4001
   silicon diodes

D5 red LED - LED-1

D6 green LED - LED-2

Q1, Q2 2N 3392 transistor

Misc PC board,
   enclosure, wire,
   hardware, sockets,
   etc.

Cut Phone Line Detector

R1, R2, R3 20 megohm
   $\frac{1}{2}$ W, 5% resistor

R4 2.2 megohm $\frac{1}{2}$ W,
   5% resistor

C1 0.47 µF 250 V mylar
   capacitor

Q1 2N3904 transistor

Q2 2N3906 transistor

Q3 IRF510 Power
   MOSFET

D1 1N914 Diode

RLY1 SPDT 6-V relay

Load Buzzer or bell

Misc PC board, wire,
   phone connectors,
   hardware, enclosure,
   etc.

9-V Power Supply

R1 240 ohm $\frac{1}{2}$ W,
   5% resistor

R2 5 K ohm potentiometer
   (PC-trim)

R3 1 K ohm $\frac{1}{4}$ W,
   5% resistor

C1 1,000 µF, 35 V
   electrolytic
   capacitor

C2 10 µF, 35 V
   electrolytic capacitor

C3 .01 µF, 35 V disc
   capacitor

D1, D2 1N4001 silicon
   diode

D3 Red LED

U1 LM317 adjustable
   regulator

T1 110 to 25 V center
   tapped transformer
   (.500 mA)

S1 SPST toggle switch

P1 110-V power plug

Misc PC board, wire, enclosure, etc.

Phone Line-in-Use Relay Controller

R1, R2, R3 3.3 megohm, $\frac{1}{2}$ W, 5% resistor

R4 10 K ohm, $\frac{1}{4}$ W, 5% resistor

D1 1N4004 silicon diode

Q1 IFR510/511 – MOSFET (Radio shack)

RLY1 12-V DC relay – 120 ohm coil

Misc PC board, wire, hardware, enclosure, load, etc.

12-V DC Power Supply

R1 10 K ohms, $\frac{1}{4}$ W, 5% resistor

R2 2.2 ohm, 1 W, 5% resistor

C1 2,200 µF, 35-V Electrolytic capacitor

C2 10 µF, 35-V electrolytic capacitor

C3 .01 µF, 35-V disc capacitor

D1, D2 1N4001 silicon diode

D3 1N4742A Zener diode

Q1 TIP31 – NPN transistor

Q2 2N3904 NPN transistor

T1 25.2-V center tapped transformer – 1.5-ampere

S1 SPST toggle – on-off switch

F1 2-amp 3AG Fuse

Misc PC board, wire, hardware, power plug, etc.

Single Phone Line Status Light

R1, R2, R4 4.7 megohm, $\frac{1}{2}$ W, 5% resistor

R3 15 K ohms, $\frac{1}{4}$ W, 5% resistor

R5 1 K ohms, $\frac{1}{4}$ W, 5% resistor

C1 10 µF, 25-V electrolytic capacitor

C2 .15 µF, 25-V polyester capacitor

D1, D2, D3, D4 1N4004 silicon diodes

D6, D7 1N4148 silicon diodes

D5 15-V Zener diode

D8 Red LED

U1 74HC14 Hex inverting Schmitt Trigger IC

B1 two "C" or "D" cells

Misc PC board, wire, IC socket, battery holders, etc.

Two-Line Phone Status Indicator

R1, R6, R15 4.7 megohm, $\frac{1}{2}$ W, 5% resistor

R2, R3, R13, R14 10 megohm, $\frac{1}{2}$ W, 5% resistor

R4, R9, R12 100 K ohms, $\frac{1}{4}$ W, 5% resistor

R5, R11 470 K ohms, $\frac{1}{4}$ W, 5% resistor

R7, R10 270 ohms, $\frac{1}{4}$ W, 5% resistor

R8 22 megohm, $\frac{1}{4}$ W,
  5% resistor

C1 10 µF, 25-V
  electrolytic capacitor

C2 .047 µF, 25-V
  mylar capacitor

D1, D2, D3, D4 1N4004
  silicon diodes

D5, D6, D7, D8 1N4004
  silicon diodes

D9, D14 1N4744 –
  15-V Zener diodes

D11, D12, D13 1N4148
  silicon diodes

U1 74HC132 – Quad
  2-input NAND Schmitt
  Trigger

OPT1, OPT2 opto-coupler
  OPI-123 OPTEK

B1 2-"C" or "D" cells

Misc PC board,
  IC socket, wire,
  hardware, enclosure,
  etc.

Smart Phone Light

R1, R2 20 K ohm, $\frac{1}{4}$ W,
  5% resistor

R3 100 K ohm, $\frac{1}{4}$ W,
  5% resistor

R4, R5 47 K ohms, $\frac{1}{4}$ W,
  5% resistor

R6 10 K ohms, $\frac{1}{4}$ W,
  5% resistor

R7 1 K ohms, $\frac{1}{4}$ W,
  5% resistor

R8 22 K ohms, $\frac{1}{4}$ W,
  5% resistor

C1 1 µF, 250-V paper
  capacitor

C2 47 µF, 35-V
  electrolytic
  capacitor

C3 1,000 µF, 35-V
  electrolytic capacitor

D1, D2 1N4007 silicon
  diode

Q1 SL100 transistor

U1, U2 MCT2E
  opto-coupler

U3 74LS123 Dual
  retriggerable
  monostable
  multivibrator

LDR light dependent
  resistor

RLY1 5-V relay

L1 lamp

Misc PC board, wire,
  IC sockets, hardware,
  enclosure, etc.

**5-V Smart Phone Light
Power Supply**

R1 1 K ohm, $\frac{1}{2}$-W,
  5% resistor

C1 1,000 µF, 35-V
  electrolytic capacitor

C2 10 µF, 35-V
  electrolytic capacitor

C3 .01 µF, 35-V
  disk capacitor

D1, D2 1N4001 silicon
  diode

D3 Red LED

U1 LM7805 – 5-V
  regulator IC

T1 110 to 25-V
  center tapped
  transformer – 1 amp

S1 SPST toggle switch –
  on-off

F1 Fuse – 1 amp

Misc PC board,
  hardware, wire, etc.

Phone Line Vigilant

R1   56 K ohms, $\frac{1}{4}$ W,
5% resistor

R2, R3, R8   100 K ohms,
$\frac{1}{4}$ W, 5% resistor

R4, R9   1 K ohms, $\frac{1}{4}$ W,
5% resistor

R5, R10   5.6 K ohms,
$\frac{1}{4}$ W, 5% resistor

R6   470 ohms, $\frac{1}{4}$ W,
5% resistor

R7, R17   10 K ohms, $\frac{1}{4}$ W,
5% resistor

R11   4.7 K ohms, $\frac{1}{4}$ W,
5% resistor

R12   220 ohms, $\frac{1}{4}$ W,
5% resistor

R13, R16   1.2 K ohms,
$\frac{1}{4}$ W, 5% resistor

R14, R15   22 K ohms,
$\frac{1}{4}$ W, 5% resistor

VR1, VR2   1 megohm
potentiometer

VR3   1 K ohm
potentiometer

C1   10 µF, 35-V
electrolytic capacitor

C2   22 µF, 35-V
electrolytic capacitor

C3   470 µF, 35-V
electrolytic capacitor

C4   1,000 µF, 35-V
electrolytic capacitor

D1, D2, D3, D4   1N4007
silicon diodes

D5, D6, D7, D8   1N4007
silicon diodes

D9   3.3-volt Zener diode

D10   9.1-volt Zener
diode

LED1   Green LED

LED2   Red LED

Q1, Q2, Q3, Q4   BC547
transistor

Q5, Q6, Q8, Q9, Q19
BC547 transistor

Q7   SL100 transistor

S1   On-Off toggle switch

Buzz   Piezo buzzer

Batt   4- 1.5-V "AA"
cells

T1   110 volt primary,
9-V secondary
@ 500 mA

Misc   PC board, wire,
hardware, modular
jack, etc.

Project 1: Phone In-Use indicator

Tired of having your phone calls interrupted by family members, roommates, or co-workers picking up the phone handset, while you are in a conversation? This Phone Line In-use Indicator circuit will show others when the phone is in use, so they will be alerted not to pickup a phone extension.

Circuit description

The Phone In-Use Indicator circuit is illustrated in Figure 14-1. When all the phones are on-hook, transistor Q2's base is turned on by a voltage divider switch circuit, which consists of R3 and R5. Note that the value for R5 causes the device to switch over at about 9 V. This can be changed to facilitate other voltage levels if desired. Transistor Q2 allows current to flow through R2 and LED2 or D6, indicating that the phone line is not in use. It also effectively grounds the base of Q1 and forces LED1 or D5 to remain off. When the voltage drops because a telephone goes off-hook, Q2 stops conducting, allowing a small flow of current from R2, D6, and R4 to Q1's base.

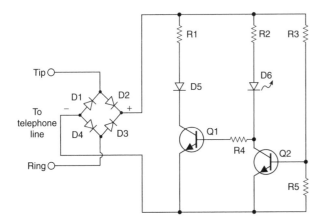

Figure 14-1 *Phone In-Use indicator circuit. Courtesy of Poptronix*

When that occurs, Q1 conducts, energizing D5, and D6 is deprived of sufficient current to light up. The bridge rectifier compensates for a possible reversal between the tip and ring phone wires, and rectifies the ring signal to power the circuit.

Circuit assembly

Before we begin constructing the Phone In-Use Indicator, you will need to locate a clean, well-lit worktable or workbench. Next we will gather a small 25–30-W pencil tipped soldering iron, a length of 60/40 tin–lead solder, and a small canister of Tip Tinner, available at your local RadioShack store; Tip Tinner is used to condition the soldering tip between solder joints. You drive the soldering tip into the compound and it cleans and prepares the soldering tip. Next grab a few small hand tools, try to locate small flat blade and Phillips screwdrivers. A pair of needle-nose pliers, a pair of tweezers along with a magnifying glass, and a pair of end-cutters and we will begin constructing the project. Now locate your schematic diagram and parts layout diagram along with all of the components needed to build the project. Once all the components are in front of you, you can check them off against the project parts list to make sure you are ready to start building the project. Now, locate Tables 14-1 and 14-2, and place them in front of you.

Table 14-1 illustrates the resistor code chart and how to read resistors, while Table 14-2 illustrates the capacitor code chart which will aid you in constructing the project.

Refer now to the resistor color chart shown in Table 14-1, which will help you to identify the resistors for the project. Note that each resistor will have either three or four color bands, which begin at one end of the resistor's body. The first colored band represents the resistor's first digit, while the second color represents the second digit of the resistor's value. The third colored

Table 14-1

Resistor color code chart

Color Band	1st Digit	2nd Digit	Multiplier	Tolerance
Black	0	0	1	
Brown	1	1	10	1%
Red	2	2	100	2%
Orange	3	3	1,000 (K)	3%
Yellow	4	4	10,000	4%
Green	5	5	100,000	
Blue	6	6	1,000,000 (M)	
Violet	7	7	10,000,000	
Gray	8	8	100,000,000	
White	9	9	1,000,000,000	
Gold			0.1	5%
Silver			0.01	10%
No color				20%

Table 14-2
Three-digit capacitor codes

pF	nF	μF	Code	pF	nF	μF	Code
1.0			**1R0**	3,900	3.9	.0039	**392**
1.2			**1R2**	4,700	4.7	.0047	**472**
1.5			**1R5**	5,600	5.6	.0056	**562**
1.8			**1R8**	6,800	6.8	.0068	**682**
2.2			**2R2**	8,200	8.2	.0082	**822**
2.7			**2R7**	10,000	10	.01	**103**
3.3			**3R3**	12,000	12	.012	**123**
3.9			**3R9**	15,000	15	.015	**153**
4.7			**4R7**	18,000	18	.018	**183**
5.6			**5R6**	22,000	22	.022	**223**
6.8			**6R8**	27,000	27	.027	**273**
8.2			**8R2**	33,000	33	.033	**333**
10			**100**	39,000	39	.039	**393**
12			**120**	47,000	47	.047	**473**
15			**150**	56,000	56	.056	**563**
18			**180**	68,000	68	.068	**683**
22			**220**	82,000	82	.082	**823**
27			**270**	100,000	100	.1	**104**
33			**330**	120,000	120	.12	**124**
39			**390**	150,000	150	.15	**154**
47			**470**	180,000	180	.18	**184**
56			**560**	220,000	220	.22	**224**
68			**680**	270,000	270	.27	**274**
82			**820**	330,000	330	.33	**334**
100		.0001	**101**	390,000	390	.39	**394**
120		.00012	**121**	470,000	470	.47	**474**
150		.00015	**151**	560,000	560	.56	**564**
180		.00018	**181**	680,000	680	.68	**684**
220		.00022	**221**	820,000	820	.82	**824**
270		.00027	**271**		1,000	1	**105**
330		.00033	**331**		1,500	1.5	**155**
390		.00039	**391**		2,200	2.2	**225**
470		.00047	**471**		2,700	2.7	**275**
560		.00056	**561**		3,300	3.3	**335**
680		.00068	**681**		4,700	4.7	**475**
820		.00082	**821**			6.8	**685**
1,000	1.0	.001	**102**			10	**106**
1,200	1.2	.0012	**122**			22	**226**
1,500	1.5	.0015	**152**			33	**336**
1,800	1.8	.0018	**182**			47	**476**

Table 14-2—cont'd

Three-digit capacitor codes

pF	nF	µF	Code	pF	nF	µF	Code
2,200	2.2	.0022	**222**			68	**686**
2,700	2.7	.0027	**272**			100	**107**
3,300	3.3	.0033	**332**			157	**157**

Code = 2 significant digits of the capacitor value + number of zeros to follow

For values below 10 pF, use "R" in place of decimal e.g. 8.2 pF = 8R2

10 pF = 100

100 pF = 101

1,000 pF = 102

22,000 pF = 223

330,000 pF = 334

1 µF = 105

band represents the resistor's multiplier value, and the fourth color band represents the resistor's tolerance value. If there is no fourth colored band then the resistor has a 20% tolerance value. If the resistor's fourth band is silver then the resistor has a 10% tolerance value and if the fourth band is gold then the resistor's tolerance value will be 5%.

Finally, we are ready to begin assembling the project, so let's get started. The prototype project was constructed on a single-sided printed circuit board or PCB. With your PC board in front of you, we can now begin populating the circuit board. We are going to place and solder the lowest height components first, the resistors and diodes. First locate our first resistor R1. Resistor R1 has a value of 3.3 K ohms and is a 5% tolerance type. Now take a look for a resistor whose first color band is orange, orange, red, and gold. From the chart you will notice that orange is represented by the digit (3), the second color band is orange, which is represented by (3). Notice that the third color is red and the multiplier is (100), so $(3)(3) \times 100 = 3,300$ ohms. Go ahead and place R1 on the circuit board and solder it in place. Now take your end-cutters and cut the excess component leads flush to the edge of the circuit board. Once you have installed the first resistor, you can move on to installing the remaining resistors and then the remaining low profile components. Solder these components in place in their respective locations on the circuit board.

Note that capacitors generally come from two major groups or types. Non-polar types, such as the ones in this project, have no polarity and can be placed in either direction on the circuit board. The other type of capacitors are called polarized types; these capacitors have polarity and have to be inserted on the board with respect to their polarity marking in order for the circuit to operate correctly. On these types of capacitors, you will observe that the components actually have a plus or minus marking or a black or white band with a plus or minus marking on the body of the capacitors

This project contains four silicon diodes. Diodes will always have some type of polarity markings, which must be observed if the circuit is going to work correctly. You will notice that each diode will have a black or white colored band at one edge of the diode's body. The colored band represents the diode's cathode lead. Check your schematic and parts layout diagrams when installing these diodes on the circuit board. Note that diodes D1 through D4 are all 1N4004 silicon diodes. Place all the diodes on the circuit board and solder them in place on the PC board. Then cut the excess component leads with your end-cutters flush to the edge of the circuit board.

The Phone In-Use Indicator utilizes two conventional transistors. Transistors are generally three lead devices, which must be installed correctly if the circuit is to work properly. Careful handling and proper orientation in the circuit is critical. A transistor will usually have a Base lead, a Collector lead, and an Emitter lead.

The base lead is generally a vertical line inside a transistor circle symbol. The collector lead is usually a slanted line facing towards the base lead and the emitter lead is also a slanted line facing the base lead, but it will have an arrow pointing towards or away from the base lead. Before installing the two transistors refer to the parts specification sheets as well as the parts layout diagram and the schematic. Be sure to orient the transistors prior to inserting them into the circuit board. If you are in doubt as to how to mount the transistors, have a knowledgeable electronics enthusiast help you. Now, insert the transistors onto the circuit board and solder them in place, remembering to cut the excess component leads with your end-cutters flush to the edge of the circuit board.

The Phone In-Use Indicator circuit uses two LEDs, which also must be oriented correctly if the circuit is to function properly. An LED has two leads, a cathode and an anode lead. The anode lead will usually be the longer of the two leads and the cathode lead will usually be the shorter lead, just under the flat side edge of the LED; this should help orient the LED on the PC board. Solder the LEDs to the PC board and then trim the excess leads.

Once the circuit board has been completed, take a short break and then we will inspect the PC board for possible "cold" solder joints and "short" circuits. Pick up the PC board with the foil side facing upwards toward you. Carefully inspect the solder joints; these should look smooth, clean, and bright. If any of the solder joints look dull, dark, or blobby, then you should remove the solder with a solder sucker or a solder wick and then resolder the joint all over again.

Now we will inspect the PC board for possible short circuits. Short circuits can be caused from small solder blobs bridging the circuit traces or from cut component leads, which can often stick to the circuit board. A sticky residue from solder will often trap component leads across the circuit traces. Look carefully for any bridging wires across circuit traces. Once this inspection is complete, we can move on to installing the circuit board into some sort of plastic enclosure.

The circuit is line powered so you will need no external power supply or batteries. Next, you will need to mount the circuit board inside of a plastic enclosure. Lay the circuit board inside of a suitable plastic case to check the fit. Locate some mounting standoffs and use

them to mount the circuit board inside of the case. Next, you will have to connect the circuit to the phone line. This can be done using a length of 4-conductor phone wire with a modular plug attached at one end of the cable. The modular plug end can then simply plug into a telephone jack, and the wire ends of the cable can connect to the circuit itself. You could also elect to hard-wire the circuit directly to the phone line without the modular plugs if desired.

Operation

Once your circuit has been mounted in an enclosure and wired to the phone line, you are now ready to test the circuit. When all phones are on-hook the green LED is lit up, indicating that the telephone is not now in use. When someone does pick up a phone in the house the red LED will light up indicating that someone has now picked up a phone or it is now in use. Your Phone In-use Indicator is now ready to serve you for many years to come. Now you will know if someone has picked up a telephone extension before you try to use the telephone.

Project 2: Cut phone line detector

The Cut Phone Line Detector is a novel circuit to detect when the phone line has been cut or broken somewhere along the line entering your home. If you have a home alarm system which dials the police or a neighbor then it is imperative that your phone line loop is always present. When the circuit detects that a phone line has been cut, it activates a MOSFET which can be used to drive a relay, motor, or local siren to alert your or your neighbors.

Circuit description

The phone line is connected to the circuit as shown in the schematic Figure 14-2. The green phone wire is normally the minus lead and the red wire is generally the plus lead. The green lead is connected to resistor R1, followed by resistor R2. The red telephone wire is

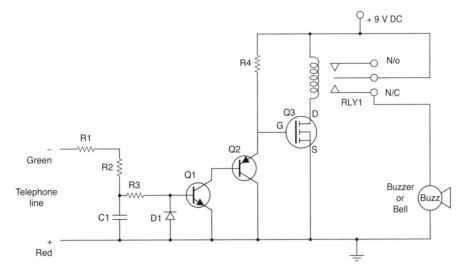

Figure 14-2 *Cut-phone detector*

connected to the ground to the circuit ground at the junction of C1, D1 the emitter of Q1, the collector of Q2 and the source of the power MOSFET at Q3. The 48-V nominal voltage on the phone line is fed to the first transistor at Q1 which is used to drive the second transistor at Q2, which in turn activates the MOSFET and pulls-in the relay at RLY1. When the phone handset is picked up, the voltage across the phone line drops to about 6 V, which is still enough to hold the relay from dropping out. But if the phone line is cut or broken, then all voltage disappears from the input of the circuit, the relay drops, and the contacts at the relay are used to activate an alarm buzzer or bell to alert you. The Cut Phone Line Detector is powered from a 9-V DC power supply, which is shown in Figure 14-3.

The 9-V power supply utilizes a center tapped 25-V transformer at T1. The two-diode rectifier is coupled to a voltage regulator chip at U1, which is an adjustable voltage regulator. The voltage output is

adjusted by trimming potentiometer R2, a 5 K pot. The output of the regulator is then fed to two capacitors, a 10 μF electrolytic and a .01 μF bypass capacitor. The regulator should be adjusted to 9 V DC before connecting it to the Cut Line Detector circuit.

Circuit assembly

Before we begin constructing the Cut Phone Line Detector, you will need to locate a clean, well-lit worktable or workbench. Next we will gather a small 25–30-W pencil tipped soldering iron, a length of 60/40 tin–lead solder, and a small canister of Tip Tinner, available at your local RadioShack store; Tip Tinner is used to condition the soldering tip between solder joints. You drive the soldering tip into the compound and it cleans and prepares the soldering tip. Next grab a few small hand tools; try to locate small flat blade and Phillips screwdrivers. A pair of needle-nose pliers, a pair of tweezers along with a magnifying glass,

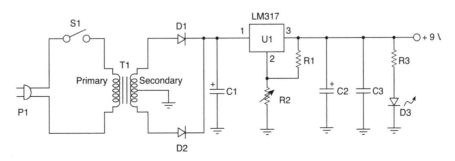

Figure 14-3 *9-V power supply circuit*

and a pair of end-cutters and we will begin constructing the project. Now locate your schematic diagram and parts layout diagram along with all of the components needed to build the project. Once all the components are in front of you, you can check them off against the project parts list to make sure you are ready to start building the project. Now, locate Tables 14-1 and 14-2, and place them in front of you.

Table 14-1 illustrates the resistor code chart and how to read resistors, while Table 14-2 illustrates the capacitor code chart which will aid you in constructing the project.

Refer now to the resistor color code chart shown in Table 14-1, which will help you to determine the proper resistors for the project. Note that each resistor will have either three or four color bands, which begin at one end of the resistor's body. The first colored band represents the resistor's first digit, while the second color represents the second digit of the resistor's value. The third colored band represents the resistor's multiplier value, and the fourth color band represents the resistor's tolerance value. If there is no fourth colored band then the resistor has a 20% tolerance value. If the resistor's fourth band is silver then the resistor has a 10% tolerance value and if the fourth band is gold then the resistor's tolerance value will be 5%.

Finally, we are ready to begin assembling the project, so let's get started. The prototype project was constructed on a single-sided printed circuit board or PCB. With your PC board in front of you, we can now begin populating the circuit board. We are going to place and solder the lowest height components, first the resistors and diodes.. First locate our first resistor R1. Resistor R1 has a value of 20 megohms and is a 5% tolerance type. Now take a look for a resistor whose first color band is red, black, blue, and gold. From the chart you will notice that red is represented by the digit (2), the second color band is black, which is represented by (0). Notice that the third color is blue and the multiplier is (1,000,000), so $(2)(0) \times 1,000,000 = 20,000,000$ ohms. Once you have identified resistor R1, go ahead and install it on the PC board, then solder it in place. Next trim the excess component lead lengths with your end-cutter. Now go ahead and locate the remaining resistors and insert them in to the circuit board at their respective locations, then solder them in place. Now solder the resistors at their locations, and remember to trim the excess

leads with your end-cutters flush to the edge of the circuit board.

Note that capacitors generally come from two major groups or types. Non-polar types, such as the ones in this project, have no polarity and can be placed in either direction on the circuit board. The other type of capacitors are called polarized types; these capacitors have polarity and have to be inserted on the board with respect to their polarity marking in order for the circuit to operate correctly. On these types of capacitors, you will observe that the components actually have a plus or minus marking or a black or white band with a plus or minus marking on the body of the capacitors.

Next refer to Table 14-2; this chart helps to determine the value of small capacitors. Many capacitors are quite small in physical size, so manufacturers have devised a three-digit code which can be placed on the capacitor instead of the actual value. The code occupies a much smaller space on the capacitor and can usually fit nicely on the capacitor body, but without the chart the builder would have a difficult time determining the capacitor's value.

Now locate the capacitor from the component pile and try and locate capacitor C1, which may be labeled .47 µF or (474); remember this is the three-digit code representing the actual value. Go ahead and place C1 on the circuit board and solder it in place. Now take your end-cutters and cut the excess component leads flush to the edge of the circuit board. Once you have installed the capacitor, you can move on to installing the remaining low profile components.

This project contains a number of silicon diodes. Diodes will always have some type of polarity markings, which must be observed if the circuit is going to work correctly. You will notice that each diode will have a black or white colored band at one edge of the diode's body. The colored band represents the diode's cathode lead. Check your schematic and parts layout diagrams when installing these diodes on the circuit board. Note, this circuit only has a single diode at D1. Place the diodes on the circuit board and solder them in place on the PC board. Then cut the excess component leads with your end-cutters.

The Cut Phone Line Detector utilizes two conventional transistors. Transistors are generally three lead devices, which must be installed correctly if the circuit is to work properly. Careful handling and proper orientation in the circuit is critical. A transistor will usually have a Base lead, a Collector lead, and an Emitter lead. The base lead is generally a vertical line inside a transistor circle symbol. The collector lead is usually a slanted line facing towards the base lead and the emitter lead is also a slanted line facing the base lead, but it will have an arrow pointing towards or away from the base lead. Pay careful attention to the pinouts of the transistors and refer to the schematic, parts layout, and component specification sheets when installing the transistors onto the circuit board. If you are uncertain of your abilities, have a knowledgeable electronics enthusiast help you. Once you have identified the transistor and MOSFET leads and can confirm the correct orientation you can install these components on the circuit board. Solder the transistors to their respective locations, then remember to cut the excess lead lengths with your wire cutters.

Now you can install the relay on to the circuit board. At some point, uou will have to decide what type of alarm device you wish to connect to the relay. The relay has both normally open and normally closed contacts. Note that when the circuit is powered, the relay remains in its rest position as shown, but as soon as the circuit is connected to the phone line the relay will pull-in and be activated. The load will have to be connected to the normally closed contacts, as shown in the schematic. Depending upon what type of relay you chose for this project, you can go ahead and mount it. If you chose a small PC board type relay you can mount this on the printed circuit board. If you chose a larger style, then you will have to mount the relay off the PC board either in a relay socket or with some sort of mounting strap to the bottom of the chassis box.

Once the circuit board has been completed, take a short break and then we will inspect the PC board for possible "cold" solder joints and "short" circuits. Pick up the PC board with the foil side facing upwards toward you. Carefully inspect the solder joints, these should look smooth, clean, and bright. If any of the solder joints look dull, dark, or blobby, then you should remove the solder with a solder sucker or a solder wick and then resolder the joint all over again.

Now we will inspect the PC board for possible short circuits. Short circuits can be caused from small solder blobs bridging the circuit traces from cut component leads, which can often stick to the circuit board.

A sticky residue from solder will often trap component leads across the circuit traces. Look carefully for any bridging wires across circuit traces. Once this inspection is complete, we can move on to installing the circuit board into some sort of plastic enclosure.

You will also need to connect the circuit to the phone line. This can be done using a length of 4-conductor phone wire with a modular plug attached at one end of the cable. The modular plug ends can then simply plug into a telephone jack. You could elect to hard-wire the circuit directly to the phone line without the modular plugs if desired. The circuit is powered from a 9-V power supply. You could elect to either build or buy your 9-V power supply. A simple solution is to utilize a 110-V to 9-V DC cube or "wall wart" power available from your local Radio Shack or electronic outlet.

Operation

Check over your circuit one more time to make sure all the components are mounted at their correct locations. Once you circuit has been mounted in an enclosure and wired to the phone line, you will be ready to test the circuit. Connect up the load or buzzer to the normally closed relay contacts. Now connect the telephone line to the circuit and finally apply power to the circuit. If all is well the buzzer will be quiet. Now disconnect the phone line from the circuit; the buzzer should sound indicating the phone line is broken or cut. If nothing happens, you will have to make sure that your power supply is working and that it is connected properly to the circuit, i.e. observe the polarity connections. If the circuit still does not work, remove the power and check to make sure all the components are installed in their proper locations and that the diodes and transistors have been installed correctly. Your Cut Phone Line Detector should give you many years of reliable service.

Project 3: Phone line-in-use relay controller

The Phone Line-in-Use Relay Controller can alert you to someone using or picking up a phone extension, or it will permit you to control a load of your choice, such as a bell, buzzer, alarm siren, or talking voice reminder!

The Phone Line In-use Controller circuit is shown in Figure 14-4. The circuit is designed to close a relay when any extension goes off-hook either by accident or intention. The circuit is pretty straightforward; the phone line is connected to points marked A and B in the diagram. The plus (+) phone lead is connected to point A, while the minus (−) phone line lead is connected to point B. A capacitor is placed across the line at C1. Note that the voltage at the gate of the MOSFET should be negative at about 1–3 V with respect to the source when the phones are on-hook. The voltage at the gate should be positive 8–10 V with respect to the source (ground) when any phone extension is off-hook. A high impedance meter is needed to measure the on-hook voltages accurately. The Phone Line In-use Controller circuit should draw less than 5 µA from the phone line. The relay (RLY1) used for controlling the external load is a 12-V DC 120 ohm coil, but almost any small 12-V DC relay should work. Power supply regulation should be +/− 2 V. The circuit provides both normally closed and normally open circuit contacts for external control of the alarm load device. The circuit is powered from a 12-V DC power supply illustrated in Figure 14-5.

Circuit assembly

Before we begin constructing the Phone Line-in-Use Controller, you will need to find a clean, well-lit worktable or workbench. Next we will gather a small 25–30-W pencil tipped soldering iron, a length of

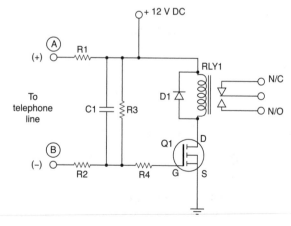

Figure 14-4 *Phone line in-use relay controller. Courtesy of Bill Bowden*

60/40 tin–lead solder, and a small canister of Tip Tinner, available at your local RadioShack store; Tip Tinner is used to condition the soldering tip between solder joints. You drive the soldering tip into the compound and it cleans and prepares the soldering tip. Next grab a few small hand tools; try to locate small flat blade and Phillips screwdrivers a pair of needle-nose pliers, a pair of tweezers along with a magnifying glass, and a pair of end-cutters and we will begin constructing the project. Now locate your schematic diagram and parts layout diagram along with all of the components needed to build the project. Once all the components are in front of you, you can check them off against the project parts list to make sure you are ready to start building the project. Now, locate Tables 14-1 and 14-2, and place them in front of you.

Table 14-1 illustrates the resistor code chart and how to read resistors, while Table 14-2 illustrates the capacitor code chart which will aid you in constructing the project.

Refer now to the resistor color code chart shown in Table 14-1, which will help you identify the resistors used in the project. Note that each resistor will have either three or four color bands, which begin at one end of the resistor's body. The first colored band represents the resistor's first digit, while the second color represents the second digit of the resistor's value. The third colored band represents the resistor's multiplier value, and the fourth color band represents the resistor's tolerance value. If there is no fourth colored band then the resistor has a 20% tolerance value. If the resistor's fourth band is silver then the resistor has a 10% tolerance value and if the fourth band is gold then the resistor's tolerance value will be 5%.

Finally, we are ready to begin assembling the project, so let's get started. The prototype project was constructed on a single-sided printed circuit board or PCB. With your PC board in front of you, we can now begin populating the circuit board. We are going to place and solder the lowest height components first, the resistors and diodes. First locate our first resistor R1. Resistor R1 has a value of 3.3 megohm and is a 5% tolerance type. Now take a look for a resistor whose first color band is orange, orange, green, and gold. From the chart you will notice that orange is represented by the digit (3); the second color band is orange, which is

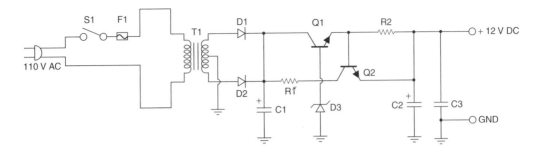

Figure 14-5 *12-V DC power supply*

represented by (3). Notice that the third color is green and the multiplier is (100,000), so (3)(3) × 100,000 = 3.3 megohms.

Locate resistors R2 and R3 and place them in their respective locations on the PC board. Now you can solder them in place and then trim the excess component leads flush to the edge of the circuit board using your end-cutters.

Note that capacitors generally come from two major groups or types. Non-polar types, such as the ones in this project, have no polarity and can be placed in either direction on the circuit board. The other type of capacitors are called polarized types; these capacitors have polarity and have to be inserted on the board with respect to their polarity marking in order for the circuit to operate correctly. On these types of capacitors, you will observe that the components actually have a plus or minus marking or a black or white band with a plus or minus marking on the body of the capacitors.

Next refer to Table 14-2. This chart helps to determine the value of small capacitors. Many capacitors are quite small in physical size, so manufacturers have devised a three-digit code which can be placed on the capacitor instead of the actual value. The code occupies a much smaller space on the capacitor and can usually fit nicely on the capacitor body, but without the chart the builder would have a difficult time determining the capacitor's value.

Now locate the capacitor C1 from the component stack, this capacitor will be labeled 0.1 μF, and will be a small size capacitor marked with (104); remember this is the three-digit code representing the actual value. Go ahead and place capacitor C1 on the circuit board and solder it in place. Now take your end-cutters and cut the excess component leads flush to the edge of the

circuit board. Once you have installed these first three components, you can move on to installing the remaining low profile components.

This project contains a single silicon diode. Diodes will always have some type of polarity markings, which must be observed if the circuit is going to work correctly. You will notice that each diode will have a black or white colored band at one edge of the diode's body. The colored band represents the diode's cathode lead. Check your schematic and parts layout diagrams when installing the diode on the circuit board. Note the diode is a 1N4004 diode. Place the diode on the circuit board and solder it in place on the PC board. Then cut the excess component leads with your end-cutters.

The Phone Line-in-Use Relay Controller utilizes a single MOSFET transistor. Transistors are generally three lead devices, which must be installed correctly if the circuit is to work properly. Careful handling and proper orientation in the circuit is critical. This type of MOSFET transistor has three leads, a Gate lead, a Drain lead and a Source lead. The gate lead is connected to resistor R4, while the drain lead is connected to the relay, and the source lead is connected to ground. Be extremely careful when installing the MOSFET since it is static sensitive and can be easily damaged by static. If you have an anti-static band, wear it while installing the MOSFET. If not, do not move around or stand on a carpet when installing the MOSFET. Be sure to refer to the component specification sheet, and the schematic and parts layout diagrams when installing the MOSFET. Once you are sure of its pin-out and orientation, you can go ahead and install it on the circuit board. Solder it in place very carefully and then remove the excess component leads with your end-cutters.

Now you can go ahead and mount the relay. If you chose a small PC board type relay, you can mount the relay on the printed circuit board. If you chose a larger conventional type relay, they you will have to use a relay socket or small strap to mount the relay to the chassis box.

Once the circuit board has been completed, take a short break and then we will inspect the PC board for possible "cold" solder joints and "short" circuits. Pick up the PC board with the foil side facing upwards toward you. Carefully inspect the solder joints; these should look smooth, clean, and bright. If any of the solder joints look dull, dark, or blobby, then you should remove the solder with a solder sucker or a solder wick and then resolder the joint all over again.

Now we will inspect the PC board for possible short circuits. Short circuits can be caused from small solder blobs bridging the circuit traces from cut component leads, which can often stick to the circuit board. A sticky residue from solder will often trap component leads across the circuit traces. Look carefully for any bridging wires across circuit traces.

You will need to locate or build a 12-V DC power supply for this project. You could elect to simply purchase a small 12-V "cube" or "wall wart" power supply at a local electronics shop, such as a 12-V DC 550 mA power supply, or you could build the power supply shown in Figure 14-5. The 12-V power supply is built around a 25.2 V transformer, which takes 110 V AC and converts it to 25 V. The center tap connects to ground, while the two secondary leads are sent to two diodes which rectify the AC to DC. The DC is first filtered by capacitor C1. A pass transistor at Q1 is current limited by the Zener diode at D3. The 12-V DC output is filtered once again by capacitors C2 and C3, which provide a clean well filtered 12 V DC supply. Once the power supply and the main Phone Line-in-Use Controller circuit have been completed, you will want to wrap some sort of case or enclosure around the circuit and power supply. Locate a suitable enclosure for the project. Center the circuit boards inside the case and use standoffs to mount the circuit board inside the case.

Once the circuit boards are mounted securely, you will have to mount the fuse holder on the rear of the chassis as well as the power switch. You will need to run the power line cord out the rear of the chassis. Finally, you will also need to connect the circuit to the phone line, and this can be done using a length of 4-conductor phone wire with a modular plug attached at one end of the cable, then you can connect the red and green wires to the phone line inputs. The modular plug ends can then simply plug into a telephone jack. You could also elect to hard-wire the circuit directly to the phone line without the modular plugs if desired.

Operation

Once your circuit has been mounted in an enclosure and wired to the phone lines, you will be ready to test the circuit. With power applied to the circuit and the phone attached, you are ready to test the circuit. Have someone pickup a phone extension and the relay should then pull-in. In the event the circuit does not work properly, you will need to remove the power and disconnect the phone line and carefully examine the circuit board for any errors. Check both circuit boards very carefully. Common construction errors include placing components in the wrong location, reading the resistor color codes incorrectly, and installing electrolytic capacitors backwards. Common mistakes also include incorrect installation of diodes and transistors and ICs. Make sure the diodes, capacitors, transistors, and ICs have been oriented correctly in the circuit. Once you have found and corrected your mistake, you can re-connect the power supply and phone line and then re-test the circuit. Also note that if the circuit does not operate once at this point, you may have to swap the telephone line lead around. The Phone Line-in-Use Relay Controller should serve you for many years to come.

Project 4: Single phone line status light

It is always handy to know quickly at a glance if a telephone is in use or if it is ringing. This circuit will allow you to shut-off your ringer while others are sleeping and watch for the light to indicate someone is calling late at night. This single line telephone status light is easy to build and is powered from 3-V batteries for six months or longer. See Figure 14-6.

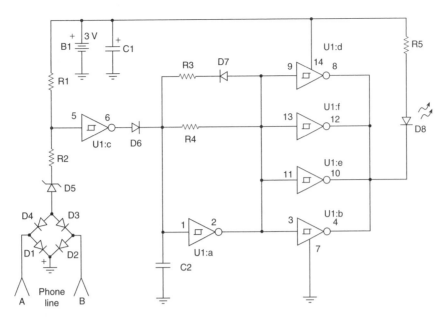

Figure 14-6 *Single line phone status light. Courtesy of Dave Johnson & Associates*

Circuit description

The sensing part of the circuit begins at the telephone line input terminals (A) and (B). The telephone line connects directly to a diode bridge circuit consisting of diodes D1 through D4. The diode bridge converts the incoming AC ringing voltage to DC and provides clean DC voltage to the Zener diode at D5. The voltage from the Zener is next passed on to a Hex Schmitt triggered inverter at U1:c. The Hex inverter cleans up the signal slightly, and the output of the inverter is then fed to a silicon diode at D6. The telephone status signal is then coupled to resistor R4, which is in parallel with inverter section U1:a. Diode D7 and series resistor R3 are used to prevent current "back-flow." The resulting status signal is then inverted once again and at the "bank" of inverters placed in parallel. The output of the inverter bank is then coupled to the indicator LED at D8. The average current when the LED is flashing is about 120 µA and the forward peak LED current is 15 mA, so the circuit does not draw very much current from the battery and therefore no power switch is utilized and the 3-V battery supply should last six months or more. The Single Line Telephone Status Light is not polarity sensitive and can be connected to the phone line without fear of damaging the circuit.

Circuit assembly

Before we begin constructing the Single Line Phone Status Indicator, you will need to locate a clean, well-lit worktable or workbench. Next we will gather a small 25–30 W pencil tipped soldering iron, a length of 60/40 tin–lead solder, and a small canister of Tip Tinner, available at your local RadioShack store; Tip Tinner is used to condition the soldering tip between solder joints. You drive the soldering tip into the compound and it cleans and prepares the soldering tip. Next grab a few small hand tools; try to locate small flat blade and Phillips screwdrivers. A pair of needle-nose pliers, a pair of tweezers along with a magnifying glass, and a pair of end-cutters and we will begin constructing the project. Now locate your schematic diagram and parts layout diagram along with all of the components needed to build the project. Once all the components are in front of you, you can check them off against the project parts list to make sure you are ready to start building the project. Now, locate Tables 14-1 and 14-2, and place them in front of you.

Table 14-1 illustrates the resistor code chart and how to read resistors, while Table 14-2 illustrates the capacitor code chart which will aid you in constructing the project.

Refer now to the resistor color code chart shown in Table 14-1, which will help you identify the resistors for the project. Note that each resistor will have either three or four color bands, which begin at one end of the resistor's body. The first colored band represents the resistor's first digit, while the second color represents the second digit of the resistor's value. The third colored band represents the resistor's multiplier value, and the fourth color band represents the resistor's tolerance value. If there is no fourth colored band then the resistor has a 20% tolerance value. If the resistor's fourth band is silver then the resistor has a 10% tolerance value and if the fourth band is gold then the resistor's tolerance value will be 5%.

Finally, we are ready to begin assembling the project, so let's get started. The prototype project was constructed on a single-sided printed circuit board or PCB. With your PC board in front of you, we can now begin populating the circuit board. We are going to place and solder the lowest height components first, the resistors and diodes. First locate our first resistor R1. Resistor R1 has a value of 4.7 megohms and is a 5% tolerance type. Now take a look for a resistor whose first color band is yellow, violet, green, and gold. From the chart you will notice that yellow is represented by the digit (4), the second color band is violet, which is represented by (7). Notice that the third color is green and the multiplier is (100,000), so $(4)(7) \times 100,000 = 4,700,000$ ohms. Go ahead and place R1 on the PC board and solder it in its respective location. Next, use your end-cutter to trim the excess component leads flush to the edge of the PC board. Now locate the remaining resistors and place them on the PC board at their proper location. Next, solder them in place, and remember to trim the excess lead lengths.

Note that capacitors generally come from two major groups or types. Non-polar types, such as the ones in this project, have no polarity and can be placed in either direction on the circuit board. The other type of capacitors are called polarized types; these capacitors have polarity and have to be inserted on the board with respect to their polarity marking in order for the circuit to operate correctly. On these types of capacitors, you will observe that the components actually have a plus or minus marking or a black or white band with a plus or minus marking on the body of the capacitors.

Next refer to Table 14-2. This chart helps to determine the value of small capacitors. Many capacitors are quite small in physical size, so manufacturers have devised a three-digit code which can be placed on the capacitor instead of the actual value. The code occupies a much smaller space on the capacitor and can usually fit nicely on the capacitor body, but without the chart the builder would have a difficult time determining the capacitor's value.

Now locate the capacitors from the component pile and try and locate capacitor C1, which is an electrolytic capacitor and will be labeled 10 µF; it should have a small black or white band with a plus or minus marking on or near the band. Be sure to observe this polarity marking when installing this capacitor on the circuit board. Next locate capacitor C2, a .15 µF capacitor, which will be a small size capacitor marked with (154). Remember, this is the three-digit code representing the actual value (refer to Table 14-2). Go ahead and place R1, C1, and C2 on the circuit board and solder them in place. Now take your end-cutters and cut the excess component leads flush to the edge of the circuit board. Once you have installed these first three components, you can move on to installing the remaining low profile components.

This project contains a number of silicon diodes as well a Zener diode. Diodes will always have some type of polarity markings, which must be observed if the circuit is going to work correctly. You will notice that each diode will have a black or white colored band at one edge of the diode's body. The colored band represents the diode's cathode lead. Check your schematic and parts layout diagrams when installing these diodes on the circuit board. Note that D5 is a Zener diode. Diodes D1 through D4 are all 1N4004 diodes which are the phone line input diodes. They all have a larger voltage/current rating since they are placed on the telephone line. The remaining diodes, D6 and D7, are small signal silicon diodes. Place all the diodes on the circuit board and solder them in place on the PC board. Then cut the excess component leads with your end-cutters flush to the edge of the circuit board.

This Single Line Status Light circuit also contains one IC at U1. ICs will also have some sort of marking to help you orient them on the PC board. Often there will be a small indented circle, a small notch, or cutout at the top end of the IC package. Pin of the device will be just to the left of the notch, or cutout. It is highly recommended that the circuit builder use IC sockets for the ICs. IC sockets are low cost insurance against a possible circuit failure at some later date. It is much easier to simply unplug an IC rather than trying to unsolder many IC pins from a PC board without causing damage to the PC board. Go ahead and install the IC on the PC board and solder it in place.

The Single Line Phone Status Light circuit uses a single indicator LED at D8, which also must be oriented correctly if the circuit is to function properly. An LED has two leads, a cathode and an anode lead. The anode lead will usually be the longer of the two leads and the cathode lead will usually be the shorter lead, just under the flat side edge of the LED; this should help orient the LED on the PC board. Solder the LEDs to the PC board and then trim the excess leads.

Once the circuit board has been completed, take a short break and then we will inspect the PC board for possible "cold" solder joints and "short" circuits. Pick up the PC board with the foil side facing upwards toward you. Carefully inspect the solder joints; these should look smooth, clean, and bright. If any of the solder joints look dull, dark, or blobby, then you should remove the solder with a solder sucker or a solder wick and then resolder the joint all over again.

Now we will inspect the PC board for possible short circuits. Short circuits can be caused from small solder blobs bridging the circuit traces from cut component leads, which can often stick to the circuit board. A sticky residue from solder will often trap component leads across the circuit traces. Look carefully for any bridging wires across circuit traces. Once this inspection is complete, we can move on to installing the circuit board into some sort of plastic enclosure.

Locate a suitable enclosure to house the circuit and the 2-cell battery holder. Choose a two "C" or "D" cell battery holder for the circuit, arrange the battery holder and circuit board within the enclosure, and mount the circuit board on standoffs within the enclosure. The battery can be hardwired directly to the circuit and the batteries should last at least 6 to 12 months with normal operation. You will also need to connect the circuit to the two phone lines. This can be done using two lengths

of 4-conductor phone wire with a modular plug attached at one end of each of the cables. Since the circuit is not sensitive to incorrect polarity, you can connect the red and green wires from each cable to each of the phone line inputs. The modular plug ends can then simply plug into a telephone jack. You could elect to hard-wire the circuit directly to the phone line without the modular plugs if desired.

Operation

Once you circuit has been mounted in an enclosure and wired to the phone line, you are now ready to test the circuit. Have someone call phone line and look for the LED to flash at D8. You can also test the in-use function by simply picking up the phone receiver and looking at the LED. Your Single Line Phone Status Light is now ready to serve you for many years to come.

Now you will know if your phone line is ringing and if someone has picked up an extension while you are on the phone. Two "C" or "D" cell batteries should last for at least 6 months.

Project 5: Two-Line Phone status indicator

Many home businesses use multiple phone lines. This circuit gives you a visual indication when a line is in operation. The two AA battery cells should provide enough power for about one year of operation. The circuit is line polarity insensitive.

The Two-Line Phone Status Indicator circuit will allow you to monitor the status of two telephone lines using a single circuit. The 2-line phone monitor is depicted in Figure 14-7. The phone line enters the circuit at two different points (A) and (B) for phone line number one and at terminals (C) and (D) for phone number two.

The first diodes bridge, on phone line #1, consists of diodes D1 through D4. The resulting line status signal is passed to resistor R2 and then through D9 which drives transistor Q1. Transistor Q1 is then used to drive the photo transistor Q2, which applies current to LED D10. LED D10 forms the "sending" unit in an opto-coupler

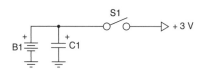

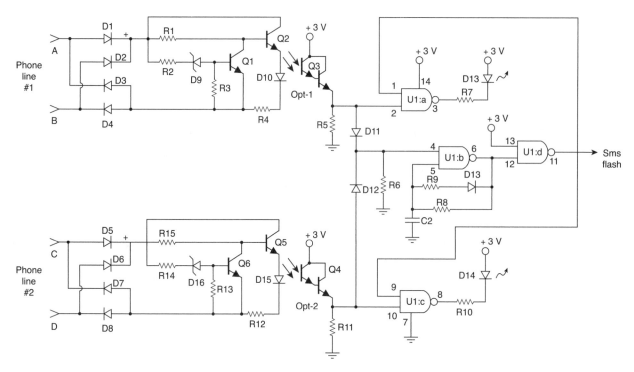

Figure 14-7 *Two-Line Phone status indicator. Courtesy of Dave Johnson & Associates*

which activates the "receive" unit in the opto-coupler. Q3 receives the isolated light signal which is passed on to D11 and U1:a. This divided signal now goes in two directions. The signal is sent to the input of a 74HC132, a Quad 2-input NAND Schmitt trigger IC, at U1:a. This NAND Schmitt Trigger is used to drive the LED at D13. LED D13 will indicate when phone line #1 is in use. Remembering the divided signal we talked about earlier, let's now go back to the junction or R5 and D11. Diode D11 directs the output from the opto-coupled ring signal from telephone line #1 and feeds it to U1:b, which is sent to U1:d. The output from U1:d is a "flash" signal which is sent back to U1:a on pin 1. This circuit allows the ring signal to flash the LED at D13 on telephone line #1.

Phone line #2 accepts the telephone line at terminals (C) and (D) and directs the line status signal through the diode bridge at D5 through D8. The diode bridge passes the line input signal on to transistor Q6 which in turn

drives the photo transistor at Q5. The isolated line #2 status signal once at photo transistor Q4 is next passed on to U1:c, which drives D14 when phone line #2 is activated. The line status signal is also split at the junction of R11 and U1:c. The signal is sent via D12 to one input of U1:b. The output of U1:b is fed directly via pin 6 on to pin12 of U1:d which provides a "flash" signal when the phone line #2 rings. This flash signal is coupled back to pin9 of U1:c in order to flash the LED at D14 when a ring signal is present on phone line #2.

Circuit assembly

Before we begin constructing the Two-Line Phone Status Indicator, you will need to find a clean, well-lit worktable or workbench. Next we will gather a small 25–30-W pencil tipped soldering iron, a length of 60/40 tin–lead solder, and a small canister of Tip Tinner, available at your local RadioShack store; Tip Tinner is used to condition the soldering tip between solder joints. You drive the soldering tip into the compound and it cleans and prepares the soldering tip. Next grab a few small hand tools; try to locate small flat blade and Phillips screwdrivers. A pair of needle-nose pliers, a pair of tweezers along with a magnifying glass, and a pair of end-cutters and we will begin constructing the project. Now locate your schematic diagram and parts layout diagram along with all of the components needed to build the project. Once all the components are in front of you, you can check them off against the project parts list to make sure you are ready to start building the project. Now, locate Tables 14-1 and 14-2, and place them in front of you.

Table 14-1 illustrates the resistor code chart and how to read resistors, while Table 14-2 illustrates the capacitor code chart which will aid you in constructing the project.

Refer now to the resistor color code chart shown in Table 14-1, which will assist you in identifying the resistors used in this project. Note that each resistor will have either three or four color bands, which begin at one end of the resistor's body. The first colored band represents the resistor's first digit, while the second color represents the second digit of the resistor's value. The third colored band represents the resistor's multiplier value, and the fourth color band represents the resistor's tolerance value. If there is no fourth colored band then the resistor has a 20% tolerance value. If the resistor's fourth band is silver then the resistor has a 10% tolerance value and if the fourth band is gold then the resistor's tolerance value will be 5%.

Finally we are ready to begin assembling the project, so let's get started. The prototype project was constructed on a single-sided printed circuit board or PCB. With your PC board in front of you, we can now begin populating the circuit board. We are going to place and solder the lowest height components first, the resistors and diodes. First, locate our first resistor R1. Resistor R1 has a value of 220 ohms and is a 5% tolerance type. Now take a look for a resistor whose first color band is yellow, violet, green, and gold. From the chart you will notice that yellow is represented by the digit (4), the second color band is violet, which is represented by (7). Notice that the third color is green and the multiplier is (100,000), so (4)(7) × 100,000 = 4,700,000 ohms.

Go ahead and place resistor R1 on the PC board and solder it in place in its respective location. Next take your end-cutters and trim the excess component leads flush to the edge of the circuit board. Next locate the remaining resistors and place them in their correct locations and solder them in place, remembering to trim the excess leads with your end-cutters.

Next refer to Table 14-2; this chart helps to determine the value of small capacitors. Many capacitors are quite small in physical size, so manufacturers have devised a three-digit code which can be placed on the capacitor instead of the actual value. The code occupies a much smaller space on the capacitor and can usually fit nicely on the capacitor body, but without the chart the builder would have a difficult time determining the capacitor's value.

Note that capacitors generally come from two major groups or types. Non-polar types, such as the ones in this project, have no polarity and can be placed in either direction on the circuit board. The other type of capacitors are called polarized types; these capacitors have polarity and have to be inserted on the board with respect to their polarity marking in order for the circuit to operate correctly. On these types of capacitors, you will observe that the components actually have a plus or minus marking or a black or white band with a plus or minus marking on the body of the capacitors.

Now locate the capacitors from the component pile and try and locate capacitor C1, which is an electrolytic capacitor will be labeled 10 µF, and should have a small black or white band with a plus or minus marking on or near the band. Be sure to observe this polarity marking when installing this capacitor on the circuit board. Next locate capacitor C2, which will be a small size capacitor marked with (473); remember this is the three-digit code representing the actual value (refer to Table 14-2). Go ahead and place C1 and C2 on the circuit board and solder them in place. Now take your end-cutters and cut the excess component leads flush to the edge of the circuit board. Once you have installed these first three components, you can move on to installing the remaining low profile components.

This project contains a number of silicon and Zener diodes. Diodes will always have some type of polarity markings, which must be observed if the circuit is going to work correctly. You will notice that each diode will have a black or white colored band at one edge of the diode's body. The colored band represents the diode's cathode lead. Check your schematic and parts layout diagrams when installing these diodes on the circuit board. Note that D9 and D16 are both Zener diodes. Diodes D1 through D8 are all 1N4004 diodes, which are the phone line input diodes. They all have a larger voltage/current rating since they are placed on the telephone line. The remaining diodes are all small signal silicon diodes. Place all the diodes on the circuit board and solder them in place on the PC board. Then cut the excess components lead with your end-cutters.

The Two-Line Phone Status Indicator utilizes four conventional transistors and two photo opto-coupler transistors. Transistors are generally three lead devices, which must be installed correctly if the circuit is to work properly. Careful handling and proper orientation in the circuit is critical. A transistor will usually have a Base lead, a Collector lead, and an Emitter lead. The base lead is generally a vertical line inside a transistor circle symbol. The collector lead is usually a slanted line facing towards the base lead and the emitter lead is also a slanted line facing the base lead but it will have an arrow pointing towards or away from the base lead.

The photo opto-coupler transistor consists of an LED and photo transistor in a small DIP package which looks like a small IC. These will often have a small notch, or cutout at the top end of the plastic package. Generally pin 1 of the device is just to the left of this notch, or cutout. Pay careful attention to the pinouts of these devices and use the schematic and parts layout and component specification sheets to install these opto-couplers onto the circuit board.

This circuit also contains an IC at U1. ICs will also have some sort of marking to help you orient them on the PC board. Often there will be a small indented circle, a small notch, or cutout at the top end of the IC package. Pin of the device will be just to the left of the notch, or cutout. It is highly recommended that the circuit builder use IC sockets for the ICs. IC sockets are low cost insurance against a possible circuit failure at some later date. It is much easier to simply unplug an IC rather than trying to unsolder many IC pins from a PC board with out causing damage to the PC board. Go ahead and install the opto-coupler and the integrated circuit on the PC board and solder them in place.

The Two-Line Phone Status Indicator circuit uses two LED, which also must be oriented correctly if the circuit is to function properly. An LED has two leads, a cathode and an anode lead. The anode lead will usually be the longer of the two leads and the cathode lead will usually be the shorter lead, just under the flat side edge of the LED, this should help orient the LED on the PC board. Solder the LEDs to the PC board and then trim the excess leads.

Once the circuit board has been completed, take a short break and then we will inspect the PC board for possible "cold" solder joints and "short" circuits. Pick up the PC board with the foil side facing upwards toward you. Carefully inspect the solder joints; these should look smooth, clean, and bright. If any of the solder joints look dull, dark, or blobby, then you should remove the solder with a solder sucker or a solder wick and then resolder the joint all over again.

Now we will inspect the PC board for possible short circuits. Short circuits can be caused from small solder blobs bridging the circuit traces from cut component leads, which can often stick to the circuit board. A sticky residue from solder will often trap component leads across the circuit traces. Look carefully for any bridging wires across circuit traces. Once this inspection is complete, we can move on to installing the circuit board into some sort of plastic enclosure.

Locate a suitable enclosure to house the circuit and the 2-cell battery holder. Choose a two "C" or "D" cell battery holder for the circuit and arrange the battery holder and circuit board within the enclosure and mount the circuit board on standoffs within the enclosure. The battery can be hardwired directly to the circuit and the batteries should last at least 6 to 12 months with normal operation. You will also need to connect the circuit to the two phone lines. This can be done using two lengths of 4-conductor phone wire with a modular plug attached at one end of each of the cables. Since the circuit is not sensitive to incorrect polarity, you can connect the red and green wires from each cable to each of the phone line inputs. The modular plug ends can then simply plug into a telephone jack. You could elect to hardwire the circuit directly to the phone line without the modular plugs if desired.

Operation

Once your circuit has been mounted in an enclosure and wired to the phone lines, you are now ready to test it. Have someone call phone line #1 first and look for the LED to flash at D13. Then have someone call phone line #2 and you should see LED D14 start to flash with the ringing. You can also test the in-use function by simply picking up each phone receiver on each line and looking at the LEDs. Your Two-Line Phone Status Indicator is now ready to serve you for many years to come. Now you will know if a phone line is ringing or if someone has picked up an extension.

Project 6: Smart phone light

The Smart Phone Light is designed to switch on a lamp when the telephone rings. If the ambient room lighting is insufficient when the phone rings, the circuit will automatically turn on a "night-light" for a period of time, so you can see where the telephone is and perhaps take notes or write down a important message. The Smart Phone Light will assist you getting your slippers on or finding a pad and paper at your night-stand. This is a very handy circuit; you may want to build a few of them and place them in different bedrooms, or give one to an older person.

Circuit description

The Smart Phone Light circuit shown in Figure 14-8 is used to switch on a lamp when the telephone rings. The circuit uses only two ICs, and it can be built quite easily. A light dependent resistance (LDR), with about 5 kilo-ohms resistance in the ambient light and greater than 100 kilo-ohms in darkness, is at the heart of the circuit. The circuit is fully isolated from the phone lines and it draws current only when the phone rings. The circuit provides automatic switching on of a lamp during darkness when the phone is kept in a place such as the bedroom. The lamp can be battery

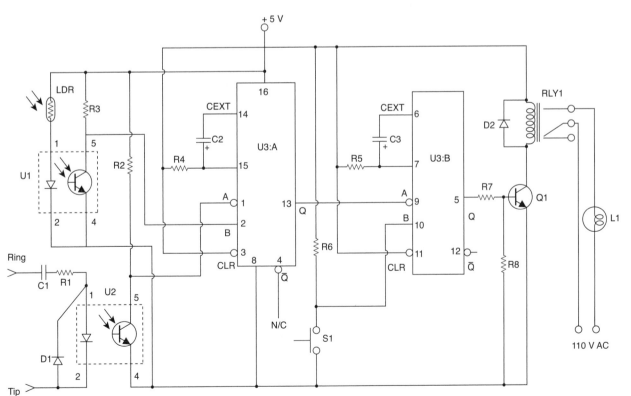

Figure 14-8 *Smart phone light project*

powered to provide light during power failure or load shedding. This avoids delay in attending to a call. The light switches off automatically after a programmable time period and it needs no attention at all. If required, the lamp lighting period can be extended by simply pressing a pushbutton switch (S1). The first part of the circuit functions as a ring detector.

When telephone is on-hook, around 48 V DC is present across the TIP and RING terminals. The diode in the opto-coupler is "off" during this condition and it draws practically no current from he telephone lines. The opto-coupler also isolates the circuit from the telephone lines. Transistor in the opto-coupler is normally "off" and a voltage of +5 V is present at the ring indicator line. When telephone rings, an AC voltage of around 70–80 V AC, which is present across the telephone lines, is used to turn on the diode inside the opto-coupler (IC2), which in turn switches on a transistor inside the opto-coupler. The voltage at its collector passes through 0-V level during ringing to trigger IC3 74LS123(A) monostable flip-flop. The other opto-coupler (IC1) is used to detect the ambient light condition. When there is sufficient light, LDR has a low resistance of about 5 kilo-ohms and the transistor inside the opto-coupler is in "on" state. When there is insufficient light available, the resistance of LDR increases to a few megohms and the transistor switches to "off" state. Thus the DC voltage present at the collector of transistor inside the opto-coupler is normally 0 V and it jumps to 5 V when there is no light or insufficient light. The 74LS123 retriggerable monostable multivibrator is used to generate a programmable pulse-width. The first monostable 74LS123(A) generates a pulse from the trigger input available during ringing, provided its pin 2 input (marked B) is logic high (i.e. during darkness). It remains high for the programmed duration and switches back to 0 V at the end of the pulse period. This high-to-low transition (trailing edge) is used to trigger the second monostable flip-flop 74LS123(B) in the same package. Output of the second monostable is used to control a relay. The lamp being controlled via the N/O contacts of the relay gets switched "on." The "on" period can be extended by simply pressing pushbutton switch S1. If nobody attends the phone, the light turns off automatically after the specific time period equal to the pulse-width of the second flip-flop. The light

sensitivity of LDR can be changed by changing resistance R2 connected at collector of the transistor in light monitor circuit. Similarly, switch-on period of the lamp can be controlled by changing capacitor C3's value in the second 74123(B) monostable circuit. The diagram shown in Figure 14-9 illustrates the timing waveforms during different circuit modes of operation.

Circuit assembly

Before we begin constructing the Smart Phone Light circuit, you will need to find a clean, well-lit worktable or workbench. Next we will gather a small 25–30-W pencil tipped soldering iron, a length of 60/40 tin–lead solder, and a small canister of Tip Tinner, available at your local RadioShack store; Tip Tinner is used to condition the soldering tip between solder joints. You drive the soldering tip into the compound and it cleans and prepares the soldering tip. Next grab a few small

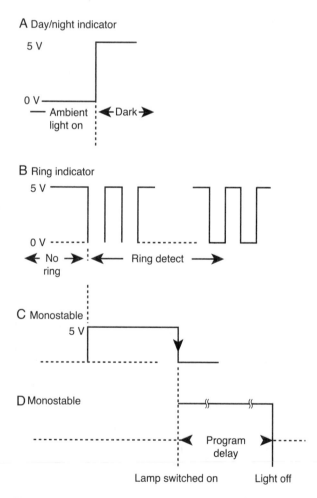

Figure 14-9 *Smart phone light waveforms*

hand tools, and try to locate small flat blade and Phillips screwdrivers. A pair of needle-nose pliers, a pair of tweezers along with a magnifying glass, and a pair of end-cutters and we will begin constructing the project. Now locate your schematic diagram and parts layout diagram along with all of the components needed to build the project. Once all the components are in front of you, you can check them off against the project parts list to make sure you are ready to start building the project. Now, locate Tables 14-1 and 14-2, and place them in front of you.

Table 14-1 illustrates the resistor code chart and how to read resistors, while Table 14-2 illustrates the capacitor code chart which will aid you in constructing the project.

Refer now to the resistor color chart in Table 14-1, which will help you to identify the resistor values for this project. Note that each resistor will have either three or four color bands, which begin at one end of the resistor's body. The first colored band represents the resistor's first digit, while the second color represents the second digit of the resistor's value. The third colored band represents the resistor's multiplier value, and the fourth color band represents the resistor's tolerance value. If there is no fourth colored band then the resistor has a 20% tolerance value. If the resistor's fourth band is silver then the resistor has a 10% tolerance value and if the fourth band is gold then the resistor's tolerance value will be 5%.

Finally we are ready to begin assembling the project, so let's get started. The prototype project was constructed on a single-sided printed circuit board or PCB; with your PC board in front of you, we can now begin populating the circuit board. We are going to place and solder the lowest height components first, the resistors and diodes. First, locate our first resistor R1. Resistor R1 has a value of 20 K ohms and is a 5% tolerance type. Now take a look for a resistor whose first color band is red, black, orange, and gold. From the chart you will notice that red is represented by the digit (2), the second color band is black, which is represented by (0). Notice that the third color is orange and the multiplier is (1,000), so (2)(0) × 1,000 = 20,000 or 20 K ohms.

Go ahead and place resistor R1 in its respective location on the PC board. Solder it in place and then trim the excess leads using your end-cutters. Cut the

excess leads flush to the edge of the circuit board. Next locate the remaining resistors, place them in their proper locations, then solder them in place. Remember to cut the excess lead lengths with your end-cutters.

Note that capacitors generally come from two major groups or types. Non-polar types, such as the ones in this project, have no polarity and can be placed in either direction on the circuit board. The other type of capacitors are called polarized types, these capacitors have polarity and have to be inserted on the board with respect to their polarity marking in order for the circuit to operate correctly. On these types of capacitors, you will observe that the components actually have a plus or minus marking or a black or white band with a plus or minus marking on the body of the capacitors.

Next refer to Table 14-2; this chart helps to determine the value of small capacitors. Many capacitors are quite small in physical size, so manufacturers have devised a three-digit code which can be placed on the capacitor instead of the actual value. The code occupies a much smaller space on the capacitor and can usually fit nicely on the capacitor body, but without the chart the builder would have a difficult time determining the capacitor's value.

Now locate the capacitors from the component pile and try and locate capacitor C1, which is an electrolytic capacitor will be labeled 1 µF, and it should be a large 250 volt paper non-electrolytic capacitor with no polarity markings. Next locate capacitor C2, which will be 47 µF electrolytic capacitor, while C3 is a 1,000 µF electrolytic capacitor. Go ahead and place R1, C1, C2, and C3 on the circuit board and solder them in place on the circuit board. Now take your end-cutters and cut the excess component leads flush to the edge of the circuit board. Once you have installed these first three components, you can move on to installing the remaining low profile components.

This project contains a two of silicon diodes. Diodes will always have some type of polarity markings, which must be observed if the circuit is going to work correctly. You will notice that each diode will have a black or white colored band at one edge of the diode's body. The colored band represents the diode's cathode lead. Check your schematic and parts layout diagrams when installing these diodes on the circuit board. Note, diodes D1 and D2 are both 1N4007 silicon diodes. Place the two diodes on the circuit board and solder them in place on the PC board. Then cut the excess components leads flush to the edge of the PC board with your end-cutters.

The Smart Phone Light utilizes one conventional transistor and two opto-couplers. Transistors are generally three lead devices, which must be installed correctly if the circuit is to work properly. Careful handling and proper orientation in the circuit is critical. A transistor will usually have a Base lead, a Collector lead, and an Emitter lead. The base lead is generally a vertical line inside a transistor circle symbol. The collector lead is usually a slanted line facing towards the base lead and the emitter lead is also a slanted line facing the base lead but it will have an arrow pointing towards or away from the base lead.

The photo opto-couplers consist of an LED and photo transistor in a small DIP package which will look like a small IC. The opto-couplers will often have a small notch, or cutout at the top end of the plastic package. Generally pin 1 of the device is just to the left of this notch, or cutout. Pay careful attention to the pinouts of these devices and use the schematic and parts layout and component specification sheets to install the opto-couplers onto the circuit board. Go ahead and install the transistor at Q1 and the opto-couplers at U1 and U2. Solder them in place and remember to trim the excess component leads.

This circuit also contains an IC at U3. ICs will also have some sort of marking to help you orient them on the PC board. Often there will be a small indented circle, a small notch, or cutout at the top end of the IC package. Pin of the device will be just to the left of the notch, or cutout. It is highly recommended that the circuit builder use IC sockets for the ICs. IC sockets are low cost insurance against a possible circuit failure at some later date. It is much easier to simply unplug an IC rather than trying to unsolder many IC pins from a PC board with out causing damage to the PC board. After installing the integrated circuit socket, you can insert the IC into its socket. The transistor, opto-couplers, and the IC must be installed properly, so as not to damage the components or other components in the circuit. If you doubt your abilities, have a knowledgeable electronics enthusiast help you install these components.

As mentioned, the heart of the Smart Phone Light circuit uses two LEDs, which also must be oriented correctly if the circuit is to function properly. The LDR does not have polarity, so it can be installed in either direction.

If you elected to use a PC board type relay at RLY1, you can go ahead and mount it on the PC board and wire it up. If you elected to use a non PC board style relay then you will have to wire it up to the circuit with separate wiring, and then mount the relay using a strap mounted to the bottom of the chassis box.

Once the circuit board has been completed, take a short break and then we will inspect the PC board for possible "cold" solder joints and "short" circuits. Pick up the PC board with the foil side facing upwards toward you. Carefully inspect the solder joints; these should look smooth, clean, and bright. If any of the solder joints look dull, dark, or blobby, then you should remove the solder with a solder sucker or a solder wick and then resolder the joint all over again.

Now we will inspect the PC board for possible short circuits. Short circuits can be caused from small solder blobs bridging the circuit traces from cut component leads, which can often stick to the circuit board. A sticky residue from solder will often trap component leads across the circuit traces. Look carefully for any bridging wires across circuit traces. Once this inspection is complete, we can move on to installing the circuit board into some sort of plastic enclosure.

If you elected to build a 5-V power supply you can refer to the schematic shown in Figure 14-10.

The power supply can be constructed on a PC board, with the transformer mounted directly on the chassis. Power is applied to the transformer primary. The secondary winding of the 25 V center tapped transformer is sent to two diodes, and then fed to the 1,000 µF capacitor at C1. The 5-V regulator is mounted on a small heat sink and isolated from the PC board. The output of the regulator is coupled to capacitor C2 followed by C3. Resistor R1 and the LED are wired in series across the 5-V power supply output.

Arrange the circuit board and power supply, if you elected to build one, within the enclosure and mount the circuit board on standoffs within the case. The power supply can be hardwired directly to the Smart Phone

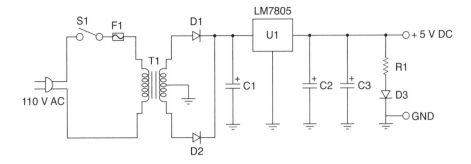

Figure 14-10 *Smart phone light 5-V power supply*

Light circuit. You will also need to connect the circuit to the phone line. This can be done using two lengths of 4-conductor phone wire with a modular plug attached at one end of each of the cables. The modular plug ends can then simply plug into a telephone jack. You could elect to hard-wire the circuit directly to the phone line without the modular plugs if desired. The load or lamp at L1 can be a small 110-V lamp if desired which is connected to the relay. A 110-V AC power source is wired in series with the lamp to the relay, and you are almost ready to try out the circuit for the first time.

Operation

Once your circuit has been mounted in an enclosure and wired to the phone lines, you will be ready to test the circuit. Connect up a 5-V power supply to the circuit. Next, connect the circuit to the phone line. Now you can press switch S1 and the lamp at L1 should light-up. If that works you can move on to testing the circuit. Turn off the lights in the room and then have someone call your phone number. As the phone rings you should see the lamp L1 light-up for the duration of the timing circuit controlled by C2/R4 and R5/C3. If nothing happens when the phone rings, you may have made a mistake in building the circuit. Disconnect the power supply and the phone line from the circuit. Carefully examine the circuit board to make sure that all the components are in their respective locations and oriented correctly. Have an electronics enthusiast recheck your wiring to make sure that you did not make any mistakes. Once your problem has been corrected you can reconnect the power supply and the phone line and your Smart Phone Light is ready to serve you.

Project 7: Telephone line vigilant

The Telephone Line Vigilant circuit will guard against misuse of your telephone lines. It monitors telephone lines round the clock and provides visual as well as audio warning when someone is using your telephone lines. The Vigilant can alert you anywhere in the house. Another advantage of using this circuit is that one comes to know of the misuse and/or snapping of the lines instantaneously on its occurrence. This enables the telephone subscriber to take necessary measures immediately.

Even when you, the subscriber, are using this telephone while the vigilant circuit is on, the buzzer beeps once every 5 s, since the vigilant circuit cannot distinguish between self-use of the subscriber lines or that of any unauthorized person. Thus to avoid unnecessary disturbance, it is advisable to install the vigilant unit away from the phone. However, if one wishes to place the unit near the telephone then switch S1 may be flipped to "off" position to switch off the buzzer. But remember to flip the switch to "on" position while replacing the handset on cradle.

Circuit description

The Telephone Line Phone Vigilant circuit is illustrated in Figure 14-11. The circuit may be divided into two parts. The first part comprises Zener D9, transistors T1 to T4 and diode D5. It is used to verify whether the telephone line loop is intact or discontinuous.

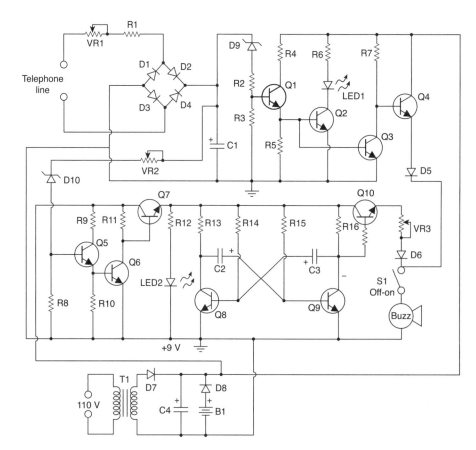

Figure 14-11 *Telephone line vigilant*

The second part comprising Zener D10 and transistors T5 to T10 is used to check whether the telephone line is in use (or misuse) or not. The Zener diode D9 (3.3 V) conducts when phone line loop is intact and not broken. Zener D9 sets control voltage for transistors T1, T2, and T3 to conduct and for T4 to cut off. As a result, green LED lights but no sound is heard from the buzzer. The Phone Vigilant circuit is insensitive to telephone line polarity at the input to the circuit; proper DC polarity is maintained across C1 due to bridge rectifier comprising diodes D1 to D4. The DC voltage developed across

capacitor C1 is used to check telephone line conditions as per Table 14-3. This circuit draws negligible current from the telephone line; thus when it is connected to the telephone line, the normal telephone operation is not affected.

When phone line loop is discontinuous, no voltage is available across capacitor C1. Thus Zener D9 and transistors T1, T2, and T3 do not conduct while T4 conducts. Now green LED extinguishes and a continuous sound is heard from the buzzer. When telephone line is all right but is not in use, Zener D10

Table 14-3

Phone line vigilant status condition chart

Number	Line Condition	Green LED	Red LED	Line Voltage	Audio Indication
1	disconnected	not lit	lit up	0 V	continuous sound
2	line inuse	lit up	lit up	9 V DC	beep-every 5 sec
3	line not in use	lit up	not lit	48 V DC	no sound

conducts as voltage across capacitor C1 is quite high. This results in conduction of transistors T5 and T6 and cutting off of transistor T7 (as collector of transistor T6 is near ground potential). Thus positive 9 V rail is not extended to the following multi-vibrator circuit built around transistors T8 and T9. Consequently, the red LED is not lit and buzzer does not sound. When phone line is in use, Zener D10 does not conduct. As a result, transistors T5 and T6 also do not conduct, while transistor T7 conducts. Now +9 V is extended to multi-vibrator circuit. This multi-vibrator is designed such that collector of transistor T9 goes high once every 5 s to forward bias transistor T10 and it conducts. Thus at every 5 s interval a beep sound is heard from buzzer. The beep sound interval can be increased or decreased by changing the value of capacitor C3, while the volume can be adjusted with the help of preset VR3.

Circuit assembly

Before we begin constructing the Telephone Line Vigilant, you will need to find a clean, well-lit worktable or workbench. Next we will gather a small 25–30-W pencil tipped soldering iron, a length of 60/40 tin–lead solder, and a small canister of Tip Tinner, available at your local RadioShack store; Tip Tinner is used to condition the soldering tip between solder joints. You drive the soldering tip into the compound and it cleans and prepares the soldering tip. Next grab a few small hand tools; try to locate small flat blade and Phillips screwdrivers. A pair of needle-nose pliers, a pair of tweezers along with a magnifying glass, and a pair of end-cutters and we will begin constructing the project. Now locate your schematic diagram and parts layout diagram along with all of the components needed to build the project. Once all the components are in front of you, you can check them off against the project parts list to make sure you are ready to start building the project. Now, locate Tables 14-1 and 14-2, and place them in front of you.

Table 14-1 illustrates the resistor code chart and how to read resistors, while Table 14-2 illustrates the capacitor code chart which will aid you in constructing the project.

Refer now to the resistor color code chart shown in Table 14-1, which will assist you in identifying the

correct resistor values for the project. Note that each resistor will have either three or four color bands, which begin at one end of the resistor's body. The first colored band represents the resistor's first digit, while the second color represens the second digit of the resistor's value. The third colored band represents the resistor's multiplier value, and the fourth color band represents the resistor's tolerance value. If there is no fourth colored band then the resistor has a 20% tolerance value. If the resistor's fourth band is silver then the resistor has a 10% tolerance value and if the fourth band is gold then the resistor's tolerance value will be 5%.

Finally we are ready to begin assembling the project, so let's get started. The prototype project was constructed on a single-sided printed circuit board or PCB. With your PC board in front of you, we can now begin populating the circuit board. We are going to place and solder the lowest height components first, the resistors and diodes. First locate our first resistor R1. Resistor R1 has a value of 56 K ohms and is a 5% tolerance type. Now take a look for a resistor whose first color band is green, blue, orange, and gold. From the chart you will notice that green is represented by the digit (5), the second color band is blue, which is represented by (6). Notice that the third color is orange and the multiplier is (1,000), so $(5)(6) \times 1,000 = 56,000$ or 56 K ohms. Once you have placed resistor R1 on the circuit board, you can go ahead and install it on the board and solder it place. Trim the excess lead length with a pair of end-cutters flush to the edge of the circuit board. Next, identify the remaining resistors and place them on the PC board and solder them in place. Remember to cut the excess lead lengths with your end-cutters.

Note that capacitors generally come from two major groups or types. Non-polar types, such as the ones in this project, have no polarity and can be placed in either direction on the circuit board. The other type of capacitors are called polarized types, these capacitors have polarity and have to be inserted on the board with respect to their polarity marking in order for the circuit to operate correctly. On these types of capacitors, you will observe that the components actually have a plus or minus marking or a black or white band with a plus or minus marking on the body of the capacitors. Electrolytic capacitors must be installed correctly in order for the circuit to work properly.

Next refer to Table 14-2, this chart helps to determine the value of small capacitors. Many capacitors are quite small in physical size, so manufacturers have devised a three-digit code which can be placed on the capacitor instead of the actual value. The code occupies a much smaller space on the capacitor and can usually fit nicely on the capacitor body, but without the chart the builder would have a difficult time determining the capacitor's value.

This project contains a number of silicon and Zener diodes. Diodes will always have some type of polarity markings, which must be observed if the circuit is going to work correctly. You will notice that each diode will have a black or white colored band at one edge of the diode's body. The colored band represents the diode's cathode lead. Check your schematic and parts layout diagrams when installing these diodes on the circuit board. Note that D9 and D10 are both Zener diodes. Diodes D1 through D4 and diodes D5 through D8 are all 1N4007 silicon diodes. Place all the diodes on the circuit board and solder them in place. Then cut the excess component leads with your end-cutters.

The Telephone Line Phone Vigilant circuit utilizes 10 conventional transistors. Transistors are generally three lead devices, which must be installed correctly if the circuit is to work properly. Careful handling and proper orientation in the circuit is critical. A transistor will usually have a Base lead, a Collector lead, and an Emitter lead. The base lead is generally a vertical line inside a transistor circle symbol. The collector lead is usually a slanted line facing towards the base lead and the emitter lead is also a slanted line facing the base lead but it will have an arrow pointing towards or away from the base lead. Transistors must be installed correctly in order for the circuit to function properly. Failure to install them correctly could possible damage the transistors as well as other components.

Refer to the schematic and the parts layout diagrams as well as the transistor specification sheets before installing the transistors onto the circuit board. Once you are certain you can identify the proper orientation of the transistors, go ahead and place them on the circuit board. Next you can solder the transistors in their respective locations and then trim the excess leads.

The Telephone Line Vigilant circuit employs two LEDs, which also must be oriented correctly if the circuit is to function properly. An LED has two leads a cathode and an anode lead. The anode lead will usually be the longer of the two leads and the cathode lead will usually be the shorter lead, just under the flat side edge of the LED. This should help orient the LED on the PC board. Solder the LEDs to the PC board and then trim the excess leads.

Now locate the three capacitors from the component pile and try and locate capacitor C1, which is an electrolytic capacitor and will be labeled 10 µF; it should have a small black or white band with a plus or minus marking on or near the band. Be sure to observe this polarity marking when installing this capacitor on the circuit board. Next locate capacitor C2, which is a 22 µF capacitor, and next locate C3, which is a 470 µF. When installing the capacitors be sure to refer to the schematic and parts layout diagrams to ensure that you install the capacitors correctly. Installing the capacitors backwards can damage them as well as other components when power is first applied to the circuit. Install the capacitors onto the circuit board and then solder them in place. Remember to trim the excess leads.

Once the circuit board has been completed, take a short break and then we will inspect the PC board for possible "cold" solder joints and "short" circuits. Pick up the PC board with the foil side facing upwards toward you. Carefully inspect the solder joints; these should look smooth, clean, and bright. If any of the solder joints look dull, dark, or blobby, then you should remove the solder with a solder sucker or a solder wick and then resolder the joint all over again.

Now we will inspect the PC board for possible short circuits. Short circuits can be caused from small solder blobs bridging the circuit traces from cut component leads, which can often stick to the circuit board. A sticky residue from solder will often trap component leads across the circuit traces. Look carefully for any bridging wires across circuit traces. Once this inspection is complete, we can move on to installing the circuit board into some sort of plastic enclosure.

Locate a suitable enclosure to house the circuit and the 4-cell "AA" battery holder. Locate a four cell battery holder for the circuit and arrange the battery holder and circuit board within the enclosure and mount the circuit board on standoffs within the enclosure. The battery can be hardwired directly to the circuit and the batteries should last at least 6 to 12 months with normal operation.

You will also need to connect the circuit to the two phone lines. This can be done using two lengths of 4-conductor phone wire with a modular plug attached at one end of each of the cables. Since the circuit is not sensitive to incorrect polarity, you can connect the red and green wires from each cable to each of the phone line inputs. The modular plug ends can then simply plug into a telephone jack. You could elect to hard-wire the circuit directly to the phone line without the modular plugs if desired.

Operation

Once your circuit has been mounted in an enclosure and wired to the phone line, you will be ready to test the circuit. With power applied to the circuit and the phone connected, observe the status LEDs when the phone is on-hook. Now pick up the phone handset and observe the LEDs and you should begin to hear the buzzer. Thus, to avoid unnecessary disturbance, it is advisable to install the vigilant unit away from the phone. However, if one wishes to place the unit near the telephone then switch S1 may be flipped to the "off" position to switch off the buzzer. But remember to flip the switch to the "on" position while replacing the handset on cradle. Various telephone line conditions and audio-visual indications available are summarized in Table 14-3. The Telephone Line Vigilant will serve you in many ways for years to come.

Chapter 15

Telephone and Phone Line Testers

Parts list

Parts Bin

Simple Telephone Line Tester

R1, R2   680 ohms, $\frac{1}{2}$ W,
  5% resistor

D1   Green LED

D2   Red LED

Misc   PC board,
  4-conductor wire,
  hardware, modular
  telephone plug, etc.

Off-Line Telephone Tester

R1   270 ohms, $\frac{1}{4}$ W,
  5% resistor

R2, R3   1.2 K ohms, $\frac{1}{4}$ W,
  5% resistor

R4   150 ohm, $\frac{1}{4}$ W,
  5% resistor

C1, C2   1,000 µF, 63-V
  electrolytic capacitor

D1, D2, D3, D4   1N4007
  silicon diodes

D5   1N5408 silicon diode

D6   Red LED

Q1   BC148 NPN transistor

Q2   BC368 NPN transistor

U1   UM66 Integrated
  circuit

T1, T2   110-V AC –
  24-V AC transformer @
  500 mA

S1   SPST Power switch

S2   Rotary switch –
  3-position – single
  pole

S3   pushbutton switch–
  normally open type

J1 $\frac{1}{8}$″ stereo headphone
  jack

Misc   PC board, IC
  socket, wire, terminal
  block, enclosure, etc.

Telephone Line Tester

R1   4.7 K ohms, $\frac{1}{4}$ W,
  5% resistor

R2   10 ohm potentiometer

R3   50 K ohm
  potentiometer

R4   100 K ohm, $\frac{1}{4}$ W,
  5% resistor

R5, R6   470 ohm, $\frac{1}{2}$ W,
  5% resistor

R7   10 K ohm, $\frac{1}{4}$ W,
  5% resistor

D1, D2, D3, D4, D5
  1N4003 silicon
  diodes

D6   Red LED

U1   UM66 Music chip IC

S1   DPDT toggle or slide
  switch

M1   0-1 mA current panel
  meter

P1 modular phone plug

Misc PC board, wire, hardware, chassis box, standoffs, etc.

Telephone Line Simulator

R1, R2 560 ohms, $\frac{1}{2}$ W, 5% resistor

R3 2.2 K ohms, $\frac{1}{4}$ W, 5% resistor

R4, R7 27 K ohms, $\frac{1}{4}$ W, 5% resistor

R5 270 K ohms, $\frac{1}{4}$ W, 5% resistor

R6 330 ohms, $\frac{1}{4}$ W, 5% resistor

R8 1.8 K ohms, $\frac{1}{4}$ W, 5% resistor

R9 220 ohms, $\frac{1}{4}$ W, 5% resistor

R10 3.9 K ohms, $\frac{1}{4}$ W, 5% resistor

C1 10 μF, 50-V electrolytic capacitor

C2 22 μF, 50-V electrolytic capacitor

C3 470 μF, 50-V electrolytic capacitor

D1 1N4001 silicon diode

D2 SCR - Silicon controlled rectifier

D3 Red LED

Q1, Q2 2N3904 NPN transistor

U1 LM317T adjustable regulator IC

T1 Transformer 110-V AC primary - 40-V AC secondary

RY1 DPDT - 24-V DC relay

RY2 SPDT - 24-V DC relay

S1, S2 SPDT toggle switch

F1 5 amp fast blow fuse

P1 110-V AC - grounded plug

Misc PC board, wire, terminal strips, enclosure, etc.

Project 1: Simple Telephone Line Tester

If you are a contractor, home builder, building new homes, or a home remodeler or even a home owner rewiring your own home or office, then you will appreciate this simple Phone Line Tester Project. The Simple Telephone Line Tester will assist you in figuring out the mysteries of telephone line wiring.

The Telephone Line Tester shown in Figure 15-1 illustrates a simple 2-LED line tester. The LEDs indicate whether a line of telephone jack is

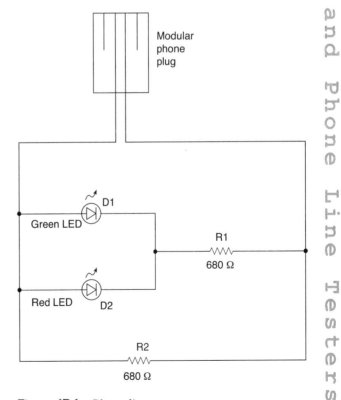

Figure 15-1 *Phone line tester*

wired correctly, has reversed polarity, or is disconnected from the telephone system.

The Simple Telephone Line Tester circuit plugs into a modular telephone jack and was designed to give a quick status check of a telephone line and jack wiring. If the green LED lights up, the jack is wired properly, and the outlet is operational. If the red LED lights up, the wiring to the jack is reversed. If neither LED lights up, then the telephone jack is not wired to the telephone system. This is a handy tester to take around the home or building site for a quick and easy test of the telephone wiring. The circuit is simple to construct and easy to use.

Circuit assembly

Before we begin constructing the Simple Phone Line Tester, you will need to locate a clean, well-lit worktable or workbench. Next we will gather a small 25–30-W pencil tipped soldering iron, a length of 60/40 tin–lead solder, and a small canister of Tip Tinner, available at your local RadioShack store; Tip Tinner is used to condition the soldering tip between solder joints. You drive the soldering tip into the compound and it cleans and prepares the soldering tip. Next grab a few small hand tools, try to locate small flat blade and Phillips screwdrivers. A pair of needle-nose pliers, a pair of tweezers along with a magnifying

glass, and a pair of end-cutters and we will begin constructing the project. Now locate your schematic diagram and parts layout diagram along with all of the components needed to build the project. Once all the components are in front of you, you can check them off against the project parts list to make sure you are ready to start building the project. Now, locate Table 15-1, and place it in front of you. Table 15-1 illustrates the resistor code chart and how to read resistors.

Each resistor will have either three or four color bands, which begin at one end of the resistor's body. The first colored band represents the resistor's first digit, while the second color represents the second digit of the resistor's value. The third colored band represents the resistor's multiplier value, and the fourth color band represents the resistor's tolerance value. If there is no fourth colored band then the resistor has a 20% tolerance value. If the resistor's fourth band is silver, then the resistor has a 10% tolerance value and if the fourth band is gold then the resistor's tolerance value will be 5%.

Finally we are ready to begin assembling the project, so let's get started. The prototype project was constructed on a single-sided printed circuit board or PCB. With your PC board in front of you, we can now begin populating the circuit board. We are going to place and solder the lowest height components first, the resistors and the diodes. First, locate our first resistor R1.

Table 15-1
Resistor color code chart

Color Band	1st Digit	2nd Digit	Multiplier	Tolerance
Black	0	0	1	
Brown	1	1	10	1%
Red	2	2	100	2%
Orange	3	3	1,000 (K)	3%
Yellow	4	4	10,000	4%
Green	5	5	100,000	
Blue	6	6	1,000,000 (M)	
Violet	7	7	10,000,000	
Gray	8	8	100,000,000	
White	9	9	1,000,000,000	
Gold			0.1	5%
Silver			0.01	10%
No color				20%

Resistor R1 has a value of 680 ohms and is a 5% tolerance type. Now take a look for a resistor whose first color band is blue, gray, brown, and gold. From the chart you will notice that blue is represented by the digit (6), the second color band is gray, which is represented by (8). Notice that the third color is brown and the multiplier is (10), so $(6)(8) \times 10 = 680$ ohms.

Go ahead and place R1 and R2 on the circuit board and solder them in place. Now take your end-cutters and cut the excess component leads flush to the edge of the circuit board. Once you have installed these components, you can move on to installing the remaining components.

The Simple Telephone Line Tester circuit utilizes two LEDs, which also must be oriented correctly if the circuit is to function properly. An LED has two leads a cathode and an anode lead. The anode lead will usually be the longer of the two leads and the cathode lead will usually be the shorter lead, just under the flat side edge of the LE; this should help orient the LED on the PC board. Solder the LEDs to the PC board and then trim the excess leads.

Once the circuit board has been completed, take a short break and then we will inspect the PC board for possible "cold" solder joints and "short" circuits. Pick up the PC board with the foil side facing upwards toward you. Carefully inspect the solder joints, these should look smooth, clean, and bright. If any of the solder joints look dull, dark, or blobby, then you should remove the solder with a solder sucker or a solder wick and then resolder the joint all over again.

Now we will inspect the PC board for possible short circuits. Short circuits can be caused from small solder blobs bridging the circuit traces or from cut component leads which can often stick to the circuit board once they have been cut. A sticky residue from solder will often trap component leads across the circuit traces. Look carefully for any bridging wires across circuit traces. Once this inspection is complete, we can move on to installing the circuit board into some sort of plastic enclosure.

Finally, you will also need to connect the circuit to the phone line. This can be done using a 6″ length of 4-conductor phone wire with a modular plug attached at one end of the cable. The modular plug can then simply plug into a telephone jack, which you will want to test.

Operation

Once your circuit has been mounted in an enclosure, you will be ready to test the circuit. Simply plug the modular jack into a telephone jack and one of the LEDs should light up immediately. If the green LED lights up then the circuit is wired correctly, but if the red LED lights up then the telephone jack is wired backwards. If neither LED lights up, then the circuit is dead and there is no voltage on the line or your jack is not wired to the active telephone line. Your Simple Telephone Line Tester is now ready to serve you for many years to come, and you will be glad you built this handy tester!

Project 2: Off-Line Telephone tester

The Off-Line Telephone Tester is a combination phone tester that will allow you test a telephone in a number of different ways. The Off-line tester can ring your test phone, it can test the side-tone audio within the phone, it can test audio from the phone to a pair of headphones. The Off-line Phone Tester can also test phone dialing by listening to the attached headphones. It also contains a music generator, which supplies music to the phone handset to test the earphone in the phone.

The multi-purpose Off-Line Telephone Tester is illustrated in the schematic at Figure 15-2. The Off-Line Telephone Tester circuit does not require any telephone line for testing a telephone instrument, so you do not have to jeopardize you phone line by testing a faulty phone set. The circuit is so simple that it can be easily assembled, even by a novice having very little knowledge of electronics.

A telephone line can be considered as a source of about 50 V DC with a source impedance of about 1 kilo-ohm. During ringing, in place of DC, an AC voltage of 70 to 80 V (at 17 to 25 Hz) is present across the telephone line. When the subscriber lifts the handset, the same is sensed by the telephone exchange and the ringing AC voltage is disconnected and DC is reconnected to the line. Lifting of the handset from the telephone cradle results in shunting of the line's two wires by low impedance of the telephone instrument.

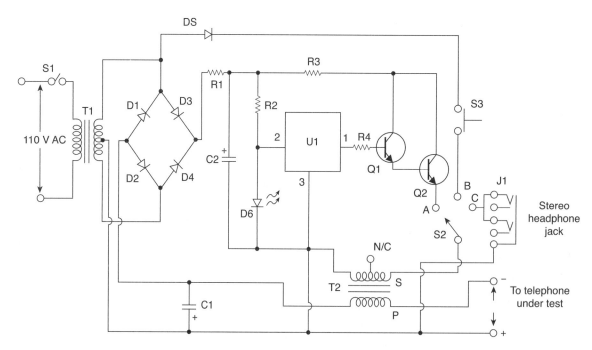

Figure 15-2   *Off-Line Telephone tester*

As a result, 50 V DC level drops to about 12 V across the telephone instrument. During conversation, the audio gets superimposed on this DC voltage. Since any DC supply can be used for testing a telephone instrument, the same is derived here from AC mains using step-down transformer T1.

The center-tap of the transformer's secondary has been used as common for the two full-wave rectifiers—one comprising diodes D1 and D2 together with smoothing capacitor C1 and the other formed by diodes D3 and D4 along with filter capacitor C2. The former supplies about 12 V for the telephone instrument through primary of transformer T2 which thus simulates a source of impedance, and a choke which blocks AC audio signals present in the secondary of transformer T2. The audio or AF signal available in secondary of T2 is sufficiently strong to directly drive a 32-ohm headset which is connected to the circuit through headphone jack J1 via rotary switch S2.

During ringing, a pulsating DC voltage from transformer T1 via rectifier diode D5, push-to-on switch S3, and contact "B" of rotary switch S2 is applied across secondary of transformer X2. The boosted voltage available across primary of transformer T2 is sufficient to drive the ringer in the telephone instrument. Please avoid pressing of switch S3 for more than a few seconds at a time to prevent damage to the circuit due to

high voltage across primary of transformer T2. The circuit also incorporates a music IC at U1, whose output is connected to secondary of transformer T2 via switch S2 after suitably boosting its output with the help of Darlington transistor pair Q1 and Q2. This output can be used to test the audio section of any telephone instrument. After having assembled the circuit satisfactorily, the following procedure may be followed for testing a telephone instrument:

Circuit assembly

Before we begin constructing the Off-Line Telephone Tester, you will need to find a clean, well-lit worktable or workbench. Next we will gather a small 25–30-W pencil tipped soldering iron, a length of 60/40 tin–lead solder, and a small canister of Tip Tinner, available at your local RadioShack store; Tip Tinner is used to condition the soldering tip between solder joints. You drive the soldering tip into the compound and it cleans and prepares the soldering tip. Next grab a few small hand tools; try and locate small flat blade and Phillips screwdrivers. A pair of needle-nose pliers, a pair of tweezers along with a magnifying glass, and a pair of end-cutters and we will begin constructing the project. Now locate your schematic diagram and parts layout diagram along with all of the components needed to build the project.

Once all the components are in front of you, you can check them off against the project parts list to make sure you are ready to start building the project. Now, locate Table 15-1 and place it in front of you. Table 15-1 illustrates the resistor code chart and how to read resistors.

Each resistor will have either three or four color bands, which begin at one end of the resistor's body. The first colored band represents the resistor's first digit, while the second color represents the second digit of the resistor's value. The third colored band represents the resistor's multiplier value, and the fourth color band represents the resistor's tolerance value. If there is no fourth colored band then the resistor has a 20% tolerance value. If the resistor's fourth band is silver then the resistor has a 10% tolerance value and if the fourth band is gold then the resistor's tolerance value will be 5%.

Finally we are ready to begin assembling the project, so let's get started. The prototype project was constructed on a single-sided printed circuit board or PCB; with your PC board in front of you, we can now begin populating the circuit board. We are going to place and solder the lowest height components first, the resistors and diodes. First, locate our first resistor R1. Resistor R1 has a value of 270 ohms and is a 5% tolerance type. Now take a look for a resistor whose first color band is red, violet, brown, and gold. From the chart you will notice that red is represented by the digit (2), the second color band is violet, which is represented by (7). Notice that the third color is brown and the multiplier is (10), so $(2)(7) \times 10 = 270$ ohms.

Go ahead and locate where resistor R1 goes on the PC and install it, next solder it in place on the PC board, and then with a pair of end-cutters trim the excess component leads flush to the edge of the circuit board. Next locate the remaining resistors and install them in their respective locations on the PC board. Solder the resistors to the board, and remember to cut the extra leads with your end-cutters.

Note that capacitors generally come from two major groups or types. Non-polar types, such as the ones in this project, have no polarity and can be placed in either direction on the circuit board. The other type of capacitors are called polarized types, these capacitors have polarity and have to be inserted on the board with respect to their polarity marking in order for the circuit

to operate correctly. On these types of capacitors, you will observe that the components actually have a plus or minus marking or a black or white band with a plus or minus marking on the body of the capacitors. Capacitors must be installed properly with respect to polarity in order for the circuit to work properly. Failure to install electrolytic capacitors correctly could cause damage to the capacitors as well as to other components in the circuit when power is first applied.

Now locate the capacitors from the component stack and locate capacitor C1, which is an electrolytic capacitor and will be labeled 1,000 μF; it should have a small black or white band with a plus or minus marking on or near the band. Be sure to observe this polarity marking when installing this capacitor on the circuit board. Next locate capacitor C2, also a 1,000 μF capacitor. Go ahead and place C1 and C2 on the circuit board and solder them in place. Now take your end-cutters and cut the excess component leads flush to the edge of the circuit board. Once you have installed these components, you can move on to installing the remaining low profile components.

The Off-Line Telephone Tester project contains five silicon diodes. Diodes, as you will remember, will always have some type of polarity markings, which must be observed if the circuit is going to work correctly. You will notice that each diode will have a black or white colored band at one edge of its body. The colored band represents the diode's cathode lead. Check your schematic and parts layout diagrams when installing these diodes on the circuit board. Note that Diodes D1 through D4 are all 1N4007 diodes, which form the diode bridge, which is connected to the transformer at T1. Diode D5 above R1 is a 1N5408 silicon diode. Place all the diodes on the circuit board in their respective locations and solder them in place. After soldering in the diodes, cut the excess component leads with your end-cutters flush to the edge of the circuit board.

The Off-Line Telephone Tester utilizes two conventional transistors. Transistors are generally three lead devices, which must be installed correctly if the circuit is to work properly. Careful handling and proper orientation in the circuit is critical. A transistor will usually have a Base lead, a Collector lead, and an Emitter lead. The base lead is generally a vertical line inside a transistor circle symbol. The collector lead is usually a slanted line facing towards the base lead and the emitter lead is also a slanted line facing the base lead but it will have an arrow pointing towards or away from the base lead. Now, identify the two transistors, and refer to the manufacturer's specification sheets as well as the schematic and parts layout diagrams when installing the two transistors. If you have trouble installing the transistors, get the help of an electronics enthusiast with some experience. Install the transistors in their respective locations and then solder them in place. Remember to trim the excess component leads with your end-cutters.

The Off-line Phone Tester circuit also contains a single integrated circuit at U1. Integrated circuits will also have some sort of marking to help you orient them on the PC board. The music generator has a non-conventional pin-out, since it only contains three pins. Refer to the manufacturer's pin-out diagram, as well as the schematic when installing this part.

The Off-line Phone Tester circuit utilizes a single LED at D6, which also must be oriented correctly if the circuit is to function properly. An LED has two leads, a cathode and an anode lead. The anode lead will usually be the longer of the two leads and the cathode lead will usually be the shorter lead, just under the flat side edge of the LED; this should help orient the LED on the PC board. You may want to leave about a ½ lead length on the LED, in order to have the LED protrude through the enclosure box. Solder the LEDs to the PC board and then trim the excess leads.

Once the circuit board has been completed, take a short break and then we will inspect the PC board for possible "cold" solder joints and "short" circuits. Pick up the PC board with the foil side facing upwards toward you. Carefully inspect the solder joints, these should look smooth, clean, and bright. If any of the solder joints look dull, dark, or blobby, then you should remove the solder with a solder sucker or a solder wick and then resolder the joint all over again.

Now we will inspect the PC board for possible short circuits. Short circuits can be caused from small solder blobs bridging the circuit traces from cut component leads which can often stick to the circuit board once they have been cut. A sticky residue from solder will often trap component leads across the circuit traces. Look carefully for any bridging wires across circuit traces.

Once this task is complete, we can move on to installing the circuit board into an aluminum enclosure. Locate a suitable enclosure to house the circuit board. Arrange the circuit board and the two transformers on the bottom of the chassis box. Now, locate four ¼″ standoffs, and mount the PC board between the standoffs and the bottom of the chassis box. Leave room for the two transformers, and try to mount them on opposite sides of the board, displaced by 90 degrees. Mount the transformers to the base of the chassis box and wire them to the circuit board.

Next, mount the rotary switch, the On—Off switch, and pushbutton switch to the front panel of the chassis box and wire them to the circuit board. The LED can be mounted on the circuit board with extra long leads so it could show through the front panel, or it could be mounted on the front directly. Connect the LED to the circuit board if not already done.

Finally, you will need to connect the Off-line Tester to the phone under test. You could elect to install a modular phone jack on the chassis box and then simply use a modular to modular phone cable to connect the circuit to the phone you wish to test. You could also elect to hard-wire a cable to the circuit and use a grommet to bring the 4-wire cable out of the chassis box. The opposite end of your cable could have a modular plug which can be quickly connected to the phone you wish to test. Once the circuit has been completed and the circuit wired, you will be ready to test the circuit and a telephone.

Circuit operation

First, connect the telephone to the terminals marked "To Telephone Under Test" and switch on power to the circuit (switch S1). Next, in order to test the ringer portion, flip switch S2 to position "B" and press S3 for a moment. You should hear the ring in the event that the ringer circuit of the telephone under test is working. Please ensure that handset is on cradle during this test.

Finally, for testing the audio section, flip switch S1 to position "C" and connect a headphone to jack J1. Pick up the telephone handset and speak into its microphone. If audio section is working satisfactorily, you should be able to hear your speech via the headphone. If you dial a number, you should be able to hear the pulse clicks or pulse tone in the headphone, depending on whether the telephone under test is functioning in pulse or tone mode. If the telephone under test has a built-in musical hold facility, on pressing the "hold" button you should be able to hear the music. Now flip switch S2 to position "A." You should be able to hear music generated by U1 through the earpiece of the handset of the telephone under test, indicating proper functioning of the AF amplifier section. Your Off-Line Telephone Tester is ready to help you and/or your family and friends to test your new found phones.

Project 3: Telephone line tester

If you build houses or remodel houses or are just re-wiring your own or trouble shooting a telephone problem for your self or friend, then you will find this Telephone Line Tester extremely helpful and a real time saver.

The Telephone Line Tester project, shown in Figure 15-3, is connected to a conventional telephone line through modular connector P1, at the right of the circuit.

The Telephone Line Tester is powered from the telephone line, so there is no external power needed for the circuit.

The Telephone Line Tester can be built into a small chassis box with the LED indicator and meter on the front panel to instantly show you the line status. The LED at D6 is connected in series with resistor D5 and R7 and will instantly inform you of the telephone line polarity.

If the green wire is the positive phone lead and the red wire the negative phone wire, the circuit is correctly wired and the LED will remain off. If the wiring is reversed, then the red indicator LED will light up.

With switch S1 set for LINE/RING, both S1(a) and S1(b) are open and the meter indicates the condition of the line voltage. Any line-voltage reading in the LINE OK range indicates a line voltage that is higher than 40 V DC. If the telephone is caused to ring, either by using a ring-back number or by dialing from another phone, the meter will indicate RING OK, and LED will pulse—indicating AC, if the ringing voltage/current is correct. The actual position of the meter's pointer depends on how many ringers are connected across the telephone line.

When S1 is closed, the voltage range of the meter is changed and a nominal load resistance of 230 ohms (R5 and R6) is connected across the line to emulate the off-hook load of the telephone. If the meter indicates LOOP OK, you can be certain that you have sufficient loop voltage for satisfactory telephone operation. If you place another load on the line, perhaps by taking an extension telephone off-hook, the meter reading will

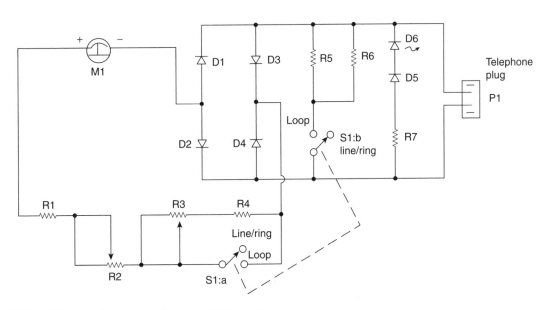

Figure 15-3 *Telephone line tester. Courtesy of Poptronix*

almost invariably drop below the LOOP OK range. If lifting the handset causes the meter reading to drop, you can at least be certain that the telephone's hook switch is working and that the repeat coil is connected to the telephone line.

Circuit assembly

Before we begin constructing the Phone Line Tester, you will need to find a clean, well-lit worktable or workbench. Next we will gather a small 25–30-W pencil tipped soldering iron, a length of 60/40 tin–lead solder, and a small canister of Tip Tinner, available at your local RadioShack store; Tip Tinner is used to condition the soldering tip between solder joints. You drive the soldering tip into the compound and it cleans and prepares the soldering tip. Next grab a few small hand tools, try to locate small flat blade and Phillips screwdrivers. A pair of needle-nose pliers, a pair of tweezers along with a magnifying glass, and a pair of end-cutters and we will begin constructing the project. Now locate your schematic diagram and parts layout diagram along with all of the components needed to build the project. Once all the components are in front of you, you can check them off against the project parts list to make sure you are ready to start building the project. Now, locate Table 15-1 and place it in front of you.

Table 15-1 illustrates the resistor code chart and how to read resistors.

Each resistor will have either three or four color bands, which begin at one end of the resistor's body. The first colored band represents the resistor's first digit, while the second color represents the second digit of the resistor's value. The third colored band represents the resistor's multiplier value, and the fourth color band represents the resistor's tolerance value. If there is no fourth colored band then the resistor has a 20% tolerance value. If the resistor's fourth band is silver then the resistor has a 10% tolerance value and if the fourth band is gold then the resistor's tolerance value will be 5%.

Finally, we are ready to begin assembling the project, so let's get started. The prototype project was constructed on a single-sided printed circuit board or PCB. With your PC board in front of you, we can now begin populating the circuit board. We are going to place and solder the lowest height components first, the resistors and diodes. First, locate our first resistor R1. Resistor R1 has a value of 4,700 ohms and is a 5% tolerance type. Now take a look for a resistor whose first color band is yellow, violet, red, and gold. From the chart you will notice that yellow is represented by the digit (4), the second color band is violet, which is represented by (7). Notice that the third color is red and the multiplier is (100), so $(4)(7) \times 100 = 4,700$ or 4.7 K ohms.

Go ahead and place R1 on the circuit board in its respective location and solder it in place. Now take your end-cutters and cut the excess component leads flush to the edge of the circuit board. Once you have installed these first three components, you can move on to installing the remaining resistors and the other low profile components. Locate the remaining resistors from the parts pile and install them in their proper location on the circuit board. Next solder them in place and remember to trim the excess lead lengths when you are finished. Finally locate and install the 50 K ohm potentiometer at R3 and the 10 K ohm potentiometer at R2. Now, solder them to the circuit board.

This project contains five silicon diodes. Diodes, you will remember, will always have some type of polarity markings, which must be observed if the circuit is going to work correctly. You will notice that each diode will have a black or white colored band at one edge of the diode's body. The colored band represents the diode's cathode lead. Check your schematic and parts layout diagrams when installing these diodes on the circuit board. Note that all five diodes are 1N4003 silicon diodes. Place all the diodes on the circuit board and solder them in place. Then trim the excess component leads with your end-cutters, cutting the leads flush to the edge of the circuit board.

The Telephone Line Tester circuit uses two LEDs, which also must be oriented correctly if the circuit is to function properly. Remember that LEDs have two leads, a cathode and an anode lead. The anode lead will usually be the longer of the two leads and the cathode lead will usually be the shorter lead, just under the flat side edge of the LED; this should help orient the LED on the PC board. Solder the LEDs to the PC board and then trim the excess leads.

Finally, you will need to install the LINE/RING—LOOP switch, at S1. The DPDT toggle switch consists of two switch sections S1(a) and S1(b), which are wired to the two different portions of the circuit.

Once the circuit board has been completed, take a short break and then we will inspect the PC board for possible "cold" solder joints and "short" circuits. Pick up the PC board with the foil side facing upwards toward you. Carefully inspect each solder joint. The solder joints should look smooth, clean, and bright. If any of the solder joints look dull, dark, or blobby, then you should remove the solder with a solder sucker or a solder wick and then resolder the joint all over again.

Now we will inspect the PC board for possible short circuits. Short circuits can be caused from small solder blobs bridging the circuit traces from cut component lead, which can often stick to the circuit board once they have been cut. A sticky residue from solder will often trap component leads across the circuit traces. Look carefully for any bridging wires across circuit traces. Once this inspection is complete, we can move on to installing the circuit board into some sort of plastic enclosure.

Locate a suitable plastic case for the project and place the circuit board in the middle of the enclosure. Use four ½″ standoffs to mount the circuit board to the base of the enclosure.

You will want to mount the meter, the two potentiometers and the LINE/RING—LOOP switch and the LED on the front panel of the enclosure.

Finally, you will need to connect the tester circuit to the phone line, this can be done using an 10″ length of 4-conductor phone wire with a modular plug attached at one end of the cable. Now you can take the four colored cable wires and connect the red and green wires to the circuit board. The modular plug can then simply plug into a telephone jack that you want to test.

Operation

Once your circuit has been mounted in an enclosure and the modular plug attached to the circuit, you will be ready to test the circuit. Your Telephone Line Tester is now ready to serve you!

The instant that the tester is connected to the telephone line, you will have an indication of the polarity, since the tester's LED polarity indicator is always connected when the tester is plugged in. If the green wire is the positive phone lead and the red wire the negative phone wire, the circuit is correctly wired and the LED will remain off. If the wiring is reversed, then the red indicator LED will light up.

With both S1(a) and S1(b) in the open position, the meter indicates the condition of the line voltage. Any line-voltage reading in the LINE OK range indicates a line voltage that is higher than 40 V DC. If the telephone line rings, either by using a ring-back number or by dialing from another phone, the meter will indicate RING OK, and LED will pulse—indicating AC, if the ringing voltage/current is correct. Note that the position of the meter's pointer depends on how many ringers are connected across the telephone line.

When switch S1 is closed, the voltage range of the meter is changed and a nominal load resistance across the line becomes 230 ohms. The two parallel resistors R5 and R6 are thus connected across the line to emulate the off-hook load of the telephone. If the meter indicates LOOP OK, you can be certain that you have sufficient loop voltage for satisfactory telephone operation.

If you place another load on the line, perhaps by taking an extension telephone off-hook, the meter reading will almost invariably drop below the LOOP OK range. If lifting the handset causes the meter reading to drop, you can at least be certain that the telephone's hook switch is working and that the repeat coil is connected to the telephone line.

The handy Telephone Line Tester will rapidly inform you of the telephone line conditions quite quickly without any guess work.

Project 4: Telephone line simulator

The Telephone Line Simulator can save you time and money when trying to determine if a telephone is working correctly or not. Perhaps you wish to wire all the rooms in your home with telephones or you want to add a number of phones in different rooms, such as your office, shop, or barn and you acquired a number of used telephones. In this situation you will want to make sure all of the phones work before you go to the trouble of installing them in the house. The Telephone Line Simulator will fit that bill nicely, since it will test a single phone or two telephones at once. It will simulate telephone on-hook as well as off-hook conditions. It will provide power to two phones at once so you can pick up both phones and listen for the side-tone while talking between phones, as well as testing the quality of each of the phones. The Telephone Line Simulator can also ring each of the telephones to test their ringers as well as answering machines (see Figure 15-4).

Circuit description

The helpful and handy Telephone Line Simulator project is depicted in Figure 15-5.

In operation, when both handsets are on-hook, resistors R1 and R2 supply power to them. When both handsets are off-hook, they transmit and receive their own audio. There are two terminal blacks which supply power to up to two different phones, through resistors R1 and R2. If switch S1 is closed, resistor R6 simulates a telephone line being plugged into J2, permitting the testing of only one phone. Now consider the situation where one handset is on-hook and one is off-hook. That causes the on-hook line to go to 24 V DC and the off-hook line to go to 7 V DC. The coil of the relay RY1 is connected across the two telephone lines, and the voltage difference between the two lines energizes it. When the contacts or RY1 are closed, C2 charges through R4; it takes about one second for C2 to charge to 12 V DC.

The 12 V DC across C2 causes a voltage controlled switch consisting of R5, R8, Q1, Q2, and SCR1 to

close, thus energizing RY2. When RY2 is energized, RY1 is removed from the circuit and a 60 Hz, 37-V p-p sine wave is placed on the telephone lines, causing the telephone to ring. Because RY1 is removed from the circuit, capacitor C2 starts discharging through R7. It takes about one second for the capacitor to discharge to about 2.4 V DC. That lower voltage level causes the voltage controlled switch to disable RY2, removing the ring voltage from the telephone lines and putting RY1 back in the circuit. If one telephone is still off-hook and one is on-hook, the cycle is repeated.

Circuit assembly

Before we begin constructing the Telephone Line Simulator, you will need to find a clean, well-lit worktable or workbench. Next we will gather a small 25–30-W pencil tipped soldering iron, a length of 60/40 tin–lead solder, and a small canister of Tip Tinner, available at your local RadioShack store; Tip Tinner is used to condition the soldering tip between solder joints. You drive the soldering tip into the compound and it cleans and prepares the soldering tip. Next grab a few small hand tools; try to locate small flat blade and Phillips screwdrivers. A pair of needle-nose pliers, a pair of tweezers along with a magnifying glass, and a pair of end-cutters and we will begin constructing the project. Now locate your schematic diagram and parts layout diagram along with all of the components needed to build the project. Once all the components are in front of you, you can check them off against the project parts list to make sure you are ready to start building the project. Now, locate Table 15-1, and place it in front of you. Table 15-1 illustrates the resistor code chart and how to read resistors.

Each resistor will have either three or four color bands, which begin at one end of the resistor's body. The first colored band represents the resistor's first digit, while the second color represents the second digit of the resistor's value. The third colored band represents the resistor's multiplier value, and the fourth color band represents the resistor's tolerance value. If there is no fourth colored band then the resistor has a 20% tolerance value. If the resistor's fourth band is silver then the resistor has a 10% tolerance value and if the fourth band is gold then the resistor's tolerance value will be 5%.

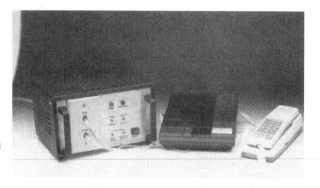

Figure 15-4 *Telephone line simulator*

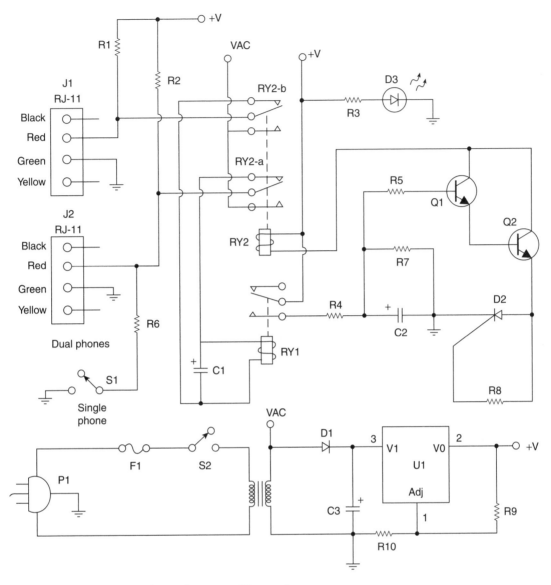

Figure 15-5 *Telephone-line simulator. Courtesy of Poptronix*

Finally, we are ready to begin assembling the project, so let's get started. The prototype project was constructed on a single-sided printed circuit board or PCB. With your PC board in front of you, we can now begin populating the circuit board. We are going to place and solder the lowest height components first, the resistors and diodes. First, locate our first resistor R1. Resistor R1 has a value of 560 ohms and is a 5% tolerance type. Now take a look for a resistor whose first color band is green, blue, brown, and gold. From the chart you will notice that green is represented by the digit (5), the second color band is blue, which is represented by (6). Notice that the third color is brown and the multiplier is (10), so (5)(6) × 10 = 560 ohms.

Once you have identified resistor R1, go ahead and install it in its respective location on the circuit board. Next, you can solder it in place and then trim the excess component lead with a pair of end-cutters. Trim the excess leads flush to the edge of the circuit board. Now, locate the remaining resistors and place them in their correct locations and then solder them in place. Remember to trim the excess component leads when finished.

Next refer to Table 15-2, this chart helps to determine value of small capacitors.

Many capacitors are quite small in physical size, so manufacturers have devised a three-digit code which

Table 15-2

Three-digit capacitor codes

pF	nF	μF	Code	pF	nF	μF	Code
1.0			**1R0**	3,900	3.9	.0039	**392**
1.2			**1R2**	4,700	4.7	.0047	**472**
1.5			**1R5**	5,600	5.6	.0056	**562**
1.8			**1R8**	6,800	6.8	.0068	**682**
2.2			**2R2**	8,200	8.2	.0082	**822**
2.7			**2R7**	10,000	10	.01	**103**
3.3			**3R3**	12,000	12	.012	**123**
3.9			**3R9**	15,000	15	.015	**153**
4.7			**4R7**	18,000	18	.018	**183**
5.6			**5R6**	22,000	22	.022	**223**
6.8			**6R8**	27,000	27	.027	**273**
8.2			**8R2**	33,000	33	.033	**333**
10			**100**	39,000	39	.039	**393**
12			**120**	47,000	47	.047	**473**
15			**150**	56,000	56	.056	**563**
18			**180**	68,000	68	.068	**683**
22			**220**	82,000	82	.082	**823**
27			**270**	100,000	100	.1	**104**
33			**330**	120,000	120	.12	**124**
39			**390**	150,000	150	.15	**154**
47			**470**	180,000	180	.18	**184**
56			**560**	220,000	220	.22	**224**
68			**680**	270,000	270	.27	**274**
82			**820**	330,000	330	.33	**334**
100		.0001	**101**	390,000	390	.39	**394**
120		.00012	**121**	470,000	470	.47	**474**
150		.00015	**151**	560,000	560	.56	**564**
180		.00018	**181**	680,000	680	.68	**684**
220		.00022	**221**	820,000	820	.82	**824**
270		.00027	**271**		1,000	1	**105**
330		.00033	**331**		1,500	1.5	**155**
390		.00039	**391**		2,200	2.2	**225**
470		.00047	**471**		2,700	2.7	**275**
560		.00056	**561**		3,300	3.3	**335**
680		.00068	**681**		4,700	4.7	**475**
820		.00082	**821**			6.8	**685**
1,000	1.0	.001	**102**			10	**106**
1,200	1.2	.0012	**122**			22	**226**
1,500	1.5	.0015	**152**			33	**336**
1,800	1.8	.0018	**182**			47	**476**

Table 15-2—cont'd

Three-digit capacitor codes

pF	nF	μF	Code	pF	nF	μF	Code
2,200	2.2	.0022	**222**			68	**686**
2,700	2.7	.0027	**272**			100	**107**
3,300	3.3	.0033	**332**			157	**157**

Code = 2 significant digits of the capacitor value + number of zeros to follow

For values below 10 pF, use "R" in place of decimal e.g. 8.2 pF = 8R2

10 pF = 100

100 pF = 101

1,000 pF = 102

22,000 pF = 223

330,000 pF = 334

1 μF = 105

can be placed on the capacitor instead of the actual value. The code occupies a much smaller space on the capacitor and can usually fit nicely on the capacitor body, but without the chart the builder would have a difficult time determining the capacitor's value.

Note that capacitors generally come from two major groups or types. Non-polar types, such as the ones in this project, have no polarity and can be placed in either direction on the circuit board. The other type of capacitors are called polarized types; these capacitors have polarity and have to be inserted on the board with respect to their polarity marking in order for the circuit to operate correctly. On these types of capacitors, you will observe that the components actually have a plus or minus marking or a black or white band with a plus or minus marking on the body of the capacitors.

Now try and locate the three electrolytic capacitors from the component pile. Capacitor C1, which is an electrolytic capacitor, will be labeled 10 μF, and it should have a small black or white band with a plus or minus marking on or near the band. Be sure to observe this polarity marking when installing the capacitors on the circuit board. Next locate capacitor C2, which will be labeled 22 μF, and locate, identify, and install capacitor C3, which will be marked 470 μF. Go ahead and solder the capacitors in place on the circuit board. Now take your end-cutters and cut the excess component leads flush to the edge of the circuit board. Once you have installed the resistors and capacitors,

you can move on to installing the remaining low profile components.

This project contains a number of single silicon diodes and an SCR or Silicon Controlled Rectifier. Diodes will always have some type of polarity markings, which must be observed if the circuit is going to work correctly. You will notice that each diode will have a black or white colored band at one edge of its body. The colored band represents the diode's cathode lead. Check your schematic and parts layout diagrams when installing the diodes on the circuit board. Note that diode D1 is a 1N4001 diode. Place the diodes on the circuit board and solder them in place on the PC board. Then cut the excess component leads with your end-cutters. The circuit also utilizes a single SCR at D2. An SCR device is a semiconductor device with three leads, a cathode, an anode and a gate lead. The SCR has an anode and cathode much the same as a diode, so the symbol for the device also has a straight horizontal line, which is the cathode, while the anode is the arrow that points to the cathode. Use the manufacturer's data sheet as well as the schematic and parts layout sheets when installing the SCR. If you are not confident that you can install the SCR without assistance, then ask a knowledgeable friend or electronics enthusiast for some help.

The Phone Line Simulator utilizes two conventional transistors at Q1 and Q2, both are 2N3904 NPN types. A transistor will usually have a Base lead, a Collector lead, and an Emitter lead. The base lead is generally a vertical line inside a transistor circle symbol. The collector lead is usually a slanted line facing towards the base lead and the emitter lead is also a slanted line facing the base lead but it will have an arrow pointing towards or away from the base lead. Identify the transistor leads and their respective location and then install the transistors on the PC board. Once this is accomplished, you can solder the transistors in place and then cut the excess component leads flush to the edge of the circuit board.

This circuit also contains a single IC regulator IC at U1. ICs will also have some sort of marking to help you orient them on the PC board. Often there will be a small indented circle, a small notch or cutout at the top end of the IC package. Pin of the device will be just to the left of the notch or cutout. It is highly recommended that the circuit builder use IC sockets for the ICs. IC sockets are low cost insurance against a possible circuit failure at some later date. It is much easier to simply unplug an IC rather than trying to unsolder many IC pins from a PC board without causing damage to the PC board. Solder the IC socket to the PC board and then insert the IC into the socket.

The Phone Line Simulator circuit utilizes a single LED at D3, which also must be oriented correctly if the circuit is to function properly. An LED has two leads, a cathode and an anode lead. The anode lead will usually be the longer of the two leads and the cathode lead will usually be the shorter, just under the flat side edge of the LED; this should help orient the LED on the PC board. You may want to leave about a ½ inch of lead length in order for the LED to extend from the PC board through to the outside of the enclosure. Solder the LED to the PC board and then trim the excess leads.

If you purchased PC board type relays, you can install them on the PC board now. Take your time to identify the correct wiring before installation. If you elected to purchase non-PC board style relays, you will have to secure them to the bottom of the chassis box and wire them to the circuit board directly using #22 gauge insulated hookup wire.

Once the circuit board has been completed, take a short break and then we will inspect the PC board for possible "cold" solder joints and "short" circuits. Pick up the PC board with the foil side facing upwards toward you. Carefully inspect the solder joints, these

should look smooth, clean, and bright. If any of the solder joints look dull, dark, or blobby, then you should remove the solder with a solder sucker or a solder wick and then resolder the joint all over again.

Now we will inspect the PC board for possible short circuits. Short circuits can be caused from small solder blobs bridging the circuit traces from cut component leads, which can often stick to the circuit board once they have been cut. A sticky residue from solder will often trap component leads across the circuit traces. Look carefully for any bridging wires across circuit traces. Once this inspection is complete, we can move on to installing the circuit board into some sort of plastic enclosure.

Locate a suitable enclosure to house the circuit board. Place the circuit board and the transformer on the bottom of the chassis box or enclosure and center them for the best fit inside the case. If you chose non-PC board relays, you will have to secure them to the bottom of the enclosure. Install four ¼″ standoffs between the circuit board and the bottom of the chassis box, in order to secure the circuit board. Mount the transformer to the base of the chassis box. Mount two 4-position terminal blocks on the front of the chassis box along with switches S1 and S2, and connect them to the circuit board. Mount the chassis mounted fuse holder on the rear of the enclosure and drill a hole to pass the power cord through the rear panel.

Finally, you will need to make up two phone test cables which go between the terminal blocks and the phones under test. Make up two 2′ lengths of 4-conductor phone wire with a modular plug attached at one end of each of the two cables. The free wire ends of the cables are wired directly to the terminal blocks, one for each phone. When you are ready to test a phone, simply plug one of the modular plugs into the phone under test.

Operation

Once your circuit has been mounted in an enclosure, you will be ready to test the circuit. Connect one of the modular phone plugs to a telephone that you wish to test at J1. Then connect up a second phone to the terminals at J2. Plug the circuit into a wall socket to connect it to a 110-V power source. Make sure you have a ½ amp fuse in the fuse holder and then toggle switch S1 to DUAL PHONE position and switch S2 to the "on" position. The LED at D3 should light up once power is applied to the circuit. With both phones on-hook, you should be able to measure 24 V across J1 and J2. Next pick up the one of the phones; you should read 24 V across the on-hook phone and 7 V across the off-hook phone line. Talk into the microphone of the off-hook phone and you should hear yourself through the side-tone. After about 30 s or so the on-hook phone should start to ring, testing the ringer circuit. You can repeat the sequence by placing both phones on-hook and picking up the opposite phone and listening to side-tone when you talk. After 30 s you should hear the other phone ringing. This handy Telephone Line Simulator will save you both time and money when trying to determine if used phones are suitable for installation in a home, office, shop, or barn. The Phone Line Simulator will serve you for many years and you may have your friends and or relatives wanting to borrow it from time to time.

Chapter 16

FM Telephone Transmitter

Parts list

Phone FM Transmitter

R1 100 ohms, $\frac{1}{4}$ W,
5% resistor

R2 33 K ohms, $\frac{1}{4}$ W,
5% resistor

R3 10 K ohms, $\frac{1}{4}$ W,
5% resistor

R4 47 K ohms, $\frac{1}{4}$ W,
5% resistor

R5 390 ohms, $\frac{1}{4}$ W,
5% resistor

C1 27 pF or 22 pF,
35 V ceramic capacitor

C2 100 nF, 35 V ceramic
capacitor (104)

C3 22 nF, 35 V ceramic
capacitor (223)

C4 1 nF, 35 V ceramic
capacitor (102)

C5 5.6 pF, 35 V
ceramic capacitor

C6, C7 47 pF, 35 V
ceramic capacitor

Q1 BC547,BC548, 2N2222,
or NTE123A transistor

Q2 ZTX320, 2N3563 or
NTE 108 transistor

D1, D2, D3, D4 1N4148
silicon diodes

L1 6 turn enameled
copper wire, wound on
$\frac{1}{8}$″ form - (see text)

L2 8 turn enameled
copper wire, wound on
$\frac{1}{8}$″ form - (see text)

L3 6 turn tinned
copper wire

Misc PC board, wire,
hardware, wire,
enclosure, etc.

CK202 – Phone
Transmitter

Kit available from:

Carls's Electronics

www.electronickits.com

Would you like to be able to amplify a phone call so everybody can hear it? Or perhaps you would like a way to record phone calls for record-keeping purposes? If either idea sounds good to you, then you might want to build the FM Telephone Transmitter described in this article (see Figure 16-1).

It is a simple yet ingenious device that connects in series with a phone line, and derives its power from the phone itself. The FM Telephone Transmitter circuit will transmit both sides of a conversation to an FM radio tuned to between 90 and 95 MHz.

The FM Telephone Transmitter circuit is built on a PC board that is so small it can easily be fitted inside the housing of a telephone base, thus making it an

Figure 16-1 *TeleXmit wireless phone bug. Courtesy of Electronickits.com*

instant pseudo-speak earphone. Keep in mind, though, that it is illegal to listen to or record a telephone conversation without informing all involved parties.

There are many legitimate reasons for wanting to broadcast a telephone call to an FM receiver. For example, when someone calls long distance, he or she does not have the time or cannot afford to stay on long, but everybody at home still wants to hear his or her voice. Or perhaps you want to record a phone call so that you have a record of some electronic banking you did or deal you made. Many other applications can be found for this interesting project.

Besides being small in size, the Telephone Transmitter is also "small" in price. Only a handful of low cost parts are needed to build the project.

Circuit description

The FM Telephone Transmitter is shown in the schematic at Figure 16-2.

The circuit is basically a radio frequency (RF) oscillator that operates around 93 MHz. Power for the circuit is derived from the full wave diode bridge, formed by diodes D1 through D4. Components C1, C8, L3, and Q1 form the main FM oscillator. In this FM transmitter circuit L1, C6, and Q2 form the power amplifier section of the unit. Audio from the telephone lines is coupled through R3 and C2 into the base of Q1 to modulate the oscillator. This is done by varying the junction

capacitance of the transistor. The junction capacitance is a function of the potential difference applied to the base of the transistor. Components R1 and C4 act as a low pass filter. Capacitor C3 forms a high frequency shunt. Coil L2 is called an RFC (radio frequency shunt.) It decouples the power and audio from the transmitter amplifier circuit itself. The coils L1 through L3 are all air-core coils, which are hand wound by the builder. Coil L1 is formed with #22 ga enameled magnet wire loosely wound on a ⅛″ drill bit, while L2 is formed with eight turns loosely wound over a ⅛″ drill bit using #22 ga enameled magnet wire. Remember to scrape off a ⅛″ of insulation in order to solder the coils to the circuit. Coil L3 is formed by wrapping six turns of tinned copper wire. The winding will need to be spaced 1 mm apart. Coil L3 is tapped between the top of the first turn of L3 and the circuit board.

Circuit assembly

Before we begin building the Telephone Transmitter project, let us first find a clean, well-lit worktable or workbench. Next we will gather a small 20–30-W pencil tipped soldering iron, a length of 60/40 tin–lead solder, and a small canister of Tip Tinner, available at your local RadioShack store, along with a few small hand tools. Try to locate small flat blade and Phillips screwdrivers. A pair of needle-nose pliers, a pair of tweezers along with a magnifying glass, and a pair of end-cutters and we will begin constructing the project.

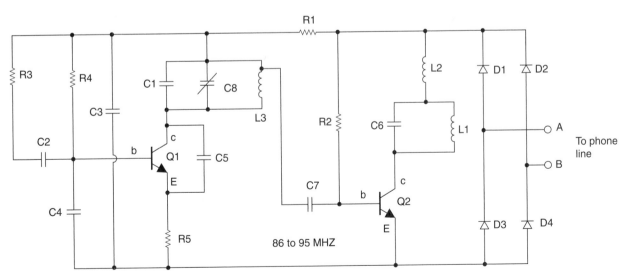

Figure 16-2 *FM telephone transmitter. Courtesy of DIY Electronics*

Locate your schematic diagram along with all of the components to build the project. Once all the components are in front of you, you can check them off against the project parts list to make sure you are ready to start building the project. Now locate Tables 16-1 and 16-2, and place them in front of you.

Table 16-1 illustrates the resistor code chart and how to read resistors, while Table 16-2 illustrates the capacitor code chart that will aid you in constructing the project.

Refer now to the resistor color code chart shown in Table 16-1, which will help you identify the resistors in this project. Note that each resistor will have either three or four color bands, which begin at one end of the resistor's body. The first colored band represents the resistor's first digit, while the second color represents the second digit of the resistor's value. The third colored band represents the resistor's multiplier value, and the fourth color band represents the resistor's tolerance value. If there is no fourth colored band then the resistor has a 20% tolerance value. If the resistor's fourth band is silver then the resistor has a 10% tolerance value and if the fourth band is gold then the resistor's tolerance value will be 5%.

Finally, we are ready to begin assembling the project, so let's get started. First, locate our first resistor R1. Resistor R1 has a value of 100 ohms and is a 5% tolerance type. Now take a look for a resistor whose

first color band is brown, black, brown, and gold. From the chart you will notice that brown is represented by the digit (1), and black is represented by (0). Notice that the third color is brown and the multiplier is (10), so $10 \times 10 = 100$ ohms. Now place resistor R1 into its proper location and solder it in place. Next, use a pair of end-cutters to trim the excess resistor leads. Cut the excess component leads flush to the edge of the circuit board. Now locate the remaining resistors and place them in their respective locations on the PC board. Solder the resistors in place and remember to trim the excess component lead with your end-cutters.

Note that capacitors generally come from two major groups or types. Non-polar types, such as the ones in this project, have no polarity and can be placed in either direction on the circuit board. The other type of capacitors are called polarized types, these capacitors have polarity and have to be inserted on the board with respect to their polarity marking in order for the circuit to operate correctly. On these types of capacitors, you will observe that the components actually have a plus or minus marking or a black or white band with a plus or minus marking on the body of the capacitors.

Next, refer to Table 16-2; this chart helps to determine the value of small capacitors. Many capacitors are quite small in physical size, so manufacturers have devised a three-digit code which

Table 16-1
Resistor color code chart

Color Band	1st Digit	2nd Digit	Multiplier	Tolerance
Black	0	0	1	
Brown	1	1	10	1%
Red	2	2	100	2%
Orange	3	3	1,000 (K)	3%
Yellow	4	4	10,000	4%
Green	5	5	100,000	
Blue	6	6	1,000,000 (M)	
Violet	7	7	10,000,000	
Gray	8	8	100,000,000	
White	9	9	1,000,000,000	
Gold			0.1	5%
Silver			0.01	10%
No color				20%

Table 16-2
Three-digit capacitor codes

pF	nF	µF	Code	pF	nF	µF	Code
1.0			**1R0**	3,900	3.9	.0039	**392**
1.2			**1R2**	4,700	4.7	.0047	**472**
1.5			**1R5**	5,600	5.6	.0056	**562**
1.8			**1R8**	6,800	6.8	.0068	**682**
2.2			**2R2**	8,200	8.2	.0082	**822**
2.7			**2R7**	10,000	10	.01	**103**
3.3			**3R3**	12,000	12	.012	**123**
3.9			**3R9**	15,000	15	.015	**153**
4.7			**4R7**	18,000	18	.018	**183**
5.6			**5R6**	22,000	22	.022	**223**
6.8			**6R8**	27,000	27	.027	**273**
8.2			**8R2**	33,000	33	.033	**333**
10			**100**	39,000	39	.039	**393**
12			**120**	47,000	47	.047	**473**
15			**150**	56,000	56	.056	**563**
18			**180**	68,000	68	.068	**683**
22			**220**	82,000	82	.082	**823**
27			**270**	100,000	100	.1	**104**
33			**330**	120,000	120	.12	**124**
39			**390**	150,000	150	.15	**154**
47			**470**	180,000	180	.18	**184**
56			**560**	220,000	220	.22	**224**
68			**680**	270,000	270	.27	**274**
82			**820**	330,000	330	.33	**334**
100		.0001	**101**	390,000	390	.39	**394**
120		.00012	**121**	470,000	470	.47	**474**
150		.00015	**151**	560,000	560	.56	**564**
180		.00018	**181**	680,000	680	.68	**684**
220		.00022	**221**	820,000	820	.82	**824**
270		.00027	**271**		1,000	1	**105**
330		.00033	**331**		1,500	1.5	**155**
390		.00039	**391**		2,200	2.2	**225**
470		.00047	**471**		2,700	2.7	**275**
560		.00056	**561**		3,300	3.3	**335**
680		.00068	**681**		4,700	4.7	**475**
820		.00082	**821**			6.8	**685**
1,000	1.0	.001	**102**			10	**106**
1,200	1.2	.0012	**122**			22	**226**
1,500	1.5	.0015	**152**			33	**336**
1,800	1.8	.0018	**182**			47	**476**

(Continued)

Table 16-2—cont'd

Three-digit capacitor codes

pF	nF	μF	Code	pF	nF	μF	Code
2,200	2.2	.0022	**222**			68	**686**
2,700	2.7	.0027	**272**			100	**107**
3,300	3.3	.0033	**332**			157	**157**

Code = 2 significant digits of the capacitor value + number of zeros to follow

For values below 10 pF, use "R" in place of decimal e.g. 8.2 pF = 8R2

10 pF = 100
100 pF = 101
1,000 pF = 102
22,000 pF = 223
330,000 pF = 334
1 μF = 105

can be placed on the capacitor instead of the actual value. The code occupies a much smaller space on the capacitor and can usually fit nicely on the capacitor body, but without the chart the builder would have a difficult time determining the capacitor's value.

Now locate the capacitors from the component pile and try and locate capacitor C1, which should be marked with either a 22 pF (223) or a 27 pF (227) value. Let's go ahead and place C1 on the circuit board and then solder it in place. Now take your end-cutters and cut the excess capacitor leads flush to the edge of the circuit board. Once you have installed C1, you can move on to installing the remaining capacitors.

The FM Phone Transmitter project contains silicon diodes at D1 through D4. These diodes also have polarity which must be observed if the circuit is going to work correctly. You will notice that each diode will have a black or white colored band at one edge of the diode's body. The colored band represents the diode's cathode lead. Check your schematic and parts layout diagrams when installing these diodes on the circuit board. Place the four diodes on the circuit board and solder them in place. Then cut the excess component leads with your end-cutters.

The Telephone Transmitter circuit also utilizes two transistors. You can substitute a number of different types (see parts list). Transistors generally have three leads, a Base lead, a Collector lead, and an Emitter lead.

Refer to the schematic and you will notice that there will be two diagonal leads pointing to a vertical line. The vertical line is the transistor's base lead and the two diagonal lines represent the collector and the emitter. The diagonal lead with the arrow on it is the emitter lead, and this should help you install the transistors onto the circuit board correctly. Go ahead and install the two transistors on the PC board and solder them in place, then follow-up by cutting the excess component leads.

When constructing the coils, note that L1 and L2 have enamel insulation lacquer on them. This must be physically removed from both ends of the coil before it can be soldered. You must use a knife to scrape off the insulation so you solder both ends of each coil to the circuit board. After constructing L3, you will need to spread out the turns of the coil about 1 mm apart. The coil turns should not touch each other or the bottom of the circuit board. You must now solder a connection or tap on the top of the first turn in the L3 coil to the pad next to the coil, closest to the junction of C1, C8, and R1. Solder a piece of wire to the top of the first turn as shown on the parts layout diagram. Then solder the other end to the pad immediately next to the L3 coil.

Next, attach a 3″ length of wire with an alligator clip on the end to the pads between the diodes marked—TO LINE. No antenna is needed. The phone line itself acts as a sufficient aerial. The telephone transmitter attaches in series with ONE of the phone wires. Note that there are two wires going to your phone,

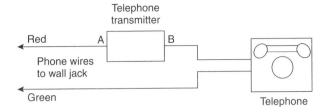

Figure 16-3 *Phone-transmitter connection*

usually a red wire and a green wire. Cut one of these wires and connect the phone transmitter in series with the phone line wire that you cut as shown in Figure 16-3.

Attach one alligator clip to one cut end and the other alligator clip to the other cut end. Take your phone off the hook and turn on an FM radio at about 93 MHz. It should be very easy to tune into the transmission. Take a portable FM receiver outside and follow the phone line; you should easily be able to hear the room noise where the transmitter is located.

Calibration

The telephone transmitter may need to be calibrated or adjusted for proper operation on the frequency of your choice. The resonant frequency of the L1-C6 amplifier circuit should be adjusted to match the resonant oscillator frequency of C1, C9-L3. If you want to try calibration you will need a frequency meter, a CRO, or just trial and error. Calibrate by moving the coils of L1 further apart. With C1 at 27 pF you will find that the transmitter tunes into the FM band in the 86–95 MHz area. With C1 at 22 pF the band is raised to about 90–95 MHz (depending on the coil spacing). If you want to move this tunable area still higher to over 100 MHz range then replace C1 by a 15 pF or 10 pF capacitor. This assumes that the on-hook voltage is about the standard 48 V. If the on-hook voltage of an extension phone network is lower, say about 39 V, C1 will have to be lower in the 15 pF to 10 pF range to be in the commercial FM band in this case. Note that you should not hold the transmitter physically in your hands if you try to do a calibration. Your own body capacitance when you touch it is more than enough to change the oscillation frequency of the whole unit. You can experiment to get greater transmission range away from the phone line by adding an aerial (about 150 cm of 26 gauge wire) to the collector of Q2.

Operation

When there is a signal on the phone line (that is, when you pick up the handset) the circuit will transmit the conversation over a distance of about 100′. In particular it will radiate from the phone line itself. It is a passive device; there is no battery. It uses the signal on the phone line for power. No aerial is needed, it feeds back the RF signal into the phone line, which radiates it in the FM band.

Note that some countries may ban any electronic device which attaches to the telephone. It is the responsibility of the purchaser to check the legal requirements for the operation of this project and to obey them.

If you simply wish to monitor a single phone, then you could just mount the Telephone Transmitter inside that particular phone case. However, if you wish to listen to all the phone calls of all the phones in the house then you will have to place the Telephone Transmitter in series with the first incoming telephone line when it arrives in your basement area. You can encase the transmitter in a small metal box, with provision for a small antenna to protrude from the case in order to radiate a signal into the air.

Troubleshooting

If your telephone transmitter does not appear to function correctly, you will need to disconnect the circuit from the phone line and carefully inspect the circuit board. It is likely that you made a soldering or parts placement mistake during construction. Pick up the circuit board with the foil side of the circuit board facing upwards toward you. Carefully inspect the circuit board against the parts placement diagram and the schematic to make sure that the proper component was inserted in the correct PC holes. You may have placed the wrong resistor at a particular location. You may have inserted a capacitor or diode backwards and this could prevent the circuit from working properly.

Once this preliminary inspection is over, you can inspect the circuit board for any possible "cold" solder joints or "short" circuits. Take a careful look at each of the PC solder joints; they should all look clean and shiny.

If any of the solder joints look dull, dark, or blobby, then you unsolder the joint, remove the solder and then resolder the joint all over again. Next, examine the PC board for any short circuits. Short circuits can be caused from two circuit traces touching each other due to a stray component lead that stuck to the PC board from solder residue, or from solder blobs bridging the solder traces. Once you have fully examined the circuit board, you can reconnect the circuit board back to the phone line and try out the phone transmitter once again.

Chapter 17

Telephone Project

Parts list

Parts Bin

Telephone Project

R1, R2, R3, R4 47 K,
  $\frac{1}{4}$ W, 5% resistor

R5, R23 10 K, $\frac{1}{4}$ W,
  5% resistor

R6, R9, R11, R21 100 K,
  $\frac{1}{4}$ W, 5% resistor

R7 2.2 K, $\frac{1}{4}$ W,
  5% resistor

R8 1 Megohm $\frac{1}{4}$ W,
  5% resistor

R10, R14 150 K, $\frac{1}{4}$ W,
  5% resistor

R12, R13, R15, R20
  3.3 K, $\frac{1}{4}$ W,
  5% resistor

R16, R17, R22 15 K,
  $\frac{1}{4}$ W, 5% resistor

R18 10 ohm, $\frac{1}{4}$ W,
  5% resistor

R19 120 ohms, $\frac{1}{4}$ W,
  5% resistor

C1 47 μF, 50 V Mylar
  (474/470)

C2 .01 μF, 50 viscap
  (.01 or 103)

C3, C4, C9 47 μF 35 V
  Electrolytic capacitor

C5 300 pF, 50 V
  Discap (300)

C6 1 μF, 50 V
  Electrolytic capacitor

C7 .04 μF, 50 V Discap
  capacitor (.04 or 403)

C8 .02 μF, 50 V Discap
  capacitor (.02 or 203)

C10, C11 30 pF, 50 V
  Discap capacitor (30)

D1, D2, D3, D4 1N4004
  or 1N4007 Silicon
  diodes

D8 55C2V7/1N5223 2.7 V
  Zener Diode

D7 55C4V7/1N5230 4.7 V
  Zener Diode

Q1, Q5 9014 C
  Transistor NPN

Q2 2N5401 Transistor
  PNP 325401

Q3 2N5551 Transistor
  NPN

Q4 9013H Transistor NPN

U1 HM9102 Integrated
  Circuit (IC)

X1 Resonator 3.58 MHz

LP1, LP2, LP3, LP4
  Neon Bulb

S1, S3 SPST Switch

S2 SPDT Hook Switch

S4 pushbutton

KB Number keypad —
  3 × 4 keypad switch
  matrix

MIC Microphone

SPK Speaker

BZ Piezo buzzer disc

J1 Modular Telephone
Jack

Misc PC board,
IC sockets, case,
telephone cord, wire,
battery clip

Model AK-700 / PT-323 K
Telephone Kit

Kit available from:

Elenco Electronics

The Telephone Project features a modern phone handset with both pulse and tone dialing, automatic redial, ringer switch, visual ring indication, and dialer. This project is also a great learning tool for electronics enthusiasts as it describes and details each telephone component and explains how they work together to form a complete useable telephone. The telephone project shown in Figure 17-1 consists of five major components: The handset, consisting of the transmitter and receiver handset section, the dialer section, the network section, and the ringer section.

Before we get started talking about the telephone project, refer to Table 17-1, which lists a phone term glossary.

Everyone knows that the telephone is used to transmit and receive voice signals, allowing two people with telephones to communicate with each other, but to be of practical value, the telephone must be connected to a

Figure 17-1 *Telephone project. Courtesy of Elenco*

Table 17-1
Telephone glossary

DTMF	Dual tone multifrequency
Electret Microphone	A microphone made up of a capacitor with a dielectric that holds a permanent electric charge
FET	Field effect transistor
Hook-Switch	A switch inside the phone which closes when the receiver is lifted
IC	Integrated circuit
FCC	Federal Communication Commission
Local Loop	The pair of wires between a telephone set and the telephone company Central Office
Mark Interval	The portion of a dial pulse during which loop current is flowing
Pulse Dialing	A dialing system in which the number of pulses on the local loop represents the number dialed
Ring Wire	The red (negative) wire of the local telephone loop
Sidetone	The sound heard from the telephone receiver due to the speaker's own voice
Space Interval	The portion of the dial pulse during which current is not flowing
Tip Wire	The green (positive) wire of the local telephone loop
Tone Dialing	A dialing system in which a combination of two frequencies on the local loop represent the number dialed
USOC	Universal Service Order Code
Zener Diode	A diode that has nearly constant voltage drop over a wide range of reverse current

switching network capable of connecting each telephone to many other telephones. To accomplish this switching, each subscriber telephone is connected to the telephone company's Central Office by two wires referred to as the Local Loop. A simplified diagram of this connection is shown in Figure 17-2.

The Tip and Ring designation of the + and − leads comes from the days of the manual switchboard. The tip of the plug the operator used to connect telephones carried the (+) lead and the ring immediately behind the tip carried the (−) lead. Now when a subscriber wishes

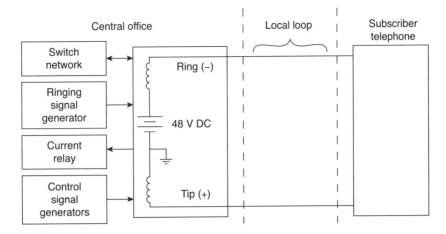

Figure 17-2 *Local telephone loop circuit*

to place a call, he or she merely picks up the telephone and a small current flows in the local loop. This current picks a relay in the Central Office (CO) indicating that service is being requested. When the CO is ready to accept the number being called, a dial tone is sent to the calling telephone. The dial pulses, or tones, then signals to the CO the number of the telephone being called. A path is then established to that telephone. This path may be a simple wire connection to a telephone connected to the same CO or it may go via wire, microwave link, or satellite to a telephone connected to a distant CO. To signal the incoming call, a ringing signal is placed on the local loop of the called telephone. The ringing signal is a 90 V AC 20 Hz signal superimposed on the 48 V DC present on the local loop. A ringing tone is also sent to the calling telephone. When the called party

picks up the telephone, voice communication is established.

A simplified diagram of a traditional rotary dial telephone phone is illustrated in Figure 17-3, and the major components of the telephone are described below.

Audio section: The handset

First we are going to discuss the main phone components or subsections. The first major components of a telephone are the transmitter or microphone and receiver or earphone.

In the older standard rotary dial phones, the transmitter or microphone consists of a metal diaphragm and a metal case, which are insulated from each other. The case is filled with carbon granules.

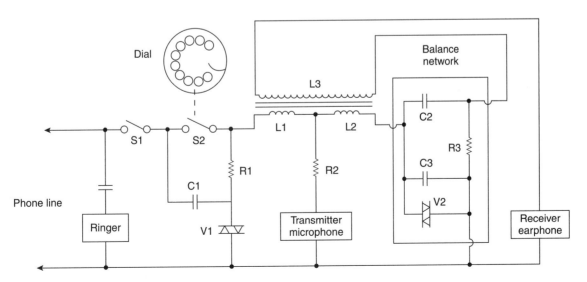

Figure 17-3 *Rotary dial telephone set*

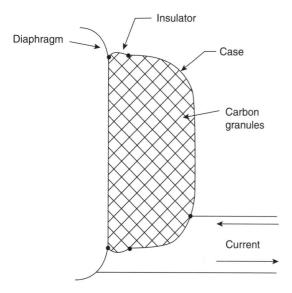

Figure 17-4 *Transmitter or microphone*

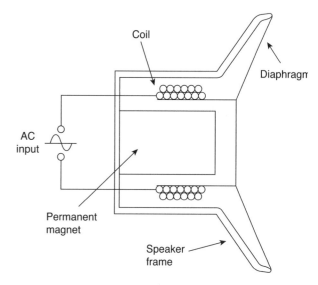

Figure 17-5 *Headphone/speaker*

When you speak into the transmitter, the sound waves of your voice strike the diaphragm and cause it to vibrate. This causes the carbon granules to compress and expand. When compressed, the resistance of the carbon granules is less than when expanded. The change of resistance causes a corresponding change in the current. The current thus varies in step with the sound waves of your voice, using a standard carbon microphone (see Figure 17-4).

In our modern electronic telephone project the transmitter or microphone consists of an electret type microphone.

There are several different types of receivers or earphone types. In principle, they work the same as the speakers in your radio and TV. The speaker consists of a small coil attached to a diaphragm. The coil is mounted over a permanent magnet. Coil current in one direction causes the coil and diaphragm to be repelled from the permanent magnet. Coil current in the other direction causes the coil and diaphragm to be attracted to the permanent. If a current of audio frequency is sent through the coil, the diaphragm vibrates and generates sound waves in step with the current. Thus, if the current from the transmitter is sent through the coil, the sound produced in the speaker or earphone will duplicate the sound striking the microphone (see Figure 17-5).

The Electronic Telephone Set schematic diagram is shown in Figure 17-6. In operation, the main Tip and

Ring telephone line input connections are fed to a diode bridge made up of diodes D1 through D4. The bridge ensures that a positive voltage is fed to the telephone circuit regardless of the polarity of the input. This is done to protect the integrated circuit and other components that may be damaged by a reversal of polarity. When the hook switch is not depressed, the telephone is ready to transmit and receive. The incoming signal from the CO is sent via R20 and C6 to the speaker amplifier made up of transistor Q5. This amplifier drives the speaker. Transmission begins when you speak into the microphone. The microphone used is an electret type. This type of microphone is made up of a capacitor with a dielectric material which holds a permanent electric charge. Sound waves striking a plate of this capacitor cause the plate to vibrate and thus generate a small voltage across the capacitor. This voltage is amplified by a Field Effect Transistor (FET) mounted inside the microphone. The signal from the microphone is then amplified and sent to the Central Office via the local loop. A portion of the signal is also sent to the speaker amplifier to provide the desired side-tone.

The induction coil/balance network

When transmitting and receiving is done over the same two wires, the problem arises that current from the transmitter flows through the receiver. The speaker then

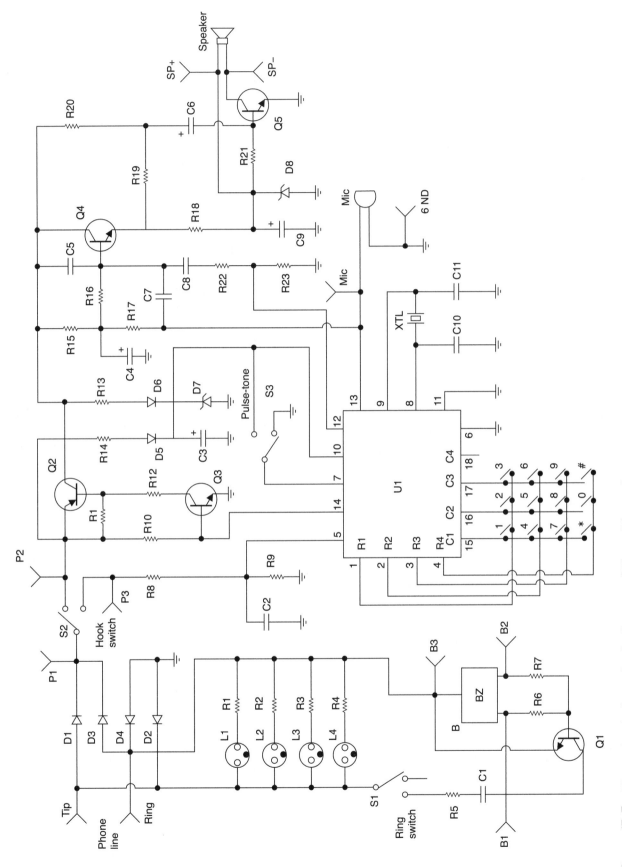

Figure 17-5 *Electronic telephone circuit. Courtesy of Elenco Electronics*

hears his or her own voice from the receiver. This is called side-tone. Too much side-tone may be objectionable to the speaker and cause them to speak too softly. A small amount of side-tone is desirable to keep the telephone from sounding dead. The induction coil and balance network limit the side-tone. The impedance of the balance network approximately matches the impedance of the local telephone line loop. Thus, about half of the current from the transmitter flows through L1 and the local loop and the other half flows through L2 and the balance network. The currents in L1 and L2 induce voltages in L3 of opposite polarity, which limits the voltage across the receiver to an acceptable level. When receiving a signal from the local loop, the currents in L1 and L2 induce voltages in L3 of the same polarity. These voltages combine to drive the receiver.

Keypad: Dialer section

In older phones, the pulse dialing is accomplished by the familiar rotary dial. The dial is rotated to the stop and then released. A spring in the dialer returns the dial to its null position. As the dial returns, the dial switch (S2) opens and closes at a fixed rate. This switch is in series with the hook switch. Opening the switch interrupts the current in the local loop. A series of current pulses is thus sent out on the local telephone line loop. The number of pulses sent corresponds to the digit dialed. Dialing "0" sends ten pulses. The dial pulses are sent at a rate of 10 pulses per second (100 ms between pulses). Each pulse consists of a mark interval (loop current) and a space interval (no loop current). In America, the mark interval is 40 ms and the space interval is 60 ms giving a mark/space ratio of 40/60. In Europe, the mark space ratio is usually 33/67.

Fortunately, the telephone CO responds to two types of dialing signals, both pulse and tone signals. Our modern electronic Telephone Project provides both pulse and tone dialing functions. The main elements of the dialer section in the Telephone Project are the keyboard and the dialer IC HM9102. These two components perform all of the functions of the rotary dial found in older telephones. The HM9102 is the tone/pulse switchable dialer with last number redial which is fabricated with CMOS technology. The HM9102 turns transistor Q3 on or off to generate dial pulses.

When the switch (S3) is in the Tone position—potential pin mode—GND and Dial will be for tone signals. When the switch (S3) is in the Pulse position—potential pin mode—VDD (4.7 V) and Dial will be used for pulse signals.

The HM9102 integrated circuit (IC) has a built-in oscillator (clock) which is used for tone generation and other timing functions. Because the frequency of dialing tones must be precise for the telephone CO to recognize them, a ceramic resonator (X) is used to accurately control the frequency of the IC's oscillator. This oscillator's frequency is 3.58 MHz, and it is divided down to lower frequencies inside the IC.

The DC voltage at the Tip and Ring inputs is fed to the dialer IC via the diode bridge, the hook switch, resistor R14, and diode D5. Since excessive voltage may damage the IC, Zener diode D7 limits the voltage to less than 4.7 V. When the phone goes on-hook, the charge on capacitor C3 provides the small current needed for the IC memory circuits to retain the last number dialed.

Tone dialing is accomplished with a keyboard of 12 keys arranged in four rows and three columns. Low frequencies of 697, 770, 852, and 941 are associated with rows R1 through R4 and high frequencies of 1,209, 1,336, and 1,477 Hz are associated with columns C1 through C3. To send each digit, two frequencies are sent to the CO simultaneously. For this reason, this method of dialing is referred to as dual tone multifrequency (DTMF). The different frequencies are generated by connecting a capacitor to different taps of a transformer to establish a resonant circuit of the correct frequency. Each of the 3 keys of row 1 is mechanically connected to switch SR1. Similarly, each of the other rows and columns are connected to their corresponding switches. Thus, pressing any key closes two switches and generates two frequencies. Pressing a 6, for example, closes switches SR2 and SC3 and generates 770 and 1,477 Hz.

Keyboard inputs C1 through C3 and R1 through R4 accept the column and row inputs from the keyboard; see Figure 17-7.

When a row and a column line are connected, the digit corresponding to that row and column is stored in the IC memory. The dial pulses start when the first button is pushed. In a conventional rotary dial, the pulses are sent as the rotor returns to its null position.

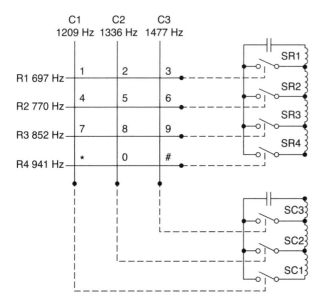

Figure 17-7 *Touch-tone dial*

This may take about one second if a zero is dialed. The next digit is entered only when the preceding digit is transmitted. With the HM9102 dialer IC, scanning of the keyboard continues while the dial pulses are being transmitted, as shown in Figure 17-8.

When the number is dialed, a memory section in the IC is loaded with each digit as it is dialed. Later, if the redial button (#) is pressed, the IC automatically dials the digits you previously entered (for Pulse Mode). For Tone Mode, button (#) – execute "#". The button "*" for Pulse Mode – execute pause (pause time 3.6 s), for Tone Mode – execute "*".

The ringer section

The telephone ringer in the Telephone Project is connected across the tip and ring inputs in series with a

capacitor, C1, to block the 48 V DC when the phone is on-hook. The diagram in Figure 17-9 illustrates the bell found in older telephone sets.

Notice that the ringer consists of a permanent magnet attached to an armature. Our modern Telephone Project circuit uses a small buzzer to simulate the bell or ringer of an old phone.

To signal an incoming call, the CO places a 90 V AC 20 Hz signal on top of the 48 V DC. If the ringer switch (S1) is closed, transistor Q1 conducts during the ½ cycle that the ring terminal is positive. The collector to emitter voltage is applied to the buzzer, causing it to change dimensions. The feedback lead is connected to the base of transistor Q1 through resistor R7, causing the circuit to oscillate at about 3 KHz. There is no oscillation during the negative portion of the 20 Hz signal. The buzzing sound is thus produced by the buzzer changing dimensions, during the positive portion of the 20 Hz ringing signal. The four neon bulbs LP1 through LP4 and resistors R1, R2, R3, and R4 are also connected across the tip and ring inputs. When the ringing signal causes the voltage across the bulbs to exceed about 100 V, the bulbs conduct, giving a visual indication of the incoming call. Install the following parts. The hammer attached to the armature thus strikes one bell and then the other to produce the ringing sound.

Hook switch

When the hook switch is open (on-hook) no current flows in the local loop, the hook-switch or HKS input of the IC is grounded through resistor R9. This enables the keyboard inputs; when the phone is on the hook, a small current flows through resistors R8 and R9;

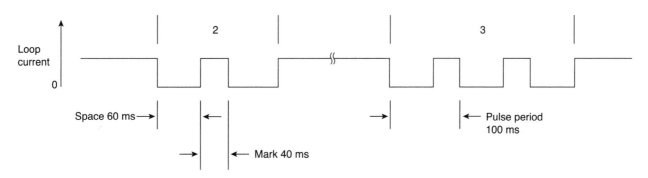

Figure 17-8 *Pulse-dialer spacing*

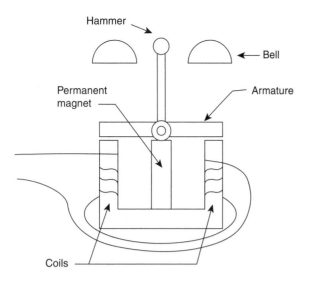

Figure 17-9 *Ringer or bell*

this biases the HKS input to about 2 V which disables the keyboard inputs. The 48 V DC from the battery in the CO appears on the tip and ring input to the telephone set. When the receiver is lifted, the hook switch closes and a current of about 20 to 120 mA flows in the local loop. The resistance of the local loop drops the voltage at the telephone to about 6 V. The current picks a relay in the CO which tells other equipment there that service is being requested. When the CO is ready to accept the number being called, a dial tone is sent to the

calling telephone. The dial tone stops when the first digit is dialed.

Phone Circuit: General assembly

Before we begin constructing the Telephone project, you will need to find a clean, well-lit worktable or workbench. Next, we will gather a small 25–30 W pencil tipped soldering iron, a length of 60/40 tin–lead solder, and a small canister of Tip Tinner, available at your local RadioShack store; Tip Tinner is used to condition the soldering tip between solder joints. You drive the soldering tip into the compound and it cleans and prepares the soldering tip. Next grab a few small hand tools; try to locate small flat blade and Phillips screwdrivers. A pair of needle-nose pliers, a pair of tweezers along with a magnifying glass, and a pair of end-cutters and we will begin constructing the project. Now locate your schematic diagram and parts layout diagram along with all of the components needed to build the project. The Telephone project was constructed on a rectangular printed circuit board

Once all the components are in front of you, you can check them off against the project parts list to make sure you are ready to start building the project. Now, locate Tables 17-2 and 17-3, and place them in front of you.

Table 17-2

Resistor color code chart

Color Band	1st Digit	2nd Digit	Multiplier	Tolerance
Black	0	0	1	
Brown	1	1	10	1%
Red	2	2	100	2%
Orange	3	3	1,000 (K)	3%
Yellow	4	4	10,000	4%
Green	5	5	100,000	
Blue	6	6	1,000,000 (M)	
Violet	7	7	10,000,000	
Gray	8	8	100,000,000	
White	9	9	1,000,000,000	
Gold			0.1	5%
Silver			0.01	10%
No color				20%

Table 17-3

Three-digit capacitor codes

pF	nF	μF	Code	pF	nF	μF	Code
1.0			**1R0**	3,900	3.9	.0039	**392**
1.2			**1R2**	4,700	4.7	.0047	**472**
1.5			**1R5**	5,600	5.6	.0056	**562**
1.8			**1R8**	6,800	6.8	.0068	**682**
2.2			**2R2**	8,200	8.2	.0082	**822**
2.7			**2R7**	10,000	10	.01	**103**
3.3			**3R3**	12,000	12	.012	**123**
3.9			**3R9**	15,000	15	.015	**153**
4.7			**4R7**	18,000	18	.018	**183**
5.6			**5R6**	22,000	22	.022	**223**
6.8			**6R8**	27,000	27	.027	**273**
8.2			**8R2**	33,000	33	.033	**333**
10			**100**	39,000	39	.039	**393**
12			**120**	47,000	47	.047	**473**
15			**150**	56,000	56	.056	**563**
18			**180**	68,000	68	.068	**683**
22			**220**	82,000	82	.082	**823**
27			**270**	100,000	100	.1	**104**
33			**330**	120,000	120	.12	**124**
39			**390**	150,000	150	.15	**154**
47			**470**	180,000	180	.18	**184**
56			**560**	220,000	220	.22	**224**
68			**680**	270,000	270	.27	**274**
82			**820**	330,000	330	.33	**334**
100		.0001	**101**	390,000	390	.39	**394**
120		.00012	**121**	470,000	470	.47	**474**
150		.00015	**151**	560,000	560	.56	**564**
180		.00018	**181**	680,000	680	.68	**684**
220		.00022	**221**	820,000	820	.82	**824**
270		.00027	**271**		1,000	1	**105**
330		.00033	**331**		1,500	1.5	**155**
390		.00039	**391**		2,200	2.2	**225**
470		.00047	**471**		2,700	2.7	**275**
560		.00056	**561**		3,300	3.3	**335**
680		.00068	**681**		4,700	4.7	**475**
820		.00082	**821**			6.8	**685**
1,000	1.0	.001	**102**			10	**106**
1,200	1.2	.0012	**122**			22	**226**
1,500	1.5	.0015	**152**			33	**336**
1,800	1.8	.0018	**182**			47	**476**

(Continued)

Table 17-3—cont'd

Three-digit capacitor codes

pF	nF	μF	Code	pF	nF	μF	Code
2,200	2.2	.0022	**222**			68	**686**
2,700	2.7	.0027	**272**			100	**107**
3,300	3.3	.0033	**332**			157	**157**

Code = 2 significant digits of the capacitor value + number of zeros to follow

For values below 10 pF, use "R" in place of decimal e.g. 8.2 pF = 8R2

10 pF = 100

100 pF = 101

1,000 pF = 102

22,000 pF = 223

330,000 pF = 334

1 μF = 105

Table 17-2 illustrates the resistor code chart and how to read resistors, while Table 17-3 illustrates the capacitor code chart which will aid you in constructing the project.

Refer now to the resistor color code chart shown in Table 17-2, which will help you identify the resistors in this project. Note that each resistor will have either three or four color bands, which begin at one end of the resistor's body. The first colored band represents the resistor's first digit, while the second color represents the second digit of the resistor's value. The third colored band represents the resistor's multiplier value, and the fourth color band represents the resistor's tolerance value. If there is no fourth colored band then the resistor has a 20% tolerance value. If the resistor's fourth band is silver then the resistor has a 10% tolerance value and if the fourth band is gold then the resistor's tolerance value will be 5%.

Finally, we are ready to begin assembling the Telephone project, so let's get started. The prototype of the Telephone project was constructed on a single-sided printed circuit board or PCB. With the PC board in front of you, we can now begin populating the circuit board. We are going to place and solder the lowest height components first, the resistors and the diodes. First, locate the first four resistors, R1 through R4. These resistors have a value of 47 K ohms and are a 5% tolerance type. Now take a look for a resistor whose first color band is yellow, violet, orange, and gold. From the chart you will notice that yellow is represented by

the digit (4), the second color band is violet, which is represented by (7). Notice that the third color is orange and the multiplier is (1,000), so (4)(7) × 1,000 = 47,000 or 47 K ohms. Go ahead and place resistor R1 in its proper location on the circuit board and solder it in place. Use a pair of end-cutters to trim the excess resistor leads. Cut the excess leads flush to the edge of the PC board. Now locate and install the remaining resistors in their proper locations; solder them in place and remember to cut the excess component leads with your end-cutters.

Note that capacitors generally come from two major groups or types. Non-polar types, such as the ones in this project, have no polarity and can be placed in either direction on the circuit board. The other type of capacitors are called polarized types; these capacitors have polarity and have to be inserted on the board with respect to their polarity marking in order for the circuit to operate correctly. On these types of capacitors, you will observe that the components actually have a plus or minus marking or a black or white band with a plus or minus marking on the body of the capacitors. Capacitors must be installed properly with respect to polarity in order for the circuit to work properly. Failure to install electrolytic capacitors correctly could cause damage to the capacitors as well as to other components in the circuit when power is first applied.

Next refer to Table 17-3. This chart helps to determine the value of small capacitors. Many capacitors are quite small in physical size, so manufacturers have devised a three-digit code which can be placed on the capacitor instead of the actual value. The code occupies a much smaller space on the capacitor and can usually fit nicely on the capacitor body, but without the chart the builder would have a difficult time determining the capacitor's value.

Now look for the capacitors from the stack of components and locate capacitor C1, which is a small capacitor labeled (561) or .47 µF. Refer to Table 17-3 to find code (561) and you will see that it represents a capacitor value of .47 µF. Go ahead and place C1 on the circuit board and solder it in place. Now take your end-cutters and cut the excess component leads flush to the edge of the circuit board. Now you can move on to installing the remaining capacitors. Locate and install the smallest physical sized capacitors first then move on to installing the larger electrolytic types. Place the capacitors in their respective locations and solder them in place on the PC board.

The Telephone Project contains six of silicon diodes and two Zener diodes. Diodes, as you will remember, will always have some type of polarity markings, which must be observed if the circuit is going to work correctly. You will notice that each diode will have a black or white colored band at one edge of the diode's body. The colored band represents the diode's cathode lead. Check your schematic and parts layout diagrams when installing these diodes on the circuit board. Place all the silicon diodes on the circuit board in their respective locations and solder them in place on the PC board. The Zener diodes are located at D7 and D8; you can find their locations, place them on the PC board, and solder them in place on the board. After soldering in the diodes and the diode bridge, remember to cut the excess component leads with your end-cutters flush to the edge of the circuit board.

The Electronic Telephone Project utilizes five conventional transistors. Transistors are generally three lead devices, which must be installed correctly if the circuit is to work properly. Careful handling and proper orientation in the circuit is critical. A transistor will usually have a Base lead, a Collector lead, and an Emitter lead. The base lead is generally a vertical line inside a transistor circle symbol. The collector lead is usually a slanted line facing towards the base lead and the emitter lead is also a slanted line facing the base lead but it will have an arrow pointing towards or away from the base lead. Now, identify the two transistors,

refer to the manufacturer's specification sheets as well as the schematic and parts layout diagrams when installing the two transistors. If you have trouble installing the transistors, get the help of an electronics enthusiast with some experience. Install the transistors in their respective locations and then solder them in place. Remember to trim the excess components leads with your end-cutter.

The Electronic Telephone Project circuit contains a single integrated circuit (IC). ICs are often static sensitive, so they must be handled with care. Use a grounded anti-static wrist strap and stay seated in one location when handling the integrated circuits. Take out a cheap insurance policy by installing IC sockets for each of the ICs. In the event of a possible circuit failure, it is much easier to simply unplug a defective IC, than trying to unsolder 14 or 16 pins from a PC board without damaging the board. ICs have to be installed correctly if the circuit is going to work properly. IC packages will have some sort of markings which will help you orient them on the PC board. An IC will have either a small indented circle, a cut-out or notch at the top end of the IC package. Pin 1 of the IC will be just to the left of the notch or cut-out. Refer to the manufacturer's pin-out diagram, as well as the schematic, when installing these parts. If you doubt you ability to correctly orient the ICs, then seek the help of a knowledgeable electronics enthusiast to help you.

The Electronic Telephone Project circuit employs four neon indicators which are installed in parallel through series resistors. The neon lamps have no polarity, so they can be installed in either direction safely. Solder the neon lamps to the PC board and then trim the excess leads. Next, locate the crystal at XTL and install it in its proper location between pins 8 and 9 of U1.

Finally, you can install the SPDT TONE-PULSE slide switch at S3, the HOOKSWITCH at S2, and the RINGER switch at S1. The HOOKSWITCH is a leaf spring type SPDT switch and the RINGER switch S1 is an SPDT slide switch. Solder the switches in to their respective locations and trim the excess leads as necessary.

Audio circuit assembly

Next, we will install the electret microphone between ground and pin 13 of U1. The microphone's (+)

marking should face pin 13 and the ground or case body connection goes to ground. Identify two wires (4″ red and 3″ gray), microphone, and microphone pad. Cut the 4″ red wire down to 2″ and the 3″ gray wire to 2″. Use the 2″ red and 2″ gray stranded wires and strip ⅛″ of insulation off of each end. Identify the microphone. Bend the wires and very carefully solder the wires to the microphone. Peel the backing off one side of the microphone pads and place them onto the microphone. Peel the second backing off of the microphone pads and thread the wires through the hole in the PC board from the solder side. Solder the red wire to the hole marked MIC on the PC board and the gray wire to the hole marked GND on the PC board. Place the microphone on the PC board on the solder side.

Locate and install the earphone/speaker between the cathode of Zener diode D8 and the collector of Q5. Identify the speaker with pad, 3″ gray wire, and the second half of the 4″ red wire (the first half was used for the microphone). Cut a 2″ piece from the 3″ gray wire and strip ⅛″ of insulation off both ends. Solder the red wire to the speaker + and the gray wire to the speaker −. Solder the red wire to SP+ on the PC board and the gray wire to SP−.

Now locate and install the piezo buzzer element. The piezo buzzer element has three leads, the outermost ring-lead connects to B3, the middle circle ring-lead connects to B1, and the inner or center lead connects to B2, as shown on the schematic diagram.

Now identify the hook-switch. The hook-switch mounts to the copper side of the PC board. Insert the switch tabs into the mounting holes and twist the tabs ⅛ of a turn as was done with the ringer switch. Solder the switch's contacts to the PC board using two pieces of discarded resistor leads and a 3″ gray stranded wire.

Keypad-dialer assembly

The keypad switch assembly is a standard Grayhill or equivalent 3 × 4 normally open push button key switch matrix. Insert the ribbon cable through the holes in the main PC board and solder. Insert the ribbon cable through the holes in the keyboard PC board. Solder the ribbon cable to the keyboard's PC board. Carefully inspect the ribbon cable solder joints. If there are many solder bridges between adjacent pads, remove them

with your soldering iron. If not already in place, snap the dial buttons onto the contact pad. Peel the backing off of the dial label and place it on the top case (make sure that the holes line up). Insert the buttons through the holes in the front plate. Be sure that the # "2" button is in the hold marked "ABC", if you are building the kit version of the project. With the contact pad lying flat on the front side, insert the keyboard's PC board under the two upper retaining tabs. Push the lower restraining tab back and snap the PC board in place. Fasten the PC board in place with the two ¼″ screws. Reconnect the speaker wires. If your kit contains extra wires, please disregard them. Insert the hook button into the top of the case. Check that there are no burrs on the sides of the button. It must move freely once installed. Cut any excess plastic off of the case. Make sure that the top of the middle post is level with or below the two outside posts.

Ringer assembly

Identify the ringer switch. The ringer switch is mounted on the copper side of the PC board. Insert the switch tabs into the holes provided and solder the three terminals. Be sure that the switch is perpendicular to the PC board. To anchor the switch in place, grasp the switch tabs with your pliers and give a ⅛ turn. Identify the telephone cord. Solder the red wire to the ring jumper and the green wire to the tip jumper. Identify the buzzer and buzzer pad. Use three 2″ wires of different color (red, orange, and gray) to connect the buzzer to the PC board. Identify the red and orange wire and cut a 2″ piece of wire off the 3″ gray wire. If needed, strip ⅛″ of insulation off of each end. Tin the buzzer with solder in the locations. Tin both ends of the three 2″ wires. Carefully solder the three 2″ wires onto the buzzer. Thread the three wires through the buzzer pad and then through the hole from the solder side of the PC board. Solder the three wires to the PC board. The gray wire goes to B1, the orange wire goes to B2, and the red wire goes to B3.

Final assembly

Place the PC board on the top case so that the ringer switch is in the ringer switch hole, the TONE-PULSE switch is in the tone/pulse hole, and the two mounting holes on the PC board line up with the mounting posts on the top case. If the PC board does not stay lined up with the two mounting posts, insert one or both of the ⅝″ screws to keep it in place. Check the following: (1) Be sure that the speaker wires do not lie between the speaker and the telephone top case. (2) Be sure that the keyboard ribbon cable is in the notch in the side of the PC board and does not extend out over the edge of the top case. (3) Make sure that no leads from the PC board touch the ¼″ screws mounting the keypad. (4) Be sure that the keyboard PC board does not hit the buzzer. Loop the telephone cord through the strain relief tabs on the bottom case. Place the bottom case over the top case. Be sure that the telephone cord is in the notch at the base of the bottom case. Snap the bottom case in place and secure it with two ⅝″ screws. The telephone number label with its cover mounts over the two ⅝″ screws. Insert the tabs at each end of the label into the slots along side the screws.

Once the Telephone Project circuit board has been completed, take a short break and when we return we will inspect the PC board for possible "cold" solder joints and "short" circuits. Pick up the PC board with the foil side facing upwards toward you. Carefully inspect the solder joints, these should look smooth, clean, and bright. If any of the solder joints look dull, dark, or blobby, then you should remove the solder with a solder sucker or a solder wick and then resolder the joint all over again. Now we will inspect the PC board for possible short circuits. Short circuits can be caused from small solder blobs bridging the circuit traces or from cut component leads which can often stick to the circuit board once they have been cut. A sticky residue from solder will often trap component leads across the circuit traces. Look carefully for any bridging wires across circuit traces.

Installing the telephone

Select a convenient place for your phone. Place the telephone cradle against the wall and mark the center of the upper lobes of the two mounting holes. Screw in the two mounting screws (not provided) to about ¼″ of the mounting surface. Slide the cradle over the two screws and then tighten the screws. The phone and cradle are very light. However, if you are mounting the cradle on

drywall, you should use drywall screws. Drywall screws are available at most hardware stores.

The procedure for connecting the telephone to the telephone service line depends on the present termination of your line. If your line terminates in a modular jack, you need only to plug in the telephone cord and the installation is complete. Modular jacks may be recognized by the distinctive shape of the outlet.

If your telephone service wiring terminates in an old type hard wired jack, you may either (1) replace it with a modular jack if your telephone is to be installed in the same place, or (2) install a modular jack in a new location and connect it to the hard wired jack with a length of 4-conductor telephone wire. Modular jacks are available in most stores where telephones are sold. To replace the old type jack: Loosen the screw from the green wire and remove the wire. Loosen each of the remaining screws and remove the wires. Unscrew the mounting screws and remove the base of the old jack. Your jack may look somewhat different, but it should have the red, green, yellow, and black wires and the four marked screws. The cover may snap or screw onto the base. Screw the modular jack to the wall. As with the cradle, if you are mounting on drywall, use drywall screws. If the telephone service wires are not already stripped and bent, strip ½″ of insulation off of each wire and bend the stripped portion into a "U". Loop the green wire around the "G" screw between the washer and the head of the screw. The loop should be in a clockwise direction. Insert the spade lug of the green wire from the modular connector under the washer and tighten the screw. Connect the red wire to the "R" screw in the same way.

If present, connect the black and yellow wires in the same way. If these are missing, merely connect the black and yellow spade lugs to the "B" and "Y" screws. This will keep them from shorting to the other wires. Place the modular jack cover on the base. Plug in the telephone line cord. To locate the modular jack in the new location: Obtain the proper length of 4-wire telephone cable. Install the modular jack, using the 4-wire telephone cable in place of the telephone service wires. Run the 4-wire telephone cable to the old type hard wired jack. Strip ½″ of insulation off the green telephone cable wire and bend the stripped portion into a "U". Loosen the screw holding the green telephone service wire. Connect the two green wires and tighten the screw. Fasten the red telephone service wire to the red telephone cable wire in the same way. Fasten the remaining wires in the same way. If the black or yellow telephone service wires are missing, fasten the telephone cable wire to the empty screw to keep it from shorting to the other wires.

If your telephone service wiring terminates in a 4-connector jack, you may replace this jack in a manner similar to replacing the hard wired jack. Be sure that the green telephone service wire is connected to the red modular jack wire and the red telephone service wire to the green modular jack wire. You may instead purchase an adapter which converts the old 4-connector jack to a modular jack. Once it has been installed, you can begin enjoying your newly completed telephone, and have great satisfaction in understanding how a telephone works that you built it yourself. Your new telephone should give you many years of trouble-free service.

Chapter 18

Talking/Musical Telephone Ringer

Parts list

Parts Bin

Talking/Musical Telephone Ringer

R1, R2   33 K ohms, $\frac{1}{4}$ W, 5% resistpor

R3   10 megohm, $\frac{1}{4}$ W, 5% resistor

R4, R5, R7, R10   10 K ohms, $\frac{1}{4}$ W, 5% resistor

R6   12 K ohms, $\frac{1}{4}$ W, 5% resistor

R8   470 K ohms, $\frac{1}{4}$ W, 5% resistor

R9   10 ohms, $\frac{1}{4}$ W, 5% resistor

R11   2 K ohms, $\frac{1}{4}$ W, 5% resistor

C1   22 µF, 250 V metal film capacitor

C2, C3   47 µF, 35 V electrolytic capacitor

C4   .1 µF, 50 V ceramic capacitor

C5   1 µF, 35 V tantalum capacitor

C6   .22 µF, 35 V polystyrene capacitor

C7   4.7 µF, 35 V tantalum capacitor

C8   22 µF, 35 V tantalum capacitor

D1   1N4758A 56 V Zener diode

D2   1N4148 silicon diode

U1   4N25 NPN opto-coupler

U2   74LS161 counter IC

U3   74LS00 Quad NAND gate

U4   ISD1016 Voice message chip

MOV   metal oxide varistor

P1   telephone line cord

S1   momentary pushbutton

S2   2-position DIP switch

S3, S4   SPDT toggle switch

SPK   8-ohm speaker

MIC   electret microphone

Misc   PC board, AC adaptor, hardware, wire, enclosure, etc.

Don't listen to that annoying bell—build our talking telephone ringer and customize the way your phone rings. Telephones ring and that's about all there is to it. Some phones buzz, beep, or chirp, but that is not very exciting. With the advent of integrated circuits (ICs) like the ISD-1016 Voice Messaging IC, you can build our talking telephone ringer and record personalized

Figure 18-1 *Telephone voice ringer (front view).*
Courtesy of Poptronix

messages that will be played whenever someone calls you; see Figure 18-1.

You will never have to listen to the same old bell again.

With the recording capabilities of the Talking/Musical Telephone Ringer you can include your own spoken messages, musical selections, your college fight song, or any other interesting sounds or phrases you can think of. You can also program several different messages that play in sequence with each ring and the possibilities are endless. The Talking/Musical Ringer does not interfere with normal telephone operation, it simply plugs into a telephone jack, and requires an external 9-V DC wall power supply for power. When someone calls you, the incoming ring signal triggers the message to be played on the project's loudspeaker.

Circuit operation

The diagram shown in Figure 18-2 is the block diagram of the Talking/Musical Telephone Ringer circuit.

When a ring signal appears across the telephone line, the ring detector sends a start signal to the message-storage chip. The counter and logic sections tell the message-storage chip which message to play based on how many times the phone rings.

Now take a look at the Talking/Musical ringer schematic diagram shown in Figure 18-3.

A 20 Hz ring signal of approximately 40 to 150 volts, present at Tip and Ring telephone line connections (green and red, respectively). This ringing signal voltage is divided down by R1, R2, and R3. Current then flows through C1 to pins 1 and 2 of U1, a 4N25 opto-coupler. The opto-coupler isolates the rest of the circuitry from the telephone line. The current flow through the input

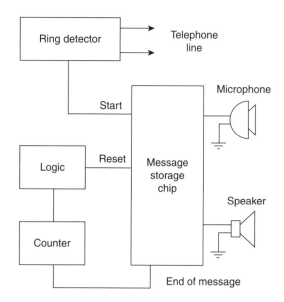

Figure 18-2 *Talking/musical ringer block diagram*
(Courtesy Poptronix)

side of U1 causes pin 5 to go low which signals a start-playback request from the ISD1016 messaging chip, U4.

The sensitivity of the ring-detection circuit is set by R6. The 12 K resistor value was obtained assuming a nominal 100-volt incoming ring signal. If the ring voltage coming in on your phone line is lower than 100 volts, you may need to increase R6 to a value between 15 K and 30 K. Metal-oxide Varistor (MOV1) protects the ringer from transient voltage spikes caused by lightning-induced current on the telephone line. Diode D1 protects the input LED of the opto-coupler from reverse voltage during the negative cycle of the AC ring signal. Diode D2 establishes a transient threshold to prevent the ringer from triggering when you pick up the phone or when dialing with a rotary telephone.

Pushbutton switch S1 has two functions: When S3 is in PLAY position, S1 acts as a test button that permits you to play back the messages under manual control. When S3 is in the RECORD position, pressing S1 enables the circuit recording function.

The operation of U2 depends on the setting of switch S4. When S4 is in the SINGLE position, one message will be played each time the phone rings (U2 has no function in this case). When S4 is in the MULTI position, U2 operates as a counter circuit that counts the number of individual messages to be played. Up to five

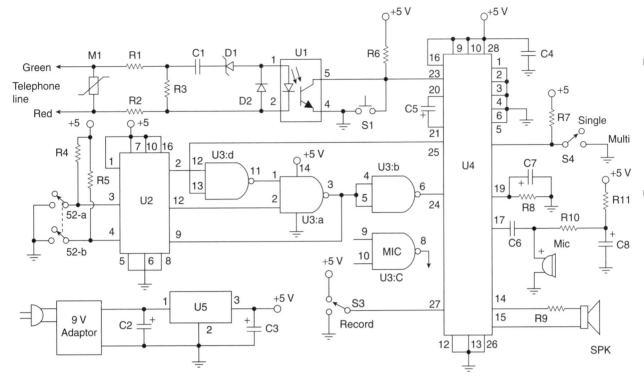

Figure 18-3 *Talking/musical ringer. Courtesy of Poptronix*

different messages can be sequenced with each successive incoming ring signal. Settings for the number of plays are shown in Table 18-1.

Quad NAND gate U3 works in conjunction with U2 to provide logic control for resetting U4 when the proper number of playback messages has been reached. The ISD1016 messaging chip (U4) stores up to l6 seconds of high-quality audio. Audio for recording is input to the chip via the electret microphone MIC, while playback occurs through the 8-ohm speaker SPK.

Power to the circuit is provided via a 9 or l2-V DC wall adapter. A 7805 regulator (U5) provides 5-V regulated power to the rest of the circuitry.

Table 18-1
Switch settings

Number of Msgs	S4	S2a	S2b
1	Single	X	X
2	Multi	L	L
3	Multi	H	L
4	Multi	L	H
5	Multi	H	H

Construction

The author chose to use a single-sided PC board for most reliable long term results.

You can make your own PC board from scratch or use pre-fab circuit board, many types of which are available from your local RadioShack store. Before we begin building the Talking/Musical Telephone Ringer project, let's first find a clean, well-lit worktable or workbench. Next we will gather a small 20–30-W pencil tipped soldering iron, a length of 60/40 tin–lead solder, a small canister of Tip Tinner, available at your local Radio Shack store, along with a few small hand tools. Try to locate small flat blade and Phillips screwdrivers. A pair of needle-nose pliers, a pair of tweezers along with a magnifying glass, and a pair of end-cutters and we will begin constructing the project. Locate your schematic diagram along with all of the components to build the project. Once all the components are in front of you, you can check them off against the project parts list to make sure you are ready to start building the project. Now, locate Tables 18-2 and 18-3, and place them in front of you.

Table 18-2 illustrates the resistor code chart and how to read resistors, while Table 18-3, illustrates the

Table 18-2

Resistor color code chart

Color Band	1st Digit	2nd Digit	Multiplier	Tolerance
Black	0	0	1	
Brown	1	1	10	1%
Red	2	2	100	2%
Orange	3	3	1,000 (K)	3%
Yellow	4	4	10,000	4%
Green	5	5	100,000	
Blue	6	6	1,000,000 (M)	
Violet	7	7	10,000,000	
Gray	8	8	100,000,000	
White	9	9	1,000,000,000	
Gold			0.1	5%
Silver			0.01	10%
No color				20%

Table 18-3

Three-digit capacitor codes

pF	nF	µF	Code	pF	nF	µF	Code
1.0			**1R0**	3,900	3.9	.0039	**392**
1.2			**1R2**	4,700	4.7	.0047	**472**
1.5			**1R5**	5,600	5.6	.0056	**562**
1.8			**1R8**	6,800	6.8	.0068	**682**
2.2			**2R2**	8,200	8.2	.0082	**822**
2.7			**2R7**	10,000	10	.01	**103**
3.3			**3R3**	12,000	12	.012	**123**
3.9			**3R9**	15,000	15	.015	**153**
4.7			**4R7**	18,000	18	.018	**183**
5.6			**5R6**	22,000	22	.022	**223**
6.8			**6R8**	27,000	27	.027	**273**
8.2			**8R2**	33,000	33	.033	**333**
10			**100**	39,000	39	.039	**393**
12			**120**	47,000	47	.047	**473**
15			**150**	56,000	56	.056	**563**
18			**180**	68,000	68	.068	**683**
22			**220**	82,000	82	.082	**823**
27			**270**	100,000	100	.1	**104**
33			**330**	120,000	120	.12	**124**
39			**390**	150,000	150	.15	**154**
47			**470**	180,000	180	.18	**184**
56			**560**	220,000	220	.22	**224**

Table 18-3—cont'd

Three-digit capacitor codes

pF	nF	µF	Code	pF	nF	µF	Code
68			**680**	270,000	270	.27	**274**
82			**820**	330,000	330	.33	**334**
100		.0001	**101**	390,000	390	.39	**394**
120		.00012	**121**	470,000	470	.47	**474**
150		.00015	**151**	560,000	560	.56	**564**
180		.00018	**181**	680,000	680	.68	**684**
220		.00022	**221**	820,000	820	.82	**824**
270		.00027	**271**		1,000	1	**105**
330		.00033	**331**		1,500	1.5	**155**
390		.00039	**391**		2,200	2.2	**225**
470		.00047	**471**		2,700	2.7	**275**
560		.00056	**561**		3,300	3.3	**335**
680		.00068	**681**		4,700	4.7	**475**
820		.00082	**821**			6.8	**685**
1,000	1.0	.001	**102**			10	**106**
1,200	1.2	.0012	**122**			22	**226**
1,500	1.5	.0015	**152**			33	**336**
1,800	1.8	.0018	**182**			47	**476**
2,200	2.2	.0022	**222**			68	**686**
2,700	2.7	.0027	**272**			100	**107**
3,300	3.3	.0033	**332**			157	**157**

Code = 2 significant digits of the capacitor value + number of zeros to follow

For values below 10 pF, use "R" in place of decimal e.g. 8.2 pF = 8R2

10 pF = 100

100 pF = 101

1,000 pF = 102

22,000 pF = 223

330,000 pF = 334

1 µF = 105

capacitor code chart which will aid you in constructing the project.

Refer now to the resistor color code chart shown in Table 18-2, which will help you identify the resistors in this project. Note that each resistor will have either three or four color bands, which begin at one end of the resistor's body. The first colored band represents the resistor's first digit, while the second color represents the second digit of the resistor's value. The third colored band represents the resistor's multiplier value, and the fourth color band represents the resistor's tolerance value. If there is no fourth colored band then the resistor has a 20% tolerance value. If the resistor's fourth band is silver then the resistor has a 10% tolerance value and if the fourth band is gold then the resistor's tolerance value will be 5%.

Finally, we are ready to begin assembling the project, so let's get started. First, locate our first resistor R1. Resistor R1 has a value of 100 ohms and is a 5% tolerance type. Now look for a resistor whose first color band is brown, black, brown, and gold. From the chart you will notice that brown is represented by the digit (1), and black is represented by (0). Notice that the third color is brown and the multiplier is (10), so

$10 \times 10 = 100$ ohms. Now place resistor R1 into its proper location and solder it in place. Next, use a pair of end-cutters to trim the excess resistor leads. Cut the excess component leads flush to the edge of the circuit board. Now locate the remaining resistors and place them in their respective locations on the PC board. Solder the resistors in place and remember to trim the excess component leads with your end-cutters.

Note that capacitors generally come from two major groups or types. Non-polar types, such as the ones in this project, have no polarity and can be placed in either direction on the circuit board. The other type of capacitors are called polarized types; these capacitors have polarity and have to be inserted on the board with respect to their polarity marking in order for the circuit to operate correctly. On these types of capacitors, you will observe that the components actually have a plus or minus marking or a black or white band with a plus or minus marking on the body of the capacitors.

Next refer to Table 18-3; this chart helps to determine the value of small capacitors. Many capacitors are quite small in physical size, so manufacturers have devised a three-digit code which can be placed on the capacitor instead of the actual value. The code occupies a much smaller space on the capacitor and can usually fit nicely on the capacitor body, but without the chart the builder would have a difficult time determining the capacitor's value.

Now locate the capacitors from the component pile and try and locate capacitor C1, which should be marked with either a 22 pF (223) or a 27 pF (227) value. Let's go ahead and place C1 on the circuit board and then solder it place. Now take your end-cutters and cut the excess capacitor leads flush to the edge of the circuit board. Once you have installed C1, you can move on to installing the remaining capacitors.

The Talking/Musical Ringer project contains a Zener diode at D1 and a single silicon diode at D2. These diodes also have polarity which must be observed if the circuit is going to work correctly. You will notice that each diode will have a black or white colored band at one edge of the diode's body. The colored band represents the diode's cathode lead. Check your schematic and parts layout diagrams when installing

these diodes on the circuit board. Place the four diodes on the circuit board and solder them in place. Then cut the excess component leads with your end-cutters.

The Talking/Musical ringer circuit contains five integrated circuits (ICs). IC U1 is an opto-coupler, while U2 is a 74LS161, U3 is a 74LS00 and U4 is the voice chip. The regulator U5 is located in the power supply section of the circuit. ICs are often static sensitive, so they must be handled with care. Use a grounded anti-static wrist strap and stay seated in one location when handling the ICs. Take out a cheap insurance policy by installing IC sockets for each of the ICs. In the event of a possible circuit failure, it is much easier to simply unplug a defective IC than trying to unsolder 14 or 16 or more pins from a PC board without damaging the board. Integrated circuits have to be installed correctly if the circuit is going to work properly. IC packages will have some sort of markings which will help you orient them on the PC board. An IC will have either a small indented circle, a cutout, or notch at the top end of the IC package. Pin 1 of the IC will be just to the left of the notch, or cutout. Refer to the manufacturer's pin-out diagram, as well as the schematic, when installing these parts. If you doubt you ability to correctly orient the ICs, then seek the help of a knowledgeable electronics enthusiast to help you. Do not forget to install wire jumpers at all locations marked "J.", at the necessary locations.

Once the circuit board has been completed, take a short break and then we will inspect the PC board for possible "cold" solder joints and "short" circuits. Pick up the PC board with the foil side facing upwards toward you. Carefully inspect the solder joints, these should look smooth, clean, and bright. If any of the solder joints look dull, dark, or blobby, then you should remove the solder with a solder sucker or a solder wick and then resolder the joint all over again. Now we will inspect the PC board for possible short circuits. Short circuits can be caused from small solder blobs bridging the circuit traces or from cut component leads which can often stick to the circuit board once they have been cut. A sticky residue from solder will often trap component leads across the circuit traces. Look carefully for any bridging wires across circuit traces.

Switches Sl, S3, and S4 were mounted on the front panel of the enclosure, while the microphone and speaker were mounted on the top of the enclosure. Any plastic or metal enclosure that will accommodate the PC board, switches, and speaker will suffice. Drill the selected enclosure to provide mounting holes for the circuit board assembly, Sl, S3, S4, SPK, MIC, and access holes for the power cord and the telephone line cord. If you are using a metal enclosure, deburr all holes and insert rubber grommets for the AC line cord and the telephone cord. If you elect to use an external 9-V "wall wart" type power supply, you will have to mount a small coaxial jack on the rear panel of the chassis box.

Use about five or six inches of insulated wire attached to each terminal to connect those components to the PC board. Once you have soldered all components in place, you can test the circuit. With U4 still not installed in its socket, plug the AC adapter into a wall outlet. Check for 5 V DC between pin 3 of U5 and ground. If you do not obtain a proper reading, make sure the 9-volt adapter is indeed delivering 9 V DC (or more) when not connected to the ringer. If it is, check the orientation of U5. Correct any wiring errors and/or replace any defective components before proceeding.

Once you obtain the proper +5-V reading from U5, disconnect power from the ringer and install U4 in its socket. Power up the ringer again and place S4 in the SINGLE position and S3 in the RECORD position. While pressing and holding Sl, speak a test message into the microphone. Then release Sl and place S3 into the PLAY position. A momentary pressing of S1 should now result in the playback of your test message. If those results are obtained, your ringer is ready to be mounted in an enclosure.

Mount the circuit-board assembly in the enclosure using ½″ standoffs. Then mount all off-board components in their respective mounting holes. Pass the AC cord and the telephone cord through their rubber grommets and tie a knot in each one, about 5 inches from the free ends inside the enclosure, to serve as strain relief for each cord. Strip the ends of each wire and tin them with solder. Then solder the free ends in their respective circuit board holes. Figure 18-4 shows the author's completed prototype installed in a metal enclosure.

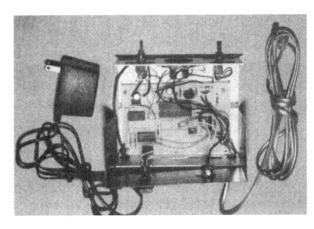

Figure 18-4 *Telephone voice ringer (inside view) (Courtesy Poptronix)*

Once the unit is complete, plug the adapter into an AC outlet and plug the telephone cord from the ringer into an unused telephone jack. If you need to attach the unit to a jack that already has a phone connected to it, you will need a duplex jack. It is a "Y" adapter that lets you plug two phones into one jack, permitting you to use the voice ringer and the telephone at the same time. You may want to shut off the telephone's internal ringer, so you can clearly hear the voice/musical ringer and its unusual message.

Some recording tips

To record multiple messages, perform the following:

1. Place S4 in the MULTI position.

2. Set S2- and S2(b) as listed in Table 18-1 for the number of messages you desire.

3. Press and hold S1 to record the first message.

4. Release Sl.

5. Press and hold S1 again to record the second message.

Table 18-1
Switch settings

Number of Msgs	S4	S2a	S2b
1	Single	X	X
2	Multi	L	L
3	Multi	H	L
4	Multi	L	H
5	Multi	H	H

6. Release S1.

7. Repeat steps 5 and 6 for each additional message.

8. Place S3 in the Play position for recording your last message.

9. Test the sequence by successively pressing S1 for each message.

Since a standard ring signal is on for 2 s and off for 2 s, you will want to make messages at least 2 s long to prevent successive triggering by one ring cycle. This can easily be accomplished, even on short messages, by holding the record button down for at least 2 s for each message.

Chapter 19

Telephone Tollsaver

Parts list

Tollsaver

R1 10 megohm, $\frac{1}{4}$ W
   5% resistor

R2, R5, R6 1 K ohm,
   $\frac{1}{4}$ W 5% resistor

R3 1 megohm, $\frac{1}{4}$ W
   5% resistor

R4 4.7 K ohm, $\frac{1}{4}$ W
   5% resistor

C1 100 µF, 35 V
   electrolytic capacitor

C2 68 µF, 35 V
   electrolytic capacitor

C3 1 µF, 35 V
   electrolytic capacitor

C4 2.2 µF, 35 V
   electrolytic
   capacitor

D1, D2, D3 1N4004
   silicon diodes

D4 1N5232B – 5.6 V
   Zener diode

D5 1N5230B – 4.7 V
   Zener diode

Q1 2N3904 transistor

Q2 BS170 – N-Channel
   MOSFET

U1 H11A5 optoislator IC

U2 HT-9212A DTMF
   dialer IC

XTL 3.579 MHz
   colorburst crystal

J1, J2 modular
   telephone jack (RJ11
   type)

PB1 pushbutton switch
   (chassis type)

S1, S2, S3, S4, S5, S6,
   S7 pushbutton
   switches (PC-mount)

Misc PC board, wire,
   IC sockets, hardware,
   plastic box, etc.

Have you checked your long distance telephone bill lately? Chances are you are paying way too much for long-distance calls, even if you participate in a calling plan. While some of those plans offer long-distance calling at 15 cents a minute or less, you might be paying a monthly fee. If you do not have a calling plan, you could be paying as much as an exorbitant 28 cents a minute for long distance!

The truth of the matter is that the cost to the consumer of long-distance calling is falling dramatically. When you consider that the actual cost to the carrier AT&T, MCI, Sprint, etc. is about 1 cent per minute, it's no wonder that competition is driving the price of telephone calls way down.

Over the past few years, "dial-around" services have sprung up all over America. These services let you make your calls at reduced rates. You simply dial a seven-digit code before the number that you are calling, bypassing your regular long-distance carrier. The access codes are easy to recognize; they usually start with the number 1010. Rates as low as 5 cents per minute with no monthly fee, no carrier switching, and no minimum calling time are presently available. These services are heavily advertised on TV and through the mail. These providers can also be found by doing a search for "dial-around service" on the Internet. If you are not taking advantage of these carriers, you are throwing your money away.

The Tollsaver project can also be used as a special purpose dialer for dialing fixed numbers on a regular basis, such as calling a taxi cab company, or the bus company or for older persons who regularly call a care-giver or relative.

The Tollsaver is a simple, easy-to-build electronic dialing device that plugs into your telephone line, and uses no batteries or AC power; see Figure 19-1.

The Tollsaver is pre-programmed by you with the seven-digit code for your selected dial around service. When you want to make a long distance call, simply pick up the telephone handset and press a pushbutton on the Tollsaver; the stored access code is dialed. You then dial the telephone number that you want to call as usual. With the Tollsaver helping you route your call through the discount services, you will save serious money from the first minute of calling.

The Tollsaver, designed for telephone systems with tone dialing, has no effect on normal telephone use. It comes into play only when the telephone handset is taken off-hook and the pushbutton is depressed. In fact, the Tollsaver may also be used as a rapid dial device for any discrete local, long distance (including the discount

code) or international. Keep in mind, however, that only one dialing sequence at a time can be stored.

Circuit description

The schematic diagram for the Tollsaver is shown in Figure 19-2. The Tollsaver is connected between a telephone and the wall jack; J1 connects to the telephone S1 and J2 goes to the wall jack. Since the telephone lines carry the DC current that powers the Tollsaver, it is important that proper polarity between the red and green wires from the telephone company be observed or the Tollsaver will not work. Note also that proper polarity is maintained between the two jacks: some older telephone equipment is as polarity sensitive as the Tollsaver. The heart of the circuit is U2, a telephone-dialer chip that has redial capability. It contains a crystal controlled oscillator circuit for generating dialing tones. Storage registers within the chip remember the pre-programmed dial-around code.

A set of four input and four output pins form a matrix for a 12- or 16-button keypad that can be scanned by U2. Since using a full-blown keypad in the Tollsaver seems like overkill, a series of discrete switches is used instead. Of those switches, S1–S7, only S1 and S2 are shown as being connected to U2 and are used instead. Those switches are wired to U2 as the digits "1" and "0" respectively—the two digits that are required to access the various dial-around services. The other switches are optional and can be wired into the circuit as any digit that you want. We will discuss that in greater detail later in the construction section of this article. For now, we will just assume that there are additional switches in the matrix along with S1 and S2 that can be used for setting the stored number in the Tollsaver.

When the Tollsaver is first connected to the telephone line, C2 (a low-leakage electrolytic capacitor) charges up through R1 to about 5 V as limited by Zener diode D4. That voltage is maintained at all times to retain the stored dial-around sequence code in U2's memory as long as the Tollsaver is connected to the telephone line and is drawing current from it. The 5 μA that the Tollsaver needs to work is well within the limits imposed by the rules concerning the switched public-telephone network.

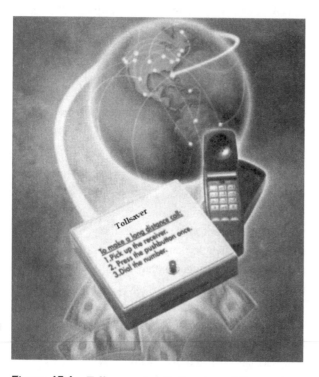

Figure 19-1 *Tollsaver circuit. Courtesy of Poptronix*

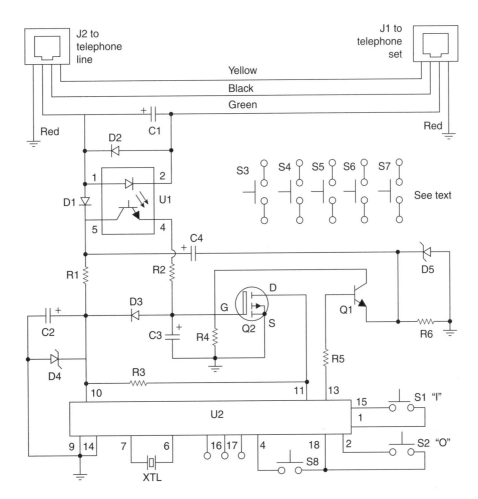

Figure 19-2 *Tollsaver. Courtesy of Poptronix*

When the telephone handset is picked up, current drawn by the phone flows through the light-emitting diode between pins 1 and 2 of Opto-isolator U1, switching on the transistor between pins 4 and 5. Current then flows through D2 and R2, charging C3 to about 5 V. That voltage is used to activate the circuit.

Once C3 is charged, Q2 turns on, providing a zero logic level to pin 1 of U2, the hook-switch sensing input. When U2 sees that input, it "wakes up" and starts scanning the switches. To dial out a long-distance access code (more later on setting that), simply press S8. That switch is wired into U2's switch matrix as a "redial the last number" switch; the stored dial-around numbers are sequentially generated and appear at pin 13 of U2.

The tones are coupled by Q1 to the telephone line through C4. Once the Moneysaver has finished dialing the access codes, you can then manually dial your call in the normal way; the call goes through the long-distance carrier of your choice at the reduced rate. When the call is completed and the handset placed back

on the phone, the voltage at pin 11 of U2 goes to a logic one. With the telephone now on-hook, U2 is ready for the next call.

The long-distance access code is programmed into the Moneysaver the first time power is applied. When any telephone is picked up, S1–S7 are used to dial the desired seven-digit access number that you want the Tollsaver to remember. After "dialing" the number with S1–S7, simply hang up the telephone; that number is saved in U2's redial register. As long as the Tollsaver is connected to a telephone line, it will remember that number.

Circuit assembly

Before we begin constructing the Tollsaver Project, you will first need to find a clean, well-lit worktable or workbench. Next we will gather a small 25–30-W pencil tipped soldering iron, a length of 60/40 tin–lead solder, and a small canister of Tip Tinner, available at your local RadioShack store; Tip Tinner is used to condition the soldering tip between solder joints. You drive the soldering tip into the compound and it cleans and prepares the soldering tip. Next grab a few small hand tools; try to locate small flat blade and Phillips screwdrivers. A pair of needle-nose pliers, a pair of tweezers along with a magnifying glass, and a pair of end-cutters and we will begin constructing the project. Now locate your schematic diagram and parts layout diagram along with all of the components needed to build the project. Once all the components are in front of you, you can check them off against the project parts list to make sure you are ready to start building the project. The Tollsaver can be constructed on a small printed circuit board.

Locate Table 19-1 and place it in front of you. Table 19-1 illustrates the resistor code chart which will help you to identify the resistors for the project. Note that each resistor will have either three or four color bands, which begin at one end of the resistor's body. The first colored band represents the resistor's first digit, while the second color represents the second digit of the resistor's value. The third colored band represents the resistor's multiplier value, and the fourth color band represents the resistor's tolerance value. If there is no fourth colored band then the resistor has a 20% tolerance value. If the resistor's fourth

band is silver then the resistor has a 10% tolerance value and if the fourth band is gold then the resistor's tolerance value will be 5%.

Finally, we are ready to begin assembling the project, so let's get started. The prototype project was constructed on a single-sided printed circuit board or PCB. With your PC board in front of you, we can now begin populating the circuit board. We are going to place and solder the lowest height components first, the resistors and diodes. First, locate our first resistor R1. Resistor R1 has a value of 10 megohms and is a ¼ watt 5% tolerance type. Now take a look for a resistor whose first color band is brown, black, green, and gold. From the chart you will notice that brown is represented by the digit (1), and again the second band is black, which is represented by (0). Notice that the third color is green and the multiplier is (100,000), so $(1)(0) \times 100,000 = 1,000,000$ or 1 megohm. Install resistor R1 in its respective location and then solder it in place on the PC board. Now take a pair of end-cutters and trim the excess resistor leads flush to the edge of the circuit board. Locate and install the remaining resistors in their correct positions on the PC board, and solder them in place. Now use your end-cutters and remove the extra lead lengths.

Note that capacitors generally come from two major groups or types. Non-polar types, such as the ones in

Table 19-1

Resistor color code chart

Color Band	1st Digit	2nd Digit	Multiplier	Tolerance
Black	0	0	1	
Brown	1	1	10	1%
Red	2	2	100	2%
Orange	3	3	1,000 (K)	3%
Yellow	4	4	10,000	4%
Green	5	5	100,000	
Blue	6	6	1,000,000 (M)	
Violet	7	7	10,000,000	
Gray	8	8	100,000,000	
White	9	9	1,000,000,000	
Gold			0.1	5%
Silver			0.01	10%
No color				20%

this project, have no polarity and can be placed in either direction on the circuit board. The other type of capacitors are called polarized types; these capacitors have polarity and have to be inserted on the board with respect to their polarity marking in order for the circuit to operate correctly. On these types of capacitors, you will observe that the components actually have a plus or minus marking or a black or white band with a plus or minus marking on the body of the capacitors.

Now locate the capacitor chart shown in Table 19-2; this will assist you installing the small capacitors which may have only a three-digit code marked on the capacitors.

Find the capacitors from the component pile and try and locate capacitor C1, which should be marked 100 μF. Now that you can read the resistor and capacitor charts, let's go ahead and place R1 and C1 on the circuit board and solder them in place. Now take your end-cutters and cut the excess component leads flush to the edge of the circuit board. Once you have installed the first two components, you can move on to installing the remaining low profile components.

Next locate and install the remaining capacitors, first installing the low profile capacitors followed by the larger electrolytic capacitors. Use good quality electrolytic or tantalum capacitors. Remember the larger

Table 19-2

Three-digit capacitor codes

pF	nF	μF	Code	pF	nF	μF	Code
1.0			**1R0**	3,900	3.9	.0039	**392**
1.2			**1R2**	4,700	4.7	.0047	**472**
1.5			**1R5**	5,600	5.6	.0056	**562**
1.8			**1R8**	6,800	6.8	.0068	**682**
2.2			**2R2**	8,200	8.2	.0082	**822**
2.7			**2R7**	10,000	10	.01	**103**
3.3			**3R3**	12,000	12	.012	**123**
3.9			**3R9**	15,000	15	.015	**153**
4.7			**4R7**	18,000	18	.018	**183**
5.6			**5R6**	22,000	22	.022	**223**
6.8			**6R8**	27,000	27	.027	**273**
8.2			**8R2**	33,000	33	.033	**333**
10			**100**	39,000	39	.039	**393**
12			**120**	47,000	47	.047	**473**
15			**150**	56,000	56	.056	**563**
18			**180**	68,000	68	.068	**683**
22			**220**	82,000	82	.082	**823**
27			**270**	100,000	100	.1	**104**
33			**330**	120,000	120	.12	**124**
39			**390**	150,000	150	.15	**154**
47			**470**	180,000	180	.18	**184**
56			**560**	220,000	220	.22	**224**
68			**680**	270,000	270	.27	**274**
82			**820**	330,000	330	.33	**334**
100		.0001	**101**	390,000	390	.39	**394**
120		.00012	**121**	470,000	470	.47	**474**
150		.00015	**151**	560,000	560	.56	**564**

(Continued)

Table 19-2—cont'd

Three-digit capacitor codes

pF	nF	μF	Code	pF	nF	μF	Code
180		.00018	**181**	680,000	680	.68	**684**
220		.00022	**221**	820,000	820	.82	**824**
270		.00027	**271**		1,000	1	**105**
330		.00033	**331**		1,500	1.5	**155**
390		.00039	**391**		2,200	2.2	**225**
470		.00047	**471**		2,700	2.7	**275**
560		.00056	**561**		3,300	3.3	**335**
680		.00068	**681**		4,700	4.7	**475**
820		.00082	**821**			6.8	**685**
1,000	1.0	.001	**102**			10	**106**
1,200	1.2	.0012	**122**			22	**226**
1,500	1.5	.0015	**152**			33	**336**
1,800	1.8	.0018	**182**			47	**476**
2,200	2.2	.0022	**222**			68	**686**
2,700	2.7	.0027	**272**			100	**107**
3,300	3.3	.0033	**332**			157	**157**

Code = 2 significant digits of the capacitor value + number of zeros to follow

For values below 10 pF, use "R" in place of decimal e.g. 8.2 pF = 8R2

10 pF = 100

100 pF = 101

1,000 pF = 102

22,000 pF = 223

330,000 pF = 334

1 μF = 105

electrolytic capacitors have polarity which must be observed if the circuit is work properly. Failure to install the electrolytic capacitors correctly can damage the component or other components when power is first applied to the circuit upon power-up. Install the capacitors in their correct positions on the PC board, and then solder them in place. Next trim the excess component leads flush to the edge of the circuit board using your end-cutters.

This project contains two silicon diodes at D2 and D3 and two Zener diodes at D4 and D5. These diodes also have polarity which must be observed if the circuit is going to work correctly. You will notice that each diode will have a black or white colored band at one edge of the diode's body. The colored band represents the diode's cathode lead. Check your schematic and parts layout diagram when installing these diodes on the circuit board. Place all the diodes on the circuit board and solder them in place. Then cut the excess component leads with your end-cutters.

The Tollsaver circuit utilizes single NPN transistors at Q1, and an FET transistor at Q2. Transistors generally have three leads, a Base lead, a Collector lead, and an Emitter lead. Refer to the schematic and you will notice that there will be two diagonal leads pointing to a vertical line. The vertical line is the transistor's base lead and the two diagonal lines represent the collector and the emitter. The diagonal lead with the arrow on it is the emitter lead, and this should help you install the transistor on to the circuit board correctly. Go ahead and install the transistor Q1 on the PC board and solder it in place. The FET transistor at Q2 also has three leads but they are labeled differently. The FET has a Drain marked (D), Source lead marked (S), and a Gate lead marked (G). FET transistors are very sensitive to static electricity, so you must handle them carefully using an anti-static wrist band which is grounded to an AC outlet. When handling the FET sit in one spot and do not move around in you chair or walk on a carpet while installing this component. Locate the FET and solder it place. Now, follow-up by cutting the excess component leads from Q1 and Q2 with your end-cutters.

The Tollsaver utilizes ICs and an opto-isolator. ICs must be handled carefully and installed properly in order for the circuit to work perfectly. Use anti-static techniques when handling the IC to avoid damage from static electricity when moving about and installing it. It is wise to install integrated circuit sockets as a cheap form of insurance against a possible circuit failure at some later point in time. Install the IC socket on the PC board and solder it in place. ICs have to be installed correctly for the circuit to work, so you must pay strict attention to inserting them correctly into their sockets. Each integrated circuit has a locator either in the form of a small indented circle or a notch, or cutout at the top end of the IC package. Pin 1 of the IC will be just to the left of the locator. The opto-isolator will be housed in a small IC-like package with six pins. Identify the pin-outs on the opto-isolator and then install it on the PC board. It is a good idea to use an integrated circuit socket for the opto-isolator as well as for the IC at U2. Do not install the IC or opto-isolator at this point, only the IC sockets.

Now is a good time to take a short break. After the break we will inspect the circuit board for any possible "cold" solder joints or "short" circuits. Take a careful look at each of the PC solder joints, they should all look clean and shiny. If any of the solder joints look dull, dark, or blobby, then you unsolder the joint, remove the solder and then resolder the joint all over again. Next, examine the PC board for any short circuits. Short circuits can be caused from two circuit traces touching each other due to a stray component lead that stuck to the PC board from solder residue or from solder blobs bridging the solder traces.

Switch wiring

As we mentioned before, although U2 is designed to use a matrix keypad, the use of such a keypad would be overkill for a project like the Tollsaver. Switches S1–S7 take the place of the keypad to keep the assembly cost down. The only switches that must be installed are S1 and S2; "1" and "0" are required to access any dial-around service.

You only need to install as many switches for S3–S7 as are needed for whatever number you would like to program into the Tollsaver. Those switches are hard-wired with short insulated jumpers so that they appear to U2 as being part of a keypad matrix. See Table 19-3 for a chart that lists the connections needed for each individual number.

Note that "1" and "0" are included in the chart for completeness; you do not need to wire those digits since S1 and S2 are already hard-wired that way.

For example, if you look closely at the photograph of the prototype, you will see that S3 and S4 are wired for "3" and "6" respectively; S5 is present but not wired. If you look at Table 19-3, you will note that both of those digits need a connection pin 3 of U2, and that is why two wires are inserted in the hole that connects to that pin. If you find yourself in a similar situation, be sure to use small-gauge wire! Switch S8 is a panel-mounted pushbutton switch that activates the Tollsaver's redial feature. It is wired to the PC board with lengths of insulated wire.

Table 19-3
Tollsaver pin-out table

Digit	UI – Pins	Connections
1	pin 1	pin 15
2	pin 2	pin 15
3	pin 3	pin 15
4	pin 1	pin 16
5	pin 2	pin 16
6	pin 3	pin 16
7	pin 1	pin 17
8	pin 2	pin 17
9	pin 3	pin 17
0	pin 2	pin 18
0	pin 1	pin 18
#	pin 3	pin 18

Final assembly

The Tollsaver was mounted in a small enclosure that has appropriate holes for the telephone jacks at J1, J2, and switch S8. As an alternative, for J2 you can substitute a length of 4-conductor telephone wire that terminates in a modular plug; that is what the author did. If you use a jack, you will need a length of telephone cord with modular jacks on each end. Center the circuit board inside the enclosure and use four plastic standoffs to mount the PC board inside the enclosure. The Tollsaver is line powered, so no power supplies are needed. Now we are ready to connect up and test the Tollsaver.

Telephone line connection

The telephone line cable must be connected to the Tollsaver circuit with the proper polarity, or it will not work. Many telephone systems use a green wire and a red wire to carry the voice and ringing signals, with green being the positive and red negative. Check your telephone line polarity with a DC voltmeter to be sure. With no telephones off-hook, there should be about 50 V DC on the line.

If you find that the wiring is backwards, you can correct that problem, if you know what you are doing. When in doubt, you can always hire a telephone line installer to make sure that the wiring is "up to snuff." If you do not want to touch the wiring in your house, you can always swap the wires from J2 to the PC board if your house is wired backwards.

With J2 connected to a telephone wall outlet, we are ready to start testing the Moneysaver. If you are a bit nervous about plugging an untested Moneysaver into the telephone line, you can simulate the telephone line on a workbench. Use a well-filtered 50-V power supply with a 1,000-ohm, two-watt resistor in series. Connect the green wire from J2 to the positive terminal; the red wire goes to the negative terminal.

Circuit testing

The only piece of test equipment that you will need is a voltmeter with a high input resistance-10 megohms

or greater. Set the meter to the 20-V DC range and connect its leads across C2. With no ICs in place, apply power to J2 and see that C2 charges up to about 5 V after a few minutes. If it does not, troubleshoot the circuit and repair the fault. Check the telephone-line polarity as well as D2-D4, R1, and C2. Be sure that C2 is a low-leakage capacitor. Disconnect the power and insert the ICs into their sockets. Make sure that the ICs have oriented correctly. Be sure that the chips are seated properly with no pins bent under them.

Connect a telephone to J1 and leave the handset on hook. Then connect the meter across C2 as before. Connect J1 into a telephone wall jack and verify once again that C2 charges up to about 5 volts after a couple of minutes. Measure the voltage across capacitor C3; it should be zero.

Pick up the telephone handset; you should hear the dial tone, and the voltage across C3 should rise to about 5 V. Press the pushbuttons on the board to dial the sequence code of your selected dial-around service. You should hear the tones in the handset, and the telephone line dial tone will be muted. Hang up the receiver and the Tollsaver is now initialized. Pick up the receiver again and listen for a dial tone. Press S8 briefly. The dial-around sequence should automatically be repeated and the dial tone silenced. For a full test, dial a long-distance number using your telephone's keypad or dial. That call will use your selected long-distance carrier.

If you need to reset U2 for any reason, disconnect the circuit from the telephone line and allow time for C2 to discharge to less than one 1 volt. That will erase the memory in U2, and it can be programmed as described above.

If everything went well, the Tollsaver is working and ready for use. If the circuit does not work, then you will have to remove it from the telephone line and recheck a few things. First make a visual check of the circuit board. Look for short circuits or solder blobs or solder bridges between PC traces. Look for cold solder joints that look dull or dark. If you see any solder joints that look this way, then remove the solder from the joint and resolder the joint over again.

Check all the ICs and make sure that they are oriented correctly, then check to see if all the diodes have been installed correctly; be sure to check all of the electrolytic capacitors to make sure that they are all installed with respect to their proper polarity. Next, check the two transistors to make sure that they have been installed correctly. If C3 does not charge up when the telephone handset is taken off the hook, then check D2, R2, and C3.

Once you know the Tollsaver is working correctly, you can mount the circuit in a small plastic box. You will need to install two RJ11 phone jacks on the case, as well as the pushbutton switches. Mount the PC board atop four ¼″ plastic standoffs, and wire the switches to the circuit board via five-inch lengths of insulated wires. Plug the telephone line into (J2) on the Tollsaver, and then plug your telephone set into the telephone jack marked (J1).

Using the tollsaver

Once the Tollsaver has been initialized as described in the testing section above, it must be left connected to the telephone line at all times to preserve the stored sequence of digits. Should it be disconnected for any length of time, you will have to reprogram U1 with the correct sequence. Use the procedure for programming described above.

With the Moneysaver properly programmed, each time you wish to make a long-distance call simply press S8 after lifting the telephone handset, then dial the telephone umber in the normal way. When you check your next month's telephone bill, you will be amazed at how much money you have saved.

Chapter 20

Infinity Bug/Transmitter Project

Parts list

Parts Bin

Infinity Bug/Transmitter

R1, R2, R8, R15, R18
22 K ohms, $\frac{1}{4}$ W,
5% resistor

R3, R13 4.7 K ohm,
$\frac{1}{4}$ W, 5% resistor

R4, R7, R10, R12
10 K ohm, $\frac{1}{4}$ W,
5% resistor

R6 15 K ohms, $\frac{1}{4}$ W,
5% resistor

R9 6.8 K ohms, $\frac{1}{4}$ W,
5% resistor

R11 120 ohms, $\frac{1}{4}$ W,
5% resistor

R14, R19 1 megohm, $\frac{1}{4}$ W,
5% resistor

R16, R20 47 K ohms,
$\frac{1}{4}$ W, 5% resistor

R17 2.2 megohm, $\frac{1}{4}$ W,
5% resistor

C1, C4, C16 10 nF,
35 V capacitor

C2, C12 100nF, 35 V

C3, C5 47 nF, 35 V

C6, C7, C9, C10, C13
10 µF, 35 V
electrolytic
capacitor

C8, C15 47 µF, 35 V
electrolytic capacitor

C11 22 nF, 35 V

C14 1 µF, 35 V
tantalum capacitor

D1, D2, D3, D4 1N4004
silicon diode

D5 8.2 volt Zener diode

D6, D7, D8, D9 1N4148
silicon diode

D10 MCR100-6-SCR

Q1, Q3, Q4 BC847
transistor

Q2, Q5 BC639 transistor

U1 HEF40106BT - Hex
inverting Schmitt
trigger IC

MIC electret microphone

Misc PC board,
IC socket, wire,
hardware, standoffs,
enclosure, etc.

A kit of parts for this
project is available
from:

Talking Electronics

www.talkingelectronics.
com/

The Infinity Bug/Transmitter is a remote telephone listening device, which can listen-in via telephone and will allow you to hear conversations from a remote location unattended; see Figure 20-1. The Infinity Bug is connected across the phone-line of a distant phone, and is line powered so no power supply is needed, since it derives its power from the phone line.

To use the Infinity Bug, the distant phone is dialed, and after having a conversation with

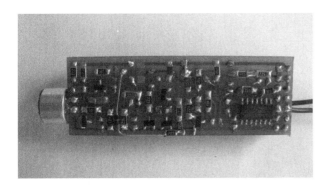

Figure 20-1 *Infinity bug circuit. Courtesy of Talking Electronics*

the person at the other end, the phone receiver will be replaced. At this point you whistle into the receiver and the Infinity Bug will pick up the line quietly and begin listening-in using its high-gain amplifier, and will pick up the audio from the remote room.

The Infinity Bug/Transmitter does not have to be next to the distant phone. It can be anywhere in the house, provided it is connected across the phone line of the distant phone. A timing circuit in the device will cancel after 3 min. To open the Bug you need to whistle again. This can be repeated any number of times as needed. If the remote phone is picked up, it cancels the Infinity Bug and goes quiet.

Circuit description

The Infinity Bug consists of a number of circuit blocks, plus components that perform very important tasks. The circuit diagram shown in Figure 20-2 illustrates the function of each of the components.

The Infinity Bug connects across the phone line and takes very little current as most of the circuit is not active when in the "waiting" state.

This Infinity Bug is a called a "leech" device as it gets its operating current from the phone line. Some phone systems detect as little as 0.5 mA, and if more than 1 mA is drawn from the line, it remains engaged. The voltage of a phone line varies from 42 to 55 V, depending on the service-provider, and the Infinity Bug will operate on this voltage range. When the Infinity Bug is connected to the line,

the "line voltage" will drop to about 36 V, due to the current drawn by the bug. This will not upset the operation of the phone system.

The diode bridge on the front-end of the circuit protects the circuit from connecting the circuit to the line backwards.

The only two sections of the circuit that draw current during quiescent conditions are the whistle-detecting circuit made up of the BC 547 and its surrounding components. This section draws less than 0.25 mA. It is frequency sensitive and will not detect the 33 Hz ringing frequency. The "timing section," of the Infinity Bug is made using the 40106 a hex Schmitt trigger IC.

When a phone (that is connected to the same line as the Infinity bug) is picked up, the line voltage drops to about 12 V and the whistle-detecting circuit does not see any voltage, as the 8.2 V Zener drops nearly all the available voltage.

Thus the circuit has a self-canceling feature. The principle of operation of the bug has been described above and when the handset is put back on the phone, the line voltage rises. This allows the person at the other end of the line to whistle. The tone produced by the whistle is passed through the 47 nF capacitor and a 4.7 K ohm resistor to the base of the BC 547 transistor. The transistor amplifies the signal to produce a 2 V p-p signal that is turned into a DC voltage by the 10 µF capacitor. This turns on the latch circuit, and the base of the BC 639 is raised to produce a voltage on the emitter of about 4.5 V. The actual voltage on the emitter of the BC 639 is determined by the 120 ohm resistor in the emitter leg of a BC 547. The BC 547 is almost fully turned on by the 10 K resistor between the collector and base. The 47 µF capacitor on the base of the BC 639 holds the base rigid to any AC signals so that the transistor operates as an emitter follower. The transistor acts as an impedance-matching stage as the phone-line has relatively low impedance while the pre-amplifier section has quite a high impedance.

Finally, the electret microphone is connected to a microphone pre-amplifier stage consisting of a BC 547 transistor, 1 M bias resistor, 4.7 K ohm load and 22 nF input capacitor. Thus the audio section consists of three stages in a very unusual arrangement.

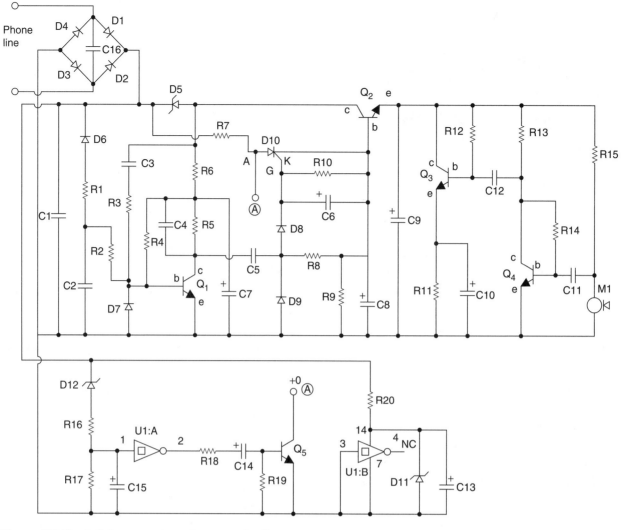

Figure 20-2 *Infinity transmitter. Courtesy of Talking Electronics.*

This arrangement is necessary as the signal has to be delivered to the same line as the supply voltage, since the circuit derives power from the phone line.

In normal operation, the Infinity Bug turns off after about 3 min due to a timing circuit connected to one gate of a hex Schmitt trigger IC. The voltage across the IC is set to 6.2 V via a Zener so the timing can be controlled, as the lower threshold at which the chip will change state is determined by the supply voltage. When the line voltage rises to 50 V, the 47 μF capacitor in the timing circuit is rapidly charged via a 47 K resistor that is connected to the rail via an 18 V Zener. This puts a maximum of 6.2 V on the 47 μF capacitor as the input of the chip cannot rise above the supply on the chip due to diodes on the input of the gate.

When the Infinity Bug is turned ON via a whistle, the line voltage falls to about 12 V and the 47 μF capacitor in the timing circuit does not see any charging voltage, as the 18 V Zener removes this voltage. The 47 μF is now slowly discharged via a 2.2 megohm resistor and when the voltage falls to below 2 V, the output of the gate goes HIGH. This HIGH is passed to the base of a BC 639 via a 22 K resistor and 1 μF electrolytic capacitor. The transistor removes voltage from the latching circuit and the Infinity Bug drops off the line. The 1 μF capacitor allows a pulse to activate the "turn-off" transistor.

The 1 megohm resistor on the base discharges the 1 μF capacitor so that it is uncharged in readiness for the circuit timing out. The 1N4148 signal diode connected to the rail of the project is connected to a 22 K resistor and 100 nF capacitor. This arrangement detects the "ring voltage" (approx 150 V) and the Zener of the signal diode allows the 100 nF capacitor to charge via the 22 K ohm resistor. All diodes have a maximum reverse voltage and the diode we have chosen has a voltage of about 110 V. This voltage can be called its "Zener voltage." Any voltage above this "Zener voltage" will pass through the diode. We have used this feature to keep the whistle-detect transistor turned on during the time when the phone is "ringing." More details on this project can be found at www.talkingelectronics.com/

Circuit assembly

Finally, we are ready to begin assembling the project, so let's get started. The prototype project was constructed on a single-sided printed circuit board or PCB. With your PC board in front of you, we can now begin populating the circuit board. The kit offered in the parts list utilizes surface mounted components in order to construct a very small circuit board. The surface-mount components are generally the first to be soldered to the board. In this project kit offered, there are through-hole components and surface-mount components. If you add the surface-mount components first, make sure you do not "fill-in" any of the holes for the through-hole components

Before we begin constructing the Infinity Bug, you will need to locate a clean, well-lit worktable or workbench. Next we will gather a small 25–30-W pencil tipped soldering iron, a length of 60/40 tin–lead solder, and a small canister of Tip Tinner, available at your local RadioShack store; Tip Tinner is used to condition the soldering tip between solder joints. You drive the soldering tip into the compound and it cleans and prepares the soldering tip. Next grab a few small hand tools, try to locate small flat blade and Phillips screwdrivers. A pair of needle-nose pliers, a pair of tweezers along with a magnifying glass, and a pair of end-cutters and we will begin constructing the project. Now locate your schematic diagram and parts layout diagram along with all of the components needed to build the project. Once all the components are in front of you, you can check them off against the project parts list to make sure you are ready to start building the project. Now, locate Tables 20-1 and 20-2, and place them in front of you.

Table 20-1 illustrates the resistor code chart and how to read resistors, while Table 20-2 illustrates the capacitor code chart which will aid you in constructing the project.

Each resistor will have either three or four color bands, which begin at one end of the resistor's body. The first colored band represents the resistor's first digit, while the second color represents the second digit of the resistor's value. The third colored band represents the resistor's multiplier value, and the fourth color band

Table 20-1
Resistor color code chart

Color Band	1st Digit	2nd Digit	Multiplier	Tolerance
Black	0	0	1	
Brown	1	1	10	1%
Red	2	2	100	2%
Orange	3	3	1,000 (K)	3%
Yellow	4	4	10,000	4%
Green	5	5	100,000	
Blue	6	6	1,000,000 (M)	
Violet	7	7	10,000,000	
Gray	8	8	100,000,000	
White	9	9	1,000,000,000	
Gold			0.1	5%
Silver			0.01	10%
No color				20%

Table 20-2
Three-digit capacitor codes

pF	nF	µF	Code	pF	nF	µF	Code
1.0			**1R0**	3,900	3.9	.0039	**392**
1.2			**1R2**	4,700	4.7	.0047	**472**
1.5			**1R5**	5,600	5.6	.0056	**562**
1.8			**1R8**	6,800	6.8	.0068	**682**
2.2			**2R2**	8,200	8.2	.0082	**822**
2.7			**2R7**	10,000	10	.01	**103**
3.3			**3R3**	12,000	12	.012	**123**
3.9			**3R9**	15,000	15	.015	**153**
4.7			**4R7**	18,000	18	.018	**183**
5.6			**5R6**	22,000	22	.022	**223**
6.8			**6R8**	27,000	27	.027	**273**
8.2			**8R2**	33,000	33	.033	**333**
10			**100**	39,000	39	.039	**393**
12			**120**	47,000	47	.047	**473**
15			**150**	56,000	56	.056	**563**
18			**180**	68,000	68	.068	**683**
22			**220**	82,000	82	.082	**823**
27			**270**	100,000	100	.1	**104**
33			**330**	120,000	120	.12	**124**
39			**390**	150,000	150	.15	**154**
47			**470**	180,000	180	.18	**184**
56			**560**	220,000	220	.22	**224**
68			**680**	270,000	270	.27	**274**
82			**820**	330,000	330	.33	**334**
100		.0001	**101**	390,000	390	.39	**394**
120		.00012	**121**	470,000	470	.47	**474**
150		.00015	**151**	560,000	560	.56	**564**
180		.00018	**181**	680,000	680	.68	**684**
220		.00022	**221**	820,000	820	.82	**824**
270		.00027	**271**		1,000	1	**105**
330		.00033	**331**		1,500	1.5	**155**
390		.00039	**391**		2,200	2.2	**225**
470		.00047	**471**		2,700	2.7	**275**
560		.00056	**561**		3,300	3.3	**335**
680		.00068	**681**		4,700	4.7	**475**
820		.00082	**821**			6.8	**685**
1,000	1.0	.001	**102**			10	**106**
1,200	1.2	.0012	**122**			22	**226**
1,500	1.5	.0015	**152**			33	**336**
1,800	1.8	.0018	**182**			47	**476**

Table 20-2—cont'd

Three-digit capacitor codes

pF	nF	μF	CODE	pF	nF	μF	Code
2,200	2.2	.0022	**222**			68	**686**
2,700	2.7	.0027	**272**			100	**107**
3,300	3.3	.0033	**332**			157	**157**

Code = 2 significant digits of the capacitor value + number of zeros to follow

For values below 10 pF, use "R" in place of decimal e.g. 8.2 pF = 8R2

10 pF = 100

100 pF = 101

1,000 pF = 102

22,000 pF = 223

330,000 pF = 334

1 μF = 105

represents the resistor's tolerance value. If there is no fourth colored band then the resistor has a 20% tolerance value. If the resistor's fourth band is silver then the resistor has a 10% tolerance value and if the fourth band is gold then the resistor's tolerance value will be 5%.

We are going to place and solder the lowest height components first, the resistors and diodes. First, locate our first resistor R1. Resistor R1 has a value of 22 K ohms and is a 5% tolerance type. Now take a look for a resistor whose first color band is red, red, orange, and gold. From the chart you will notice that red is represented by the digit (2), the second color band is red, which is represented by (2). Notice that the third color is orange and the multiplier is (1,000), so $(2)(2) \times 1,000 = 22,000$ or 22 K ohms.

Go ahead and place R1 on the circuit board and solder it in place. Now take your end-cutters and cut the excess component leads flush to the edge of the circuit board. Once you have installed the first resistor, you can move on to installing the remaining ones. Solder them in place in their respective locations on the circuit board, and follow up by trimming the excess component leads with you end-cutters.

Next refer to Table 20-2; this chart helps to determine the value of small capacitors. Many capacitors are quite small in physical size, so manufacturers have devised a three-digit code which can be placed on the capacitor instead of the actual value. The code occupies a much smaller space on the capacitor and can usually fit nicely

on the capacitor body, but without the chart the builder would have a difficult time determining the capacitor's value.

Note that capacitors generally come from two major groups or types. Non-polar types, such as the ones in this project, have no polarity and can be placed in either direction on the circuit board. The other type of capacitors are called polarized types, these capacitors have polarity and have to be inserted on the board with respect to their polarity marking in order for the circuit to operate correctly. On these types of capacitors, you will observe that the components actually have a plus or minus marking or a black or white band with a plus or minus marking on the body of the capacitors.

Now, we will install the capacitors for the project. The kit offered uses surface-mount capacitors that are not identified and are contained in a strip for this project in the following order: Do not remove any of the capacitors until they are required. Only remove one at a time and solder it to the board. First, locate capacitor C1, a 10 nF capacitor. Refer to Table 20-2 to find the three-digit code for the 10nf capacitor. Once identified, go ahead and install C1 and solder it in place in its proper location. Next locate the remaining small or low profile capacitors and install them in their respective locations on the PC board. Solder these capacitors to the PC board and then remember to cut the excess component leads. Finally locate the electrolytic capacitors. The electrolytic capacitors are generally larger capacitors and they will have polarity marking on the side of the component. Look for a white or black band with a plus or minus marking next to the colored band. Once you have identified the electrolytic capacitors, refer to the schematic and parts layout diagrams to ensure that you install the capacitors correctly. Electrolytic capacitors installed backwards can damage the component and possibly damage the circuit when power is first applied to the circuit.

This project contains four silicon diodes, three Zener diodes, and an SCR or silicon controller rectifier. Diodes always have some type of polarity markings, which must be observed if the circuit is going to work correctly. You will notice that each diode will have a black or white colored band at one edge of the diode's body. The colored band represents the diode's cathode lead. Check your schematic and parts layout diagrams when installing these diodes on the circuit board. Note that the diodes are marked "A6." The rectifier bridge diodes D1 through D4 are all 1N4004 silicon diodes types. There are also three voltage/current diodes at D7, D8 and D9. Next locate and install the Zener diodes. Note that all of the Zener diodes have different voltage cut-off voltages, so pay particular attention, to install the correct Zener in its exact location. Finally, locate the single SCR. The SCR will have three leads, an Anode, a Cathode and a Gate lead. Check the component specification sheet and then refer to both the schematic and parts layout diagrams when installing the SCR. Once all the diodes and the SCR have been placed in their respective locations on the PC board, you can then solder them in place. Finally, trim the excess component leads with your end-cutter flush to the edge of the circuit board.

The Infinity Transmitter/Bug utilizes five conventional NPN type transistors. Transistors are generally three lead devices, which must be installed correctly if the circuit is to work properly. Careful handling and proper orientation in the circuit is critical. A transistor will usually have a Base lead, a Collector lead, and an Emitter lead. The base lead is generally a vertical line inside a transistor circle symbol. The collector lead is usually a slanted line facing towards the base lead and the emitter lead is also a slanted line facing the base lead but it will have an arrow pointing

towards or away from the base lead. In the kit offered, the transistors are marked "1K" or "1G" Before installing the two transistors refer to the parts specification sheets as well as the parts layout diagram and the schematic. Be sure to orient the transistors prior to inserting them into the circuit board. If you are in doubt as to how to mount the transistors, have a knowledgeable electronics enthusiasts help you. Now, insert the transistors on to the circuit board and solder them in place. Remember to cut the excess component leads with your end-cutters, flush to the edge of the circuit board.

The Infinity Transmitter/Bug uses a single IC at U1. Prior to installing the IC, install an IC socket. An IC socket is low cost insurance in the event of a circuit failure at some point in time. It is much easier to simply unplug an IC from its socket and replace it rather than trying unsolder 14 or 16 pins. ICs have to be installed correctly if the circuit is to work properly. You must orient the IC before inserting it into the socket. Take a look at the IC package. At the top end of the IC package, you will notice either a small indented circle, a cutout, or notch. This notch or cutout is the IC locator. Pin 1 of the IC will be just to the left of the notch or cutout. Once certain of the orientation, install the IC in its socket.

Finally, locate and install the electret microphone at M1; it is installed at the junction of R14/R15 and C11. The case of the electret microphone goes to the "-" on the board. This is the negative rail and runs along the bottom of the board. Solder the microphone in place and remember to trim the excess lead lengths.

Once the circuit board has been completed, take a short break and then we will inspect the PC board for possible "cold" solder joints and "short" circuits. Pick up the PC board with the foil side facing upwards toward you. Carefully inspect the solder joints, these should look smooth, clean, and bright. If any of the solder joints look dull, dark, or blobby, then you should remove the solder with a solder sucker or a solder wick and then resolder the joint all over again.

Now we will inspect the PC board for possible short circuits. Short circuits can be caused from small solder blobs bridging the circuit traces or from cut component leads which can often stick to the circuit board once they have been cut. A sticky residue from solder will often trap component leads across the circuit traces. Look carefully for any bridging wires across circuit traces. Once this inspection is complete, we can move on to installing the circuit board into some sort of plastic enclosure.

The circuit is line powered so you will need no external power supply or batteries. Now you will have to mount the circuit board inside of a plastic enclosure. Lay the circuit board inside of a suitable plastic case, to check the fit. Locate some mounting standoffs and use them to mount the circuit board inside of the case. Next, you will have to connect the circuit to the phone line; this can be done using a length of 4-conductor phone wire with a modular plug attached at one end of the cable. The modular plug end can then simply plug into a telephone jack, and wire ends of the cable can connect to the circuit itself. You could also elect to hard-wire the circuit directly to the phone line without the modular plugs if desired.

Testing the infinity bug

To efficiently test the Infinity Bug you will need two phone lines, but if they are not available, the next best solution is to build an Infinity Bug Test Circuit. The 470 ohm resistors are current-limit resistors and the LEDs indicate when the phone and Infinity Bug are on the line.

The Infinity Bug can be tested with the following Test Circuit, as shown in Figure 20-3.

Connect the Infinity Bug to the test circuit shown, and pick up the left-hand phone. Whistle into the mouthpiece and the Bug will pick up the line and you will be able to hear the ticking of a clock in the room. Pick up the second phone and the Bug will drop off the line. Replace the second phone and whistle. The Infinity Bug will pick up the line again. Keep listening to the Bug and it will drop off after about 2–3 min. Whistle again and the Infinity Bug will pick up the line. Repeat this again and if everything works perfectly, the Infinity Bug is fully tested.

What to do if the circuit does not work. The first thing to do is measure the current taken by the Infinity Bug when it is connected to the line. It should be

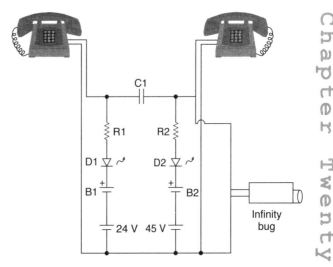

Figure 20-3 *Infinity bug test circuit. Courtesy of Talking Electronics*

less than 0.5 mA and only the whistle-detecting stage should be active. If you decide to measure the amplitude of the signal produced by the whistle-detecting stage at the collector of the transistor, you will need an oscilloscope. The measured voltage should be about 2.5 V p-p. The DC voltage across the 10 µF electrolytic capacitor that provides a turn-on voltage for the latching circuit will rise to 0.6 V and the Infinity Bug will turn on.

When the Infinity Bug is turned on, the emitter voltage on the BC 639 transistor will be about 4.5 V. This voltage is created by the value of the 120 ohm resistor as it is the load resistor for the project. The 4.5 V becomes the rail voltage for the audio stages and the only way to test the audio stages is with an oscilloscope. Since the Infinity Bug is connected to a current-limited power supply (the 470 ohm resistor in the test power supply provides current limiting) you can use a jumper on the project with fear of damaging anything.

To test the audio stages, connect a jumper between point "A" on the board and the base of the BC 639 amplifier transistor. You will be able to hear the faintest sounds in a room. If audio is not detected, measure the voltage on the collector of the first audio transistor. It should be about 2 V. The second transistor should have about 4.5 V on the collector and 4 V on the emitter. The most common fault will be a faulty connection to one of the surface-mount components or a damaged transistor. You will need an oscilloscope to detect the passage of audio through each of the stages.

There are a number of critical components in the circuit and the audio section is very important. The current taken by the audio section causes the rail voltage to drop to about 4.2 V when the bug is active. This voltage is determined mainly by the 120 ohm resistor in the emitter of the second stage. The transistor is turned on by the 10 K resistor between the base and collector and the transistor is actually a common-emitter device, although it appears to be an emitter-follower. If the 120 ohm resistor is increased, the voltage on the section rises and creates a feedback squeal. If the 22 K ohm load resistor for the microphone is decreased, a squeal is also created. If the microphone is removed, a squeal is also created. The circuit is somewhat complex as we are creating an audio signal on the same lines that are supplying the voltage to the circuit.

In the event that the circuit does not function at all, then you will have to disconnect the Infinity Transmitter from the phone and carefully inspect the circuit board to make sure you have not installed some components incorrectly. Commonly, electrolytic capacitors, diodes, or transistors are installed backwards or incorrectly causing the circuit to fail. Once you have detected the problem, you can go ahead and reconnect the Infinity Transmitter to the phone line and retest the circuit.

Chapter 21

Radio Phone Patch

Parts list

Radio Phone Patch

R1 10 K ohms, $\frac{1}{4}$ W,
   5% resistor

R2 100 ohm, $\frac{1}{4}$ W,
   5% resistor

R3 10 K potentiometer

R4 10 ohm, $\frac{1}{4}$ W,
   5% resistor

R5 1 K ohm
   potentiometer

R6, R12, R13 10 K ohm,
   $\frac{1}{4}$ W, 5% resistor

R7, R11 47 K ohms,
   $\frac{1}{4}$ W, 5% resistor

R8, R9 100 K ohms,
   $\frac{1}{4}$ W, 5% resistor

R10 100k ohms, $\frac{1}{4}$ W,
   5% resistor

C1, C8 10 μF, 35 V
   tantalum capacitor

C2, C6, C11 2.2 μF, 35 V
   tantalum capacitor

C3 .01 μF, 35 V disc
   capacitor

C4 1,000 pF, 15 V
   tantalum capacitor

C5, C12 .22 μF, 35 V
   tantalum capacitor

C7 100 μF, 15 V
   tantalum capacitor

C9, C10 .1 μF, 35 V
   ceramic capacitor

Q1 2N3904 NPN
   transistor

U1 LM324 Op-amp IC

U2 LM386 audio
   amplifier IC

T1 900 to 900 or
   600 to 600 ohm – mini
   transformer (Microtran
   T2110)

S1, S3 SPDT toggle
   switch

S2 SPST toggle switch

SPK 4-8 ohm speaker

B1 9-volt transistor
   radio battery

P1 4-pin telephone plug

Misc PC board, wire,
   hardware, battery
   holder/clip,
   enclosure, connectors.

For many years amateur radio enthusiasts have taken advantage of the telephone to extend their voice communications beyond the physical location of his/her transmitter and receiver. Such an arrangement is called a Telephone Patch, and the Federal Communications Commission has authorized this procedure to be used within the Amateur Radio and the Citizens band radio frequencies. As with all radio communications, strict rules specified by the FCC must be followed, and the telephone patch equipment interfaced with the telephone network must meet the requirements of the FCC Rules and Regulations. The Radio Phone Patch described here is a simple and easy to build circuit which will provide the means to connect and operate an amateur radio or CB radio transceiver with the telephone network (see Figure 21-1).

Figure 21-1 *Telephone phone patch circuit. Courtesy of Cengage Learning*

The Radio Phone Patch circuit operates from a common 9-V transistor battery, and should have an operating life equivalent to that provided by any of the small handheld radios. An additional feature of this telephone patch is a monitoring circuit which permits the radio operator to hear both sides of the telephone conversation.

Circuit description

Take a look at the schematic shown in Figure 21-2. The Radio Phone Patch consists of two sections of a quad operational amplifier chip, an audio power amplifier, and an isolation transformer which connects the amplifier circuitry to the telephone line. The primary of T1 is connected to the telephone line through series capacitor C2. This provides DC blocking so that only audio frequencies are impressed upon the transformer winding. The circuit composed of R1, R2, C1, and Q1 is a holding circuit which provides a DC path for the telephone line current, which is necessary to hold the line in operation when the telephone set is on the hook. The AC impedance of this circuit is very high, due to the action of C1 which does not permit any audio frequencies to be impressed upon the base of Q1. This prevents the holding circuit from loading down the line, and preserves the impedance matching action of T1.

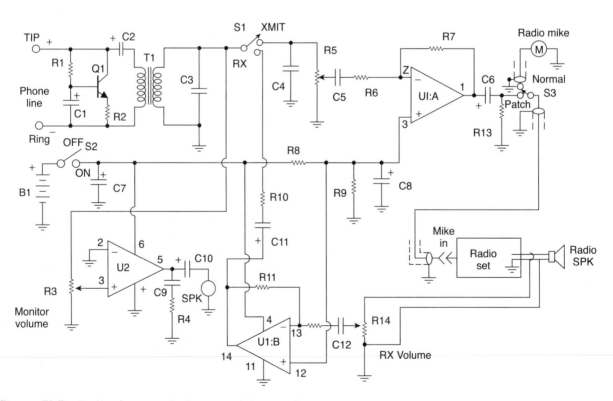

Figure 21-2 *Radio phone patch. Courtesy of Cengage Learning*

When the Radio Phone Patch is operated in transmit mode, S1 connects the transformer winding to operational amplifier U1 through the transmit volume control, R5. U1(A) has a gain of about 5, which provides sufficient drive to the microphone input of the radio transceiver through the NORMAL/PATCH switch. Thus, any audio signals present on the telephone line will be fed to the microphone input circuit of the radio transceiver when the circuit is operated in transmit mode.

During receiving mode, S1 is thrown to the receive position. Audio information received by the radio transceiver and fed to its speaker is also impressed upon the input of operational amplifier U1(B) through the receive volume control, R14. U1(B) has a gain of about 5, and drives the transformer winding through R10. Thus, in receive mode, any audio signals received by the radio transceiver will be impressed upon the telephone line through T1.

The use of a monitoring audio amplifier in the Radio Phone Patch permits hanging up the phone once the call is initiated. The monitor amplifier will reproduce both sides of the conversation and its volume can be set to a satisfactory level by means of the monitor volume control R3. U2 is a low power amplifier which will drive a small speaker.

Power for the circuit is provided by a common 9-V transistor radio battery. Current drain on the battery will depend upon the level of volume provided by the monitor speaker, and will be in the 20-mA range at low volumes.

Circuit assembly

Before we begin constructing the Radio Phone Patch, you will need to locate a clean, well-lit worktable or workbench. Next we will gather a small 25–30-W pencil tipped soldering iron, a length of 60/40 tin–lead solder, and a small canister of Tip Tinner, available at your local RadioShack store; Tip Tinner is used to condition the soldering tip between solder joints. You drive the soldering tip into the compound and it cleans and prepares the soldering tip. Next grab a few small hand tools; try to locate small flat blade and Phillips screwdrivers. A pair of needle-nose pliers, a pair of tweezers along with a magnifying glass, and a pair of end-cutters and we will begin constructing the project. Now locate your schematic diagram and parts layout diagram along with all of the components needed to build the project.

The circuitry for the Radio Phone Patch is shown in Figure 21-2 and was constructed on a printed circuit board measuring 2-¾″ × 4-¾″ and may be inserted into the cabinet of a radio transceiver or can be housed in a separate chassis box placed next to your radio transceiver

Once all the components are in front of you, you can check them off against the project parts list to make sure you are ready to start building the project. Now, locate Tables 21-1 and 21-2, and place them in front of you.

Table 21-1 illustrates the resistor code chart and how to read resistors, while Table 21-2 illustrates the capacitor code chart which will aid you in constructing the project.

Refer now to Table 21-1, which introduces the resistor color chart that will help you identify resistors. Each resistor will have either three or four color bands, which begin at one end of the resistor's body. The first colored band represents the resistor's first digit, while the second color represents the second digit of the resistor's value. The third colored band represents the resistor's multiplier value, and the fourth color band represents the resistor's tolerance value. If there is no fourth colored band then the resistor has a 20% tolerance value. If the resistor's fourth band is silver then the resistor has a 10% tolerance value and if the fourth band is gold then the resistor's tolerance value will be 5%.

Finally, we are ready to begin assembling the project, so let's get started. The prototype was constructed on a single-sided printed circuit board or PCB. With your PC board in front of you, we can now begin populating the circuit board. We are going to place and solder the lowest height components first, the resistors and diodes. First, locate our first resistor R1. Resistor R1 has a value of 10 K ohms and is a 5% tolerance type. Now take a look for a resistor whose first color band is brown, black, orange, and gold. From the chart you will notice that brown is represented by the digit (1), the second color band is black, which is represented by (0). Notice that the third color is orange and the multiplier is (1,000), so (1) (0) × 1,000 = 10,000 or 10 K ohms.

Go ahead and locate where resistor R1 goes on the PC and install it, next solder it in place on the PC board, and then with a pair of end-cutters trim the excess component leads flush to the edge of the circuit board. Next locate the remaining resistors and install them in their respective locations on the main controller

Table 21-1
Resistor color code chart

Color Band	1st Digit	2nd Digit	Multiplier	Tolerance
Black	0	0	1	
Brown	1	1	10	1%
Red	2	2	100	2%
Orange	3	3	1,000 (K)	3%
Yellow	4	4	10,000	4%
Green	5	5	100,000	
Blue	6	6	1,000,000 (M)	
Violet	7	7	10,000,000	
Gray	8	8	100,000,000	
White	9	9	1,000,000,000	
Gold			0.1	5%
Silver			0.01	10%
No color				20%

Table 21-2
Three-digit capacitor codes

pF	nF	µF	Code	pF	nF	µF	Code
1.0			**1R0**	3,900	3.9	.0039	**392**
1.2			**1R2**	4,700	4.7	.0047	**472**
1.5			**1R5**	5,600	5.6	.0056	**562**
1.8			**1R8**	6,800	6.8	.0068	**682**
2.2			**2R2**	8,200	8.2	.0082	**822**
2.7			**2R7**	10,000	10	.01	**103**
3.3			**3R3**	12,000	12	.012	**123**
3.9			**3R9**	15,000	15	.015	**153**
4.7			**4R7**	18,000	18	.018	**183**
5.6			**5R6**	22,000	22	.022	**223**
6.8			**6R8**	27,000	27	.027	**273**
8.2			**8R2**	33,000	33	.033	**333**
10			**100**	39,000	39	.039	**393**
12			**120**	47,000	47	.047	**473**
15			**150**	56,000	56	.056	**563**
18			**180**	68,000	68	.068	**683**
22			**220**	82,000	82	.082	**823**
27			**270**	100,000	100	.1	**104**
33			**330**	120,000	120	.12	**124**
39			**390**	150,000	150	.15	**154**
47			**470**	180,000	180	.18	**184**

Table 21-2—cont'd

Three-digit capacitor codes

pF	nF	µF	Code	pF	nF	µF	Code
56			**560**	220,000	220	.22	**224**
68			**680**	270,000	270	.27	**274**
82			**820**	330,000	330	.33	**334**
100		.0001	**101**	390,000	390	.39	**394**
120		.00012	**121**	470,000	470	.47	**474**
150		.00015	**151**	560,000	560	.56	**564**
180		.00018	**181**	680,000	680	.68	**684**
220		.00022	**221**	820,000	820	.82	**824**
270		.00027	**271**		1,000	1	**105**
330		.00033	**331**		1,500	1.5	**155**
390		.00039	**391**		2,200	2.2	**225**
470		.00047	**471**		2,700	2.7	**275**
560		.00056	**561**		3,300	3.3	**335**
680		.00068	**681**		4,700	4.7	**475**
820		.00082	**821**			6.8	**685**
1,000	1.0	.001	**102**			10	**106**
1,200	1.2	.0012	**122**			22	**226**
1,500	1.5	.0015	**152**			33	**336**
1,800	1.8	.0018	**182**			47	**476**
2,200	2.2	.0022	**222**			68	**686**
2,700	2.7	.0027	**272**			100	**107**
3,300	3.3	.0033	**332**			157	**157**

Code = 2 significant digits of the capacitor value + number of zeros to follow

For values below 10 pF, use "R" in place of decimal e.g. 8.2 pF = 8R2

10 pF = 100

100 pF = 101

1,000 pF = 102

22,000 pF = 223

330,000 pF = 334

1 µF = 105

PC board. Solder the resistors to the board, and remember to cut the extra lead with your end-cutters.

Next refer to Table 21-2; this chart helps to determine the value of small capacitors. Many capacitors are quite small in physical size, so manufacturers have devised a three-digit code which can be placed on the capacitor instead of the actual value. The code occupies a much smaller space on the capacitor and can usually fit nicely on the capacitor body, but without the chart the builder would have a difficult time determining the capacitor's value.

Note that capacitors generally come from two major groups or types. Non-polar types, such as the ones in this project, have no polarity and can be placed in either direction on the circuit board. The other type of capacitors are called polarized types, these capacitors have polarity and have to be inserted on the board with respect to their polarity marking in order for the circuit to operate correctly. On these types of capacitors, you will observe that the components actually have a plus or minus marking or a black or white band with a plus or minus marking on the body of the capacitors. Capacitors must be installed properly with respect to polarity in order for the circuit to work properly. Failure to install electrolytic capacitors correctly could cause damage to the capacitor as well as to other components in the circuit when power is first applied.

Now look for the capacitors from the component stack and locate capacitor C1, labeled 10 μF. Go ahead and place C1 on the circuit board and solder it in place. Now take your end-cutters and cut the excess component leads flush to the edge of the circuit board. Note that there are a number of small capacitors in the circuit that may not be labeled with their actual value, so you will have to check Table 21-2, to learn how to use the three-digit capacitor code. Capacitor C4, for example is a .001 μF capacitor but you may have to look for a three-digit code marked (102). There are a number of electrolytic capacitors in the circuit and placing them in the circuit with respect to their proper polarity is essential for proper operation of the circuit. Once you have installed these components, you can move on to installing the remaining capacitors.

The Radio Phone Patch utilizes a single conventional transistor at Q1. Transistors are generally three lead devices, which must be installed correctly if the circuit is to work properly. Careful handling and proper orientation in the circuit is critical. A transistor will usually have a Base lead, a Collector lead, and an Emitter lead. The base lead is generally a vertical line inside a transistor circle symbol. The collector lead is usually a slanted line facing towards the base lead and the emitter lead is also a slanted line facing the base lead but it will have an arrow pointing towards or away from the base lead. Now, identify the two transistors, referring to the manufacturer's specification sheets as well as the schematic and parts layout diagrams when installing the two transistors. If you have trouble installing the transistors, get the help of an electronics enthusiast with some experience. Install the transistors in their respective locations and then solder them in place. Remember to trim the excess component leads with your end-cutters.

The Radio Phone Patch circuit also contains two ICs. ICs are often static sensitive, so they must be handled with care. Use a grounded anti-static wrist strap and stay seated in one location when handling the ICs. Take out a cheap insurance policy by installing IC sockets for each of the ICs. In the event of a possible circuit failure, it is much easier to simply unplug a defective IC than trying to unsolder 14 or 16 pins from a PC board without damaging the board. ICs have to be installed correctly if the circuit is going to work properly. IC packages will have some sort of markings which will help you orient them on the PC board. An IC will have either a small indented circle, a cut-out, or notch at the top end of the IC package. Pin 1 of the IC will be just to the left of the notch or cut-out. Refer to the manufacturer's pin-out diagram, as well as the schematic, when installing these parts. Accidental reversal of pin insertion can cause the integration to fail. If you doubt your ability to correctly orient the ICs, then seek the help of a knowledgeable electronics enthusiast.

Next, locate transformer T1 and place it on the circuit board. Any small isolation transformer with 900 to 900 or 600 to 600 ohms will work fine for this circuit. Use the transformer tabs to hold the transformer in place then solder the transformer leads to the PC board, remembering to trim the excess leads.

Once the main Radio Phone Patch circuit board has been completed, take a short break and then we will inspect the PC board for possible "cold" solder joints and "short" circuits. Pick up the PC board with the foil side facing upwards toward you. Carefully inspect the solder joints, these should look smooth, clean, and bright. If any of the solder joints look dull, dark, or blobby, then you should remove the solder with a solder sucker or a solder wick and then resolder the joint all over again. Now we will inspect the PC board for possible short circuits. Short circuits can be caused from small solder blobs bridging the circuit traces or from cut component leads which can often stick to the circuit board once they have been cut. A sticky residue from solder will often trap component leads across the circuit traces. Look carefully for any bridging wires across circuit traces.

Once this inspection is complete, we can move on to installing the circuit board into an aluminum enclosure. Locate a suitable enclosure to house the circuit board.

Arrange the circuit board and the transformer on the bottom of the chassis box. Now, locate four ¼″ standoffs, and mount the PC board between the standoffs and the bottom of the chassis box. If you selected a conventional relay you will have to mount it on the chassis and wire it to the PC board. If you selected a non-PC board transformer then you will have to mount the transformer to the base of the chassis box and wire it to the circuit board. Locate a 9-V battery holder and mount it to the bottom of the chassis box. You could also elect to use a small "wall wart" 9-V power supply to power the circuit if desired.

Operating controls for the circuit R3, R5, R14, S1, and S3 are shown mounted on the printed circuit board. This method of construction is left as an option to the builder. You may want to locate these controls on the front of the panel on the enclosure, so you may elect to use standard chassis mounted controls instead.

Next you will have to install an RJ11 phone jack or a 2-position terminal strip on the front panel of the chassis to bring the phone into the chassis box. Do not forget to install a power switch; this is shown as S2, and can be mounted on the front panel of the chassis box. Finally, you will have to mount the NORMAL/PATCH switch (S3) on the front panel, as well as TRANSMIT/RECEIVE switch (S1) to the front panel of the enclosure.

Installation

When connecting the Radio Phone Patch to your radio transceiver, follow the directions shown in the schematic diagram. There are two connections which must be made to the radio transceiver, not including ground wires. These are the microphone input jack and the speaker terminals. You may wish to purchase a microphone plug and jack which match those components on your radio transceiver. This will permit you to make the connections to the radio without disturbing either the original microphone cable or the connections to the jack within the radio. Use shielded microphone wire for the microphone connections. Note that when S3 is set to the normal position, the microphone will be connected to the radio transceiver in the usual manner, permitting normal operation. When making the connection to the speaker of the radio transceiver, you may use a twisted pair of light gauge stranded wires. Be sure to connect the grounded side of

the speaker to the ground terminal of the printed circuit board. Using two wires of different color will help in avoiding misconnections.

The connections to the telephone line should be made using a standard 4-prong telephone plug, or the new modular connector which is now standard equipment with many telephone companies. Either of these types of connectors is readily available from electronics supply houses, as well as those outlets which sell telephones to the general public. Before making a connection between the printed circuit and telephone plug, check the polarity of the telephone line with a DC voltmeter. Once you have done this you will be able to properly connect the Radio Phone Patch to the telephone line. Note: Do not connect the Radio Phone Patch to the telephone line unless you are placing it in operation. If the unit is left connected to the line it will prevent normal operation of the telephone.

Operation

The three operating volume controls in the Radio Phone Patch can be adjusted while the unit is in operation. As a start, set each control to mid-position. If your telephone is the type which uses a rotary dial you will have to dial the number you are calling before you connect the Radio Telephone Patch to the telephone line. If your telephone line is equipped for Touch Tone dialing, you can connect the Radio Phone Patch to the telephone line either before you dial, or after. Once you have dialed the number and connected the Radio Phone Patch to the telephone line, you can hang up the telephone whenever it is no longer needed for communication. The holding circuit of the Radio Phone Patch will keep the telephone line in operation.

Set the NORMAL/PATCH switch, S3, to the Patch position. The only other operating switch will be the TRANSMIT/RECEIVE switch, S1, which will have to be operated manually as you transmit the telephone conversation and receive the reply from the radio transceiver. Since your radio transceiver also has a Transmit/Receive button on the microphone, you will have to operate these switches in unison.

If you have a modulation indicator on your radio transceiver, you can use it to give you an approximate indication of the proper setting of the transmit volume control, R5. Otherwise you will have to rely on information provided by the remote party receiving the radio broadcast as to whether or not the level of modulation is sufficient.

Before adjusting the receive volume control, R14, set your radio transceiver's volume control for a comfortable volume setting. Use the squelch control to eliminate receiver noise when no signal is present. By speaking into your telephone handset, you can ask the party at the other end of the telephone line if the received volume from the radio transceiver is sufficient. This will enable you to adjust R14 for proper volume on the telephone line signal.

Once the receive and transmit volume controls are set to a satisfactory volume level for both parties, you can hang up your telephone and use the monitor feature of Radio Phone Patch to hear both sides of the conversation as you operate the receive/transmit switches. Set the monitor volume control, R3, for a comfortable volume setting. After you have used your Radio Phone Patch for the first time, you can leave the three volume control settings as they are. The next time you operate the unit they will probably need little or no readjustment. When the telephone patch conversation is

terminated, be sure to disconnect the Radio Phone Patch from the telephone line. Failure to do this will result in an inoperative telephone.

Before placing the Radio Phone Patch on the air, read and obey the following rules which are spelled out under FCC rules: (a) You may connect your radio transmitter to a telephone if you comply with all of the following: (1) You, or someone authorized to operate under your license, must be present at your radio station and must: (i)Manually make the connection (the connection must not be made by remote control); (ii) Supervise the operation of the transmitter during the connection; (iii) Listen to each communication during the connection; and (iv) Stop all communications if there are operations in violation of these rules. (2) Each communication during the telephone connection must comply with all of these rules. (3) You must obey any restriction that the telephone company places on the connection of a radio transmitter to a telephone. (b) The radio transmitter you connect to a telephone must not be shared with any other radio station. (c) If you connect your radio transmitter to a telephone, you must use a phone patch device which has been registered with the FCC.

Chapter 22

Telephone Phone Intercom

Parts list

Parts Bin

Telephone Phone Intercom

R1   600 ohms, $\frac{1}{4}$ W,
  5% resistor

R2, R3   100 K ohm, $\frac{1}{4}$ W,
  5% resistor

R4   300 K ohm, $\frac{1}{4}$ W,
  5% resistor

R5, R6, R10, R11, R13
  56 K ohms, $\frac{1}{4}$ W,
  5% resistor

R7   10 ohms, $\frac{1}{4}$ W,
  5% resistor

R8   220 K ohm, $\frac{1}{4}$ W,
  5% resistor

R9   10 K ohm, $\frac{1}{4}$ W,
  5% resistor

R12   560 ohm, $\frac{1}{4}$ W,
  5% resistor

C1   .01 μF, 200 volts
  ceramic disc capacitor

C2, C4, C5, C6, C7
  .01 μF, 50 volts
  ceramic disc capacitor

C3   .1 μF, 35 V ceramic
  capacitor

C8   22 μF, 16 V
  tantalum

C7   .1 μF, 50 V ceramic
  disc capacitor

C9   1,000 μF, 16 V
  electrolytic capacitor

C10   10 μF, 35 V
  electrolytic capacitor

U1   Teltone 8870-01
  Touch Tone Decoder
  (C.P. Claire Co.)

U2   MC14049 hex
  inverting buffer

U3   MC14013B dual D-type
  flip-flop

U4   LM555N timer

U5   LM7805 – 5-V
  regulator

Q1, Q2, Q3   MPSA14
  transistor

D3, 34, 37, 38   1N4148
  silicon diodes

31, D2, D5   1N4004
  silicon diode

LED1   red LED

BR1   silicon diode
  bridge

BZ   Piezo buzzer

XTL   3.58 MHz crystal

RY1   4PDT relay 12 V,
  1-amp coil –
  communications relay

T1   600/600 ohm audio
  isolation transformer

J1, J2   RJ11 phone jacks

Misc   PC board, IC
  sockets, wire,
  12 V- 500 mA
  plug-in transformer,
  enclosure, etc.

Alert Module

R1 2.2 megohm, $\frac{1}{4}$ W, 5%

R2, R4, R8 220 K ohm, $\frac{1}{4}$ w, 5%

R3 10 megohms, $\frac{1}{4}$ W, 5%

R5 56 K ohm, $\frac{1}{4}$ W, 5%

R6 33 K ohm, $\frac{1}{4}$ W, 5%

R7 680 K ohm, $\frac{1}{4}$ W, 5%

R9, R10 100 K ohm, $\frac{1}{4}$ W, 5%

R11 300 K ohm, $\frac{1}{4}$ W, 5%

C1 .01 µF, 500 V ceramic disc capacitor

C2 1 µF, 16 V electrolytic capacitor

C3 .1 µF, 35 V ceramic capacitor

U1 Teltone 8870-01 Touch Tone decoder (C.P. Claire Co.)

Q1, Q2, Q4 MPSA14 transistor

Q3 2N5401 transistor

D1, D2, D3 1N4004 silicon diode

D4, D5 1N4148 silicon diode

XTL 3.58 MHz color burst crystal

BZ piezo buzzer

B1 9-volt battery

PL1 4-wire modular phone plug

Misc PC board, IC sockets, wire, battery clip/holder, enclosure, etc.

Use your home telephones as a home intercom system. It is quick and easy to add an intercom feature to your existing home telephones. Intercoms have been around for many years, providing a valuable tool in communications for home and industry. Unfortunately these systems either require added hardware or hours of labor installing wires. Responding to demand, many manufacturers have incorporated the intercom as an added feature in their telephones. However, even replacing your existing equipment is an expense that usually outweighs the justification.

The home phone intercom project, shown in Figure 22-1, may be used concurrently with any touch-tone phone system, and it provides features that make it practical, easy to use, and inexpensive. Because it connects to your existing telephone equipment, there are no unsightly boxes to clutter up your desk.

How it works

To engage the intercom at any time, all you have to do is pick up any phone and press the "#" key that causes one or more alert nodules to sound an alarm, signaling other people in the home to pick up a phone. If you answer a call that comes in for someone else, pressing the "#" key will place the call on hold and the alarm will sound on the alert modules, signaling someone else in the home to pick up the phone. That someone else may then release the call on hold by pressing the

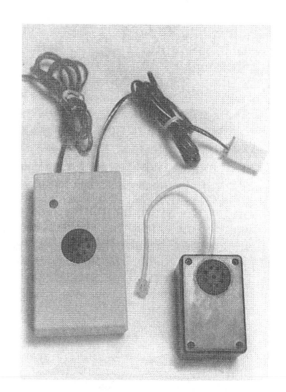

Figure 22-1 *Phone intercom components. Courtesy of Poptronix*

"#" key or talk to you in private before answering the call. Since the system is voice operated (VOX), the intercom will disengage and return to normal operation after approximately 30 s.

The home phone intercom takes advantage of a device called a "network interface." Installed in most newer homes over the past 10 years; despite the complicated name it is simply a connector box that separates the outside phone line from your internal wiring (see Figure 22-2).

Since a network interface is used by the phone company to determine whether problems are internal or external, a substantial premium service charge may be imposed if you do not have one. Therefore, it is highly recommended that one be installed even if not for this project.

Theory of operation

Glance at the Intercom Controller board in Figure 22-3 and at the schematic depicted in Figure 22-4.

In the standby mode, relay RY1 is not energized and the only connection to the phone line is the coupling-capacitor C1. The Teltone 8870-01 (U1) is a telephone tone decoder whose BCD output is dependent on which tones are present at pin 9 (see decoder diagram in Figure 22-5).

The two most significant bits (pins 16 and 17) will only be high during a "#" key depression. The high on pin 7 of U4 and pin 4 of U2 allow Q3 to turn on,

thereby energizing the audible alarm. The alarm will remain on for as long as the "#" key is depressed.

Pin 6 of U2 is now low which sends a clock pulse to U3. That transfers the high at U3 pin 2 to the output at pin 1, which energizes the relay through Q1, R5, and U2. The trigger input of U4, also being low, starts the VOX-timer U4. The clock pulse to U3 is delayed slightly by R7 and C2 until the output of U4 pin at 3 has enough time to remove the reset signal at U3 pin 4. When the "#" is released, pin 11 of U3 goes high and toggles the output of U3 pin 13 to a low state, preventing the alarm to sound during the next depression of the "#" key (manual turn off).

If conversation continues, U4 is prevented from time-out through C7 and Q2. If conversation stops for more than 30 s, C4 charges to the threshold voltage at U4 pin 6 and pin 3 return to a low state. That places a high on the reset pin (pin 4) and the set pin (pin 8) of U3, which turns off RY1, returning the system to the standby mode. Any calls in process are held by maintaining central office loop current through R1, while in intercom mode.

Alert module

An Alert module is placed at every room near the room's telephone. The Alert module is illustrated in Figure 22-6 and in the schematic at Figure 22-7; it plugs into any home phone jack, and can be powered by either a 7.2-V rechargeable Ni-Cd or 9-V alkaline battery.

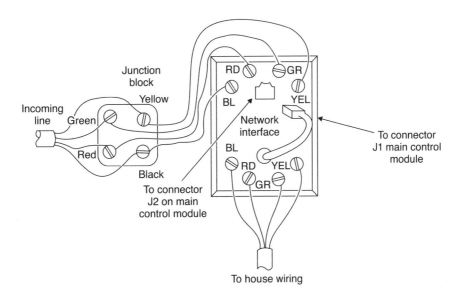

Figure 22-2 *Network interface*

Figure 22-3 *Telephone intercom main circuit board. Courtesy of Poptronix*

However the circuit is only active during an off-hook condition to conserve the battery during normal operation. (There are approximately 50 V on the phone line). That allows base current to flow through R1 which turns on Q1 and holds Q2 in an off state preventing power to U1 and the audible alarm. When the line voltage falls, as evident in an off-hook condition, Q2 turns on, thereby placing U1 in the standby mode.

Telephone tones are decoded by U1 as previously discussed. A "#" key activation will activate the alarm. Diode D1 is important in that it protects U1 and the rest of the semiconductors from damage when the AC ringing voltage is present on the phone line. If you are using a rechargeable battery charging current is supplied through R3 during on-hook conditions.

Circuit assembly

Before we begin constructing the Telephone Phone Intercom, you will need to locate a clean, well-lit worktable or workbench. Next we will gather a small 25–30-W pencil tipped soldering iron, a length of 60/40 tin–lead solder, and a small canister of Tip Tinner, available at your local RadioShack store; Tip Tinner is used to condition the soldering tip between solder joints. You drive the soldering tip into the compound and it cleans and prepares the soldering tip. Next grab a few small hand tools, try to locate small flat blade and Phillips screwdrivers. A pair of needle-nose pliers, a pair of tweezers along with a magnifying glass, and a pair of end-cutters and we will begin constructing the project.

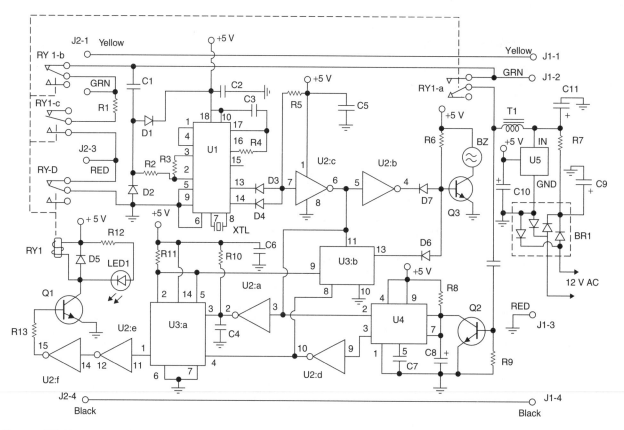

Figure 22-4 *Phone intercom main controller schematic. Courtesy of Poptronix*

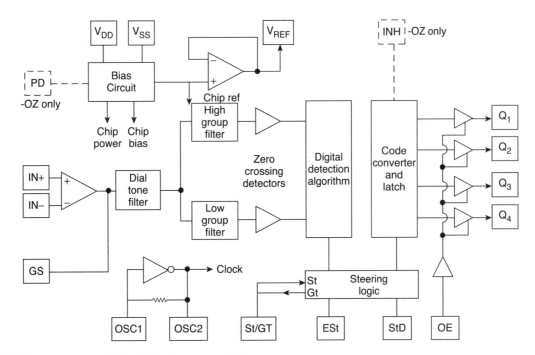

Figure 22-5 *Teltone M-8870-01/02. Courtesy of Teletone*

Figure 22-6 *Telephone intercom remote ring board. Courtesy of Poptronix*

Now locate your schematic diagram and parts layout diagram along with all of the components needed to build the project.

The Home Phone Intercom controller is built on one PC board, and the alert modules are built on separate boards for each telephone location. Determine the number of alert modules you will need including the master control unit. The modules are loud enough to cover approximately 1,000 square feet each, even when placed behind furniture. It is recommended that one module be installed in a central location on each phone of your house.

Once all the components are in front of you, you can check them off against the project parts list to make sure you are ready to start building the project. Now, locate Tables 22-1 and 22-2, and place them in front of you.

Table 22-1 illustrates the resistor code chart and how to read resistors, while Table 22-2 illustrates the capacitor code chart which will aid you in constructing the project.

Finally, we are ready to begin assembling the project, so let's get started. The prototype project was constructed on a single-sided printed circuit board or PCB. With your PC board in front of you, we can now begin populating the circuit board. We are going to place and solder the lowest height components first, such as the resistors and the diodes.

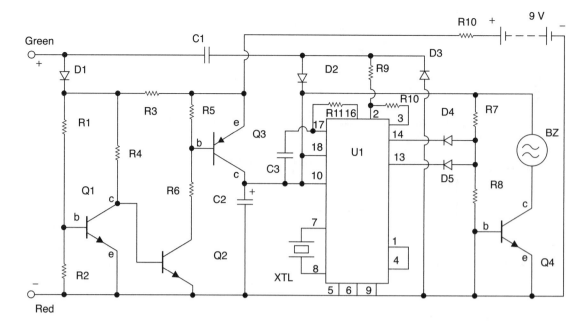

Figure 22-7 *Alert module. Courtesy of Poptronix*

Refer now to Table 22-1 which depicts the resistor color codes; this chart will help you identify each resistor before you mount them on the PC board. Note that each resistor will have either three or four color bands, which begin at one end of the resistor's body. The first colored band represents the resistor's first digit, while the second color represents the second digit of the resistor's value.

The third colored band represents the resistor's multiplier value, and the fourth color band represents the resistor's tolerance value. If there is no fourth colored band then the resistor has a 20% tolerance value. If the resistor's fourth band is silver then the resistor has a 10% tolerance value and if the fourth band is gold then the resistor's tolerance value will be 5%.

Table 22-1

Resistor color code chart

Color Band	Ist Digit	2nd Digit	Multiplier	Tolerance
Black	0	0	1	
Brown	1	1	10	1%
Red	2	2	100	2%
Orange	3	3	1,000 (K)	3%
Yellow	4	4	10,000	4%
Green	5	5	100,000	
Blue	6	6	1,000,000 (M)	
Violet	7	7	10,000,000	
Gray	8	8	100,000,000	
White	9	9	1,000,000,000	
Gold			0.1	5%
Silver			0.01	10%
No color				20%

Table 22-2

Three-digit capacitor codes

pF	nF	μF	Code	pF	nF	μF	Code
1.0			**1R0**	3,900	3.9	.0039	**392**
1.2			**1R2**	4,700	4.7	.0047	**472**
1.5			**1R5**	5,600	5.6	.0056	**562**
1.8			**1R8**	6,800	6.8	.0068	**682**
2.2			**2R2**	8,200	8.2	.0082	**822**
2.7			**2R7**	10,000	10	.01	**103**
3.3			**3R3**	12,000	12	.012	**123**
3.9			**3R9**	15,000	15	.015	**153**
4.7			**4R7**	18,000	18	.018	**183**
5.6			**5R6**	22,000	22	.022	**223**
6.8			**6R8**	27,000	27	.027	**273**
8.2			**8R2**	33,000	33	.033	**333**
10			**100**	39,000	39	.039	**393**
12			**120**	47,000	47	.047	**473**
15			**150**	56,000	56	.056	**563**
18			**180**	68,000	68	.068	**683**
22			**220**	82,000	82	.082	**823**
27			**270**	100,000	100	.1	**104**
33			**330**	120,000	120	.12	**124**
39			**390**	150,000	150	.15	**154**
47			**470**	180,000	180	.18	**184**
56			**560**	220,000	220	.22	**224**
68			**680**	270,000	270	.27	**274**
82			**820**	330,000	330	.33	**334**
100		.0001	**101**	390,000	390	.39	**394**
120		.00012	**121**	470,000	470	.47	**474**
150		.00015	**151**	560,000	560	.56	**564**
180		.00018	**181**	680,000	680	.68	**684**
220		.00022	**221**	820,000	820	.82	**824**
270		.00027	**271**		1,000	1	**105**
330		.00033	**331**		1,500	1.5	**155**
390		.00039	**391**		2,200	2.2	**225**
470		.00047	**471**		2,700	2.7	**275**
560		.00056	**561**		3,300	3.3	**335**
680		.00068	**681**		4,700	4.7	**475**
820		.00082	**821**			6.8	**685**
1,000	1.0	.001	**102**			10	**106**
1,200	1.2	.0012	**122**			22	**226**
1,500	1.5	.0015	**152**			33	**336**
1,800	1.8	.0018	**182**			47	**476**
2,200	2.2	.0022	**222**			68	**686**

(Continued)

Table 22-2—cont'd

Three-digit capacitor codes

pF	nF	µF	Code	pF	nF	µF	Code
2,700	2.7	.0027	**272**			100	**107**
3,300	3.3	.0033	**332**			157	**157**

Code = 2 significant digits of the capacitor value + number of zeros to follow

For values below 10 pF, use "R" in place of decimal e.g. 8.2 pF = 8R2

10 pF = 100

100 pF = 101

1,000 pF = 102

22,000 pF = 223

330,000 pF = 334

1 µF = 105

Now, locate our first resistor R1. Resistor R1 has a value of 600 ohms and is a 5% tolerance type. Now take a look for a resistor whose first color band is blue, black, brown, and gold. From the chart you will notice that blue is represented by the digit (6), the second color band is black, which is represented by (0). Notice that the third color is brown and the multiplier is (10), so (6) (0) × 10 = 600 ohms.

Go ahead and locate where resistor R1 goes on the PC and install it, next solder it in place, and then with a pair of end-cutters trim the excess component leads flush to the edge of the circuit board. Next locate the remaining resistors and install them in their respective locations on the main controller PC board. Solder the resistors to the board, and remember to cut the extra lead with your end-cutters.

Note that capacitors generally come from two major groups or types. Non-polar types, such as the ones in this project, have no polarity and can be placed in either direction on the circuit board. The other type of capacitors are called polarized types, these capacitors have polarity and have to be inserted on the board with respect to their polarity marking in order for the circuit to operate correctly. On these types of capacitors, you will observe that the components actually have a plus or minus marking or a black or white band with a plus or minus marking on the body of the capacitors. Capacitors must be installed properly with respect to polarity in order for the circuit to work properly. Failure to install electrolytic capacitors correctly could cause damage to the capacitor as well as to other components in the circuit when power is first applied.

Many capacitors are quite small in physical size, so manufacturers have devised a three-digit code which can be placed on the capacitor instead of the actual value. The code occupies a much smaller space on the capacitor and can usually fit nicely on the capacitor body, but without the chart the builder would have a difficult time determining the capacitor's value.

Now look for the capacitors from the component stack and locate capacitors C1 and C2, they are small capacitors labeled (103). Refer to the Table 22-2 to find code (103) and you will see that it represents a capacitor value of .01 µF. Go ahead and place C1 and C2 on the circuit board and solder them in place on the circuit board. Now take your end-cutters and cut the excess component leads flush to the edge of the circuit board. Once you have installed these components, you can move on to installing the remaining capacitors.

The main Telephone Intercom board contains five silicon diodes. Diodes, as you will remember, will always have some type of polarity markings, which must be observed if the circuit is going to work correctly. You will notice that each diode will have a black or white colored band at one edge of its body. The colored band represents the diode's cathode lead. Check your schematic and parts layout diagrams when installing these diodes on the circuit board. Place all the diodes on the circuit board in their respective locations and solder them in place. Diode bridge BR1 on the right side of the schematic can be installed now. After soldering in the diodes and the diode bridge, remember to cut the excess component leads with your end-cutters, flush to the edge of the circuit board.

The Telephone Intercom main board utilizes three conventional transistors. Transistors are generally three lead devices, which must be installed correctly if the circuit is to work properly. Careful handling and proper orientation in the circuit is critical. A transistor will usually have a Base lead, a Collector lead, and an Emitter lead. The base lead is generally a vertical line inside a transistor circle symbol. The collector lead is usually a slanted line facing towards the base lead and the emitter lead is also a slanted line facing the base lead but it will have an arrow pointing towards or away from the base lead. Now, identify the two transistors, referring to the manufacturer's specification sheets as well as the schematic and parts layout diagrams when installing the two transistors. If you have trouble installing the transistors, get the help of an electronics enthusiast with some experience. Install the transistors in their respective locations and then solder them in place. Remember to trim the excess component leads with your end-cutters.

The main Intercom Controller circuit also contains five ICs. ICs are often static sensitive, so they must be handled with care. Use a grounded anti-static wrist strap and stay seated in one location when handling the integrated circuits. Take out a cheap insurance policy by installing IC sockets for each of the ICs. In the event of a possible circuit failure, it is much easier to simply unplug a defective IC, than trying to unsolder 14 or 16 pins from a PC board without damaging the board. ICs have to be installed correctly if the circuit is going to work properly. IC packages will have some sort of markings which will help you orient them on the PC board. An IC will have either a small indented circle, a cutout, or notch at the top end of the IC package. Pin 1 of the IC will be just to the left of the notch, or cutout. Refer to the manufacturer's pin-out diagram, as well as the schematic when installing these parts. If you doubt your ability to correctly orient the ICs, then seek the help of a knowledgeable electronics enthusiast.

The Intercom Controller circuit utilizes a single LED at LED1, which also must be oriented correctly if the circuit is to function properly. An LED has two leads, a cathode and an anode lead. The anode lead will usually be the longer of the two leads and the cathode lead will usually be the shorter lead, just under the flat side edge of the LED; this should help orient the LED on the PC board. You may want to lead about ½ lead length on the

LED, in order to have the LED protrude through the enclosure box. Solder the LED to the PC board and then trim the excess leads.

Now, locate the crystal at XTL and install it in its proper location between pins 7 and 8 of U1. Next, locate the piezo buzzer at BZ which is connected to the collector of Q3. Locate and install the mini transformer at T1, which is connected to the input of U5. Finally go ahead and install the 4PDT communications relay. If you purchased a PC board relay you can install it on the PC board, but if you purchased a conventional relay you will have to mount it on the chassis and wire it to the circuit board.

Once the main Intercom Controller circuit board has been completed, take a short break and then we will inspect the PC board for possible "cold" solder joints and "short" circuits. Pick up the PC board with the foil side facing upwards toward you. Carefully inspect the solder joints, these should look smooth, clean, and bright. If any of the solder joints look dull, dark, or blobby, then you should remove the solder with a solder sucker or a solder wick and then resolder the joint all over again. Now we will inspect the PC board for possible short circuits. Short circuits can be caused from small solder blobs bridging the circuit traces or from cut component leads which can often stick to the circuit board once they have been cut. A sticky residue from solder will often trap component leads across the circuit traces. Look carefully for any bridging wires across circuit traces.

Once the inspection is complete, we can move on to installing the circuit board into an aluminum enclosure. Locate a suitable enclosure to house the circuit board. Arrange the circuit board and the transformer on the bottom of the chassis box. Now, locate four ¼ inch standoffs, and mount the PC board between the standoffs and the bottom of the chassis box. If you selected a convention relay you will have to mount it on the chassis and wire it to the PC board. Mount the transformer to the base of the chassis box and wire it to the circuit board.

Mount a coaxial power jack on the rear panel of the chassis box for the AC power supply at BR1's input. The 12 V AC power supply could be a low cost "wall wart" type power supply. Mount the status LED on the front panel if desired. Next you will have to install an RJ11 phone jack or a 2-position terminal strip on the front panel of the chassis to bring the phone into the chassis box. Now you are ready to connect the main controller to the phone line, but before we do, you will need to construct the desired number of Alert Modules.

Alert module construction

Next refer to the Alert Module schematic shown in Figure 22-7. First, install all the resistors on the PC board then solder them in place and trim the excess component leads with you end-cutters. Next install all of the capacitors on the Alert Module and then solder them in place on the PC board. Next install the four silicon diodes, remembering they are polarity sensitive, so care must be taken to install them correctly on the PC. Now locate the four transistors and install them on the PC board. Refer to the schematic, parts layout diagram, and manufacturer's specification sheets when installing them. Transistors must be installed properly the first time, so that they are damage free when power is first applied to the circuit. Now install the piezo buzzer at BZ; it is connected between the collector of Q4 and resistor R7.

Finally, install an IC socket for the touch-tone decoder at U1. Locate the touch-tone decoder chip, and insert it into the socket, paying particular attention to the positioning notch when inserting the IC into its socket, in order to avoid damage to the IC upon power up.

The values for resistors R3 and R10 depend on what type of battery you are using. Use l0 megohms for R3 and 82 ohms for R10 for a 9-V alkaline battery. If a rechargeable battery is used, change the value of R3 to 82 K and R10 to l0 ohms.

The Alert Modules will need to be placed in a suitable enclosure next to a phone jack in each desired room installation. A small plastic box can be used to mount each of the Alert Modules. Make sure you have enough room for the 9-V battery and the buzzer. You will want to either install a modular phone jack on the enclosure or you will need to make up a cable with a modular phone plug between the circuit and the phone line for each of the Alert Modules. Once the Alert Modules are installed in their cases, with batteries connected, you will be ready to connect each one to the phone line and test the modules.

Installation and check out

Determine where your network interface is by locating the area where the phone line enters the house. In some cases, the device is mounted on the outside. It is a wall box with a short wire loop connecting to a modular jack. Refer back to Figure 22-2 on how to install a network interface if it is not already present. During the next few steps, your phone system will be inoperative until installation is complete.

Disconnect the short wire from the jack on the network interface. Connect that wire to J1 on the main control module. Make sure all phones on the same extension are on-hook, and connect the controller to a 12-V AC source. Pick up a telephone receiver and press the "#" key. The alarm will sound and the LED should be on. You will also be able hear yourself talk through the handset. Hang up the phone. The LED should remain on for approximately 30 s before turning off.

Connect the remaining wire from the main control module to the jack on the network interface. Pick up the receiver again and initiate a call to determine normal operation. If you are unable to dial out, the red and green wires (tip and ring) have been reversed somewhere in the system. Remember that positive phone line voltage must be present at J2 pin 2. It may be necessary to toggle the "#" key once or twice to get everything going when the system is first installed or after a power failure. Install the alert modules and determine the correct polarity by measuring a positive voltage at the anode of D1. Reverse the wires in the module if it turns out to be necessary. If a rechargeable battery is used, you should allow it to charge for at least 24 hours before activating the system.

Now, you can test each of the Alert Modules in each location and you are ready to begin using your new Telephone Intercom system. The Telephone Intercom will become your favorite phone accessory, saving your voice and many footsteps, or having to yell or walk around the house trying to let someone know a phone call is for them. The Telephone Intercom will serve you for many years as a faithful servant around the house.

Chapter 23

Speaker-Phone Project

Parts list

Parts Bin

Speaker-Phone

R1, R6, R14, R20
10 K ohm, $\frac{1}{4}$ W,
5% resistor

R2 1 K ohm, $\frac{1}{4}$ W,
5% resistor

R3, R9, R13, R19
4.7 K ohm, $\frac{1}{4}$ W,
5% resistor

R4 alternate R10
270 ohm, $\frac{1}{4}$ W,
5% resistor

R5 68 K ohm, $\frac{1}{4}$ W,
5% resistor

R7, R12 47 K ohm,
$\frac{1}{4}$ W, 5% resistor

R8 820 ohm, $\frac{1}{4}$ W,
5% resistor

R10 100 ohm, $\frac{1}{4}$ W,
5% resistor

R11, R15, R16, R17
100 K ohm, $\frac{1}{4}$ W,
5% resistor

R18 10 K potentiometer
(PC board type)

C1 100 pF disc, 50 V
(100 or 101)

C9, C10, C13, C14 .01 µF
disc, 50 V (103)

C3, C18 .05 or .047 µF
disc, 50 V (473)

C28 .1 µF disc, 50 V
(104)

C2, C4, C6, C7, C8,
C17, C22, C23 .47 µF,
50 V electrolytic
capacitor

C5, C11, C15, C19, C20,
C21, C29 2.2 µF,
50 V electrolytic
capacitor

C16, C26 47 µF, 50 V
electrolytic capacitor

C24 220 µF, 50 V
electrolytic capacitor

C12, C25, C27 470 µF
or 330 µF, 50 V
electrolytic capacitor

U1 MC34118 28-pin DIP
Speaker-phone IC

U2 LM386 8-pin DIP
audio amplifier IC

D5 6.2 volt zener diode

D1, D2, D3, D4, D6,
D7 1N4002 black
epoxy rectifier
diode

MK1 miniature
microphone element

T1, T2 1,000-to-8 ohm
audio transformer

S1, S2 PC mount DPDT
push-switch

J1 modular RJ-11
telephone line jack

J2 subminiature
    phone/speaker jack

SPK 8-ohm-3″ speaker

Misc PC board, IC
    sockets, wire, battery
    holder, battery clip,
    enclosure, hardware, etc

Kit available from:

Ramsey Electronics

590 Fishers Station Drive

Victor, NY 14564

585-924-4560

The Speaker-Phone Project efficiently replaces a cluster of add-on phone gadgets to give you what you really want in the shop or office: The ability to answer incoming calls and continue what you are doing while talking from anywhere in the room.

The Speaker-Phone Project is great for family and conference calling, or you can connect two speaker-phones together for a completely hands-free intercom; this is ideal for linking the shop with the rest of the house. Imagine how handy it would be to simply answer back from anywhere in your shop without stopping what you are doing. It's like linking two rooms together. The great features of this project include the circuit: (1) Needs no battery when connected to the phone line. (2) Can be simply connected to the phone line by itself without using an extra telephone. For intercom use, a regular 9-V battery powers the unit. It includes built-in electret microphone; just add your choice of speaker.

Figure 23-1 *Speaker-phone. Courtesy of Ramsey Electronics*

The Speaker-Phone Project shown in Figure 23-1 gives you full, state-of-the-art speaker-phone technology.

First, it is a complete stand alone telephone for incoming calls. Use some other phone or dialer right along with it only if you have a frequent need to initiate those unique Speaker-phone-style calls. Otherwise, you can use your speaker-phone to answer all incoming calls, handling them while you work or play.

A pair of speaker-phone units can make a great hands-free intercom system for which you will imagine dozens of practical applications. You will figure out how to set them up so they can serve as intercoms and convenient speaker-phones.

How does a speaker-phone work?

You must be able to turn the speaker up to normal room volume, and the microphone circuit must be sensitive enough for you to speak from anywhere in a small room. How would you do it? Remember: No switches, just two wires, and no squealing "feedback"! To put it very simply, the speaker-phone is "smart": it knows when to be idle, when to transmit, when to receive, when to mute, when to attenuate, and what to do about background noise. It even recognizes the difference between steady background noise and the phone line's dial tone. The Speaker-phone block diagram, illustrated in Figure 23-2, was designed around a high-tech Motorola MC65118 Speaker-phone IC.

Motorola explains the MC34118 this way: The fundamental difference between the operation of a speaker-phone and a handset is that of half-duplex versus full-duplex.

The handset is full duplex, since conversation can occur in both directions (transmit and receive) simultaneously. A Speaker-phone has higher gain levels in both paths, and attempting to converse full duplex results in oscillatory problems due to the loop that exists within the system. The loop is formed by the receive and transmit paths, the hybrid, and the acoustic coupling (speaker to microphone.)

The most practical solution used to date is to design the speaker-phone to function in a half-duplex mode, i.e. only one person speaks at a time, while the other listens. To achieve this requires a circuit which can detect who is talking, switch on the appropriate path

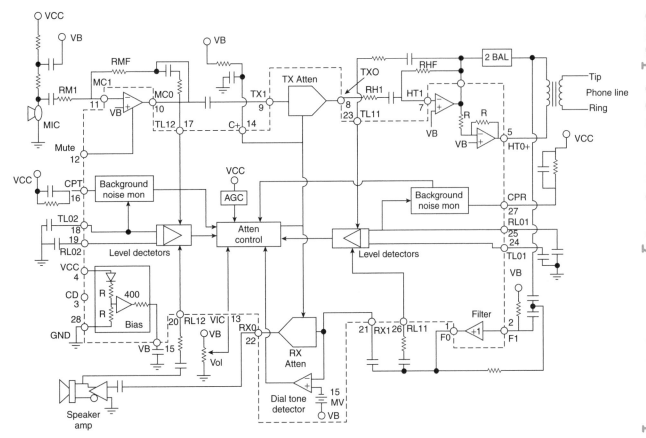

Figure 23-2 *MC34118 Block diagram. Courtesy of Motorola Inc*

(transmit or receive), and switch off (attenuate) the other path. In this way, the loop gain is maintained less than unity. When the talkers exchange function, the circuit must quickly detect this, and switch the circuit appropriately. By providing speech level detectors, the circuit operates in hands-free mode, eliminating the need for a push-to-talk switch. The MC34118 provides the necessary level detectors, attenuators, and switching control for a properly operating Speaker-phone. The Speaker-phone IC also provides background noise monitors which make the circuit insensitive to room as well as line noise, hybrid amplifiers for interfacing to Tip and Ring, i.e. the phone line connections.

Circuit description

The Speaker-phone schematic diagram, in Figure 23-3 also provides a complete diagram of the MC34118's functions. How self-evident the circuit functions seem to you depends on your level of electronics experimenting or learning. Here are the highlights of what makes the SP1 tick. S1(A) and a jumper from B to A connect the

circuit to the phone line's two wires. There is nothing hi-tech about calling these two wires "tip" and "ring." That is our inheritance from manual switchboard days. Tip and ring referred to parts of the plug and patch cord sets handled by the operator. As mentioned, S1(A) performs the same function as a conventional telephone "hook" switch. As soon as S1 is pressed in, the central office equipment is able to sense that the Speaker-phone is connected to the line or "loop."

The phone line carries both audio energy (AC) and DC voltage. T1 and T2 couple the audio signals to U1's hybrid amplifiers (pins 5 and 6). In telephone terminology, "hybrid" is a circuit that divides a single transmission channel into two, one for each direction. Diodes D1-D4 are arranged to make the input insensitive to DC polarity; the circuit gets correct DC polarity no matter which way the phone line is connected! IC U2 is a self-contained audio amplifier capable of outputs up to 400 mW, quite sufficient to drive quality speakers to room volume. The LM386 audio amplifier is connected directly to a speaker for conversation listening.

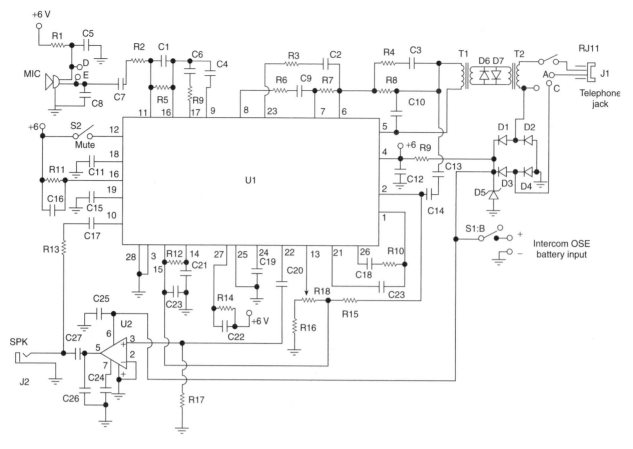

Figure 23-3 *Speaker-phone circuit. Courtesy of Ramsey Electronics*

Switch S2 simply activates the microphone amplifier muting feature built into the Speaker-phone IC. Pin 12 is ordinarily low (0 voltage) for normal operation. The microphone input circuitry is set up to accommodate both self-powered microphones and those which require a small DC voltage for proper operation. Hands-free intercom operation is the same as half-duplex Speaker-phone operation except that a source of DC voltage is required. This is made easy by S1(B) and a jumper from A to C. Perhaps the easiest overview of the circuit would be the following pin by pin notes on the MC34118 (see Table 23-1).

Circuit assembly

Before we begin constructing the Speaker-phone, you will need to locate a clean, well-lit worktable or workbench. Next we will gather a small 25–30-W pencil tipped soldering iron, a length of 60/40 tin–lead solder, and a small canister of Tip Tinner, available at your local RadioShack store; Tip Tinner is used to condition the soldering tip between solder joints. You drive the soldering tip into the compound and it cleans and prepares the soldering tip. Next grab a few small hand tools; try to locate small flat blade and Phillips screwdrivers. A pair of needle-nose pliers, a pair of tweezers along with a magnifying glass, and a pair of end-cutters and we will begin constructing the project. Now locate your schematic diagram and parts layout diagram along with all of the components needed to build the project.

Once all the components are in front of you, you can check them off against the project parts list to make sure you are ready to start building the project. Now, locate Tables 23-2 and 23-3, and place them in front of you. Table 23-2 illustrates the resistor code chart and how to read resistors, while Table 23-3 illustrates the capacitor code chart which will aid you in constructing the project.

Finally, we are ready to begin assembling the project, so let's get started. The prototype was constructed on a single-sided printed circuit board or PCB. With your

Table 23-1

Speaker-phone IC pin descriptions

IC Pin Descriptions

1. Filter output
2. Filter input
3. Chip disable (not used in SP1design)
4. Vcc (DC supply voltage)
5. Hybrid amplifier output #2
6. Hybrid amplifier output #1
7. Input to hybrid amplifier #1
8. TX attenuator output
9. TX attenuator input
10. Microphone amplifier output
11. Microphone amplifier input
12. Mute input (mike amplifier muted by + DC at pin 12)
13. Volume control input
14. Response time to switch between transmit and receive established by R15 and C21
15. Supplies bias to volume control, filtered by C24
16. Time constant for TX BNM set by R11 and C16
17. Input to TX level detector on mike/speaker side
18. Output of TX level detector and input to TX BNM
19. Output of RX level detector on mike/speaker side
20. Input to RX level detector on mike/speaker side
21. Input to RX attenuator and dial tone detector
22. Output of RX attenuator
23. Input to transmit level detector on line side
24. Output of transmit level detector on line side
25. Output of RX level detector on line side, also input to RX BNM
26. Input to RX level detector on line side
27. Time constant for RX BNM set by R16, C26
28. Ground for entire IC.

PC board in front of you, we can now begin populating the circuit board. Now refer to the resistor color code chart shown in Table 23-2. You will note that each resistor will have either three or four color bands, which begin at one end of the resistor's body. The first colored band represents the resistor's first digit, while the second color represents the second digit of the resistor's value. The third colored band represents the resistor's multiplier value, and the fourth color band represents the resistor's tolerance value. If there is no fourth colored band then the resistor has a 20% tolerance value. If the resistor's fourth band is silver then the resistor has a 10% tolerance value and if the fourth band is gold then the resistor's tolerance value will be 5%.

We are going to place and solder the lowest height components first, the resistors and diodes. First, locate our first resistor R1. Resistor R1 has a value of 10 K ohms and is a 5% tolerance type. Now take a look for a resistor whose first color band is brown, black, orange, and gold. From the chart you will notice that brown is represented by the digit (1), the second color band is black, which is represented by (0). Notice that the third color is orange and the multiplier is (1,000), so (1) (0) × 1,000 = 10,000 or 10 K ohms.

Go ahead and locate where resistor R1 goes on the PC and install it, next solder it in place on the PC board, and then with a pair of end-cutters trim the excess component leads flush to the edge of the circuit board. Next locate the remaining resistors and install them in their respective locations on the main controller PC board. Solder the resistors to the board, and remember to cut the extra lead with your end-cutters.

Note that capacitors generally come from two major groups or types. Non-polar types, such as the ones in this project, have no polarity and can be placed in either direction on the circuit board. The other type of capacitors are called polarized types, these capacitors have polarity and have to be inserted on the board with respect to their polarity marking in order for the circuit to operate correctly. On these types of capacitors, you will observe that the components actually have a plus or minus marking or a black or white band with a plus or minus marking on the body of the capacitors. Capacitors must be installed properly with respect to polarity in order for the circuit to work properly. Failure to install electrolytic capacitors correctly could cause damage to the capacitor as well as to other components in the circuit when power is first applied.

Many capacitors are quite small in physical size, so manufacturers have devised a three-digit code which can be placed on the capacitor instead of the actual value. The code occupies a much smaller space on the capacitor and can usually fit nicely on the capacitor body, but without the chart the builder would have a difficult time determining the capacitor's value.

Now look for the lowest profiles capacitors from the component stack and locate capacitor C1, labeled (100)

Table 23-2
Resistor color code chart

Color Band	1st Digit	2nd Digit	Multiplier	Tolerance
Black	0	0	1	
Brown	1	1	10	1%
Red	2	2	100	2%
Orange	3	3	1,000 (K)	3%
Yellow	4	4	10,000	4%
Green	5	5	100,000	
Blue	6	6	1,000,000 (M)	
Violet	7	7	10,000,000	
Gray	8	8	100,000,000	
White	9	9	1,000,000,000	
Gold			0.1	5%
Silver			0.01	10%
No color				20%

Table 23-3
Three-digit capacitor codes

pF	nF	µF	Code	pF	nF	µF	Code
1.0			**1R0**	3,900	3.9	.0039	**392**
1.2			**1R2**	4,700	4.7	.0047	**472**
1.5			**1R5**	5,600	5.6	.0056	**562**
1.8			**1R8**	6,800	6.8	.0068	**682**
2.2			**2R2**	8,200	8.2	.0082	**822**
2.7			**2R7**	10,000	10	.01	**103**
3.3			**3R3**	12,000	12	.012	**123**
3.9			**3R9**	15,000	15	.015	**153**
4.7			**4R7**	18,000	18	.018	**183**
5.6			**5R6**	22,000	22	.022	**223**
6.8			**6R8**	27,000	27	.027	**273**
8.2			**8R2**	33,000	33	.033	**333**
10			**100**	39,000	39	.039	**393**
12			**120**	47,000	47	.047	**473**
15			**150**	56,000	56	.056	**563**
18			**180**	68,000	68	.068	**683**
22			**220**	82,000	82	.082	**823**
27			**270**	100,000	100	.1	**104**
33			**330**	120,000	120	.12	**124**
39			**390**	150,000	150	.15	**154**
47			**470**	180,000	180	.18	**184**

Table 23-3—cont'd

Three-digit capacitor codes

pF	nF	μF	Code	pF	nF	μF	Code
56			**560**	220,000	220	.22	**224**
68			**680**	270,000	270	.27	**274**
82			**820**	330,000	330	.33	**334**
100		.0001	**101**	390,000	390	.39	**394**
120		.00012	**121**	470,000	470	.47	**474**
150		.00015	**151**	560,000	560	.56	**564**
180		.00018	**181**	680,000	680	.68	**684**
220		.00022	**221**	820,000	820	.82	**824**
270		.00027	**271**		1,000	1	**105**
330		.00033	**331**		1,500	1.5	**155**
390		.00039	**391**		2,200	2.2	**225**
470		.00047	**471**		2,700	2.7	**275**
560		.00056	**561**		3,300	3.3	**335**
680		.00068	**681**		4,700	4.7	**475**
820		.00082	**821**			6.8	**685**
1,000	1.0	.001	**102**			10	**106**
1,200	1.2	.0012	**122**			22	**226**
1,500	1.5	.0015	**152**			33	**336**
1,800	1.8	.0018	**182**			47	**476**
2,200	2.2	.0022	**222**			68	**686**
2,700	2.7	.0027	**272**			100	**107**
3,300	3.3	.0033	**332**			157	**157**

Code = 2 significant digits of the capacitor value + number of zeros to follow

For values below 10 pF, use "R" in place of decimal e.g. 8.2 pF = 8R2

10 pF = 100
100 pF = 101
1,000 pF = 102
22,000 pF = 223
330,000 pF = 334
1 μF = 105

or 100 pF. Note that there are a number of small capacitors in the circuit that may not be labeled with their actual value, so you will have to check Table 23-2 to learn how to use the three-digit capacitor code. Go ahead and place C1 on the circuit board and solder it in place. Now take your end-cutters and cut the excess component leads flush to the edge of the circuit board. Next, locate the remaining low profile capacitors and install them on the printed circuit board, then solder them in place. Remember to trim the extra component leads.

There are a number of electrolytic capacitors in the circuit and placing them in the circuit with respect to their proper polarity is essential for proper operation of the circuit. Once you have located the larger electrolytic capacitors install them in their respective locations and solder them in place. Finally trim the excess component leads.

The Speaker-Phone Project contains a number of silicon diodes and a Zener diode. Diodes, as you will remember, always have some type of polarity markings, which must be observed if the circuit is going to work correctly. You will notice that each diode will have a black or white colored band at one edge of its body. The colored band represents the diode's cathode lead. Check your schematic and parts layout diagrams when installing these diodes on the circuit board. Place all the diodes on the circuit board in their respective locations and solder them in place. The diode bridge on the right side of the schematic can be installed now. After soldering the diodes and the diode bridge in place in their respective locations, remember to cut the excess component leads with your end-cutters, flush to the edge of the circuit board. Note, if you plan to using the project as a telephone speaker-phone, install Zener diode D5 (small gray body with black band). The banded end marks the cathode and must be oriented as illustrated. If you plan to use your speaker-phone only as a duplex intercom, do not install D5. Place it in a safe place for possible future use.

The main speaker-phone circuit also contains two ICs. The main speaker-phone chip at U1 is the 28 pin Motorola MC34118 chip. IC number U2 is an LM386 audio amplifier chip. ICs are often static sensitive, so they must be handled with care. Use a grounded anti-static wrist strap and stay seated in one location when handling the ICs. Take out a cheap insurance policy by installing IC sockets for each of the ICs. In the event of a possible circuit failure, it is much easier to simply unplug a defective IC, than trying to unsolder 14 or more pins from a PC board without damaging the board. Integrated circuits have to be installed correctly if the circuit is going to work properly. IC packages will have some sort of markings which will help you orient them on the PC board. An IC will have either a small indented circle, a cutout, or notch at the top end of the IC package. Pin 1 of the IC will be just to the left of the notch or cutout. Refer to the manufacturer's pin-out diagram, as well as the schematic when installing these parts. Accidental reversal of pin insertion can cause the integration to fail. If you doubt your ability to correctly orient the ICs, then seek the help of a knowledgeable electronics enthusiast.

Next, locate transformers T1 and T2 and place them on the circuit board. Both transformers are miniature 100 to 8 ohm types. Examine the audio transformers, T1

and T2. Notice the red marking on one side of the top insulator on each unit. These marked sides must be pointed toward the center of the board, as shown by the dark sides for T1 and T2 on the PC board drawing. For each transformer (they are identical units), a total of six solder connections should be made. The outer two tabs are soldered to common ground for mechanical rigidity. Install both transformers then solder the leads to the PC board, remembering to trim the excess leads.

Next locate and install the electret microphone MK1. The electret microphone will have either two or three leads. The ground lead is often obvious and is connected to the body of the microphone. Observe the correct polarity, then go ahead and install the microphone on the PC board. Now, locate the On-off switch at S1 and the Mute switch at S2; the prototype project used DPST pushbutton switches, but you could elect to use toggle switches for these if you cannot locate the pushbutton switches. Finally, locate and install the subminiature phone jack at J1, the modular phone jack at J1, and the potentiometer at R18 on the edges of the circuit board. Solder these components to the circuit board.

Once the main Speaker-phone circuit board has been completed, take a short break and then we will inspect the PC board for possible "cold" solder joints and "short" circuits. Pick up the PC board with the foil side facing upwards toward you. Carefully inspect the solder joints, these should look smooth, clean, and bright. If any of the solder joints look dull, dark, or blobby, then you should remove the solder with a solder sucker or a solder wick and then resolder the joint all over again. Now we will inspect the PC board for possible short circuits. Short circuits can be caused from small solder blobs bridging the circuit traces of from cut component leads which can often stick to the circuit board once they have been cut. A sticky residue from solder will often trap component leads across the circuit traces. Look carefully for any bridging wires across circuit traces.

Once this inspection is complete, we can move on to installing the circuit board into an aluminum enclosure. Locate a suitable enclosure to house the circuit board. Arrange the circuit board and the transformer on the bottom of the chassis box. Now, locate four 1/4" standoffs, and mount the PC board between the standoffs and the bottom of the chassis box. If you selected a conventional relay you will have to mount it on the chassis and wire it to the PC board. If you selected a non-PC board transformer then you will have to mount the transformer to the base of the chassis box and wire it to the circuit board. Locate a 9-V battery holder and mount it to the bottom of the plastic box and solder the battery clip wire to the circuit board.

Speaker-phone setup and use—phone mode

Connect up the telephone line to the Speaker-phone circuit, install the 9-V battery, plug in an 8-ohm speaker to the output jack J2, and you will be ready to test the speaker-phone for the first time. First, set both S1 and S2 to their "out" or off position. Set the volume control to a mid-range point. Next, test the modular phone cord you intend to use with the speaker-phone, even if it is brand new. Do this simply by trying it on another telephone. Then, connect the cord between the speaker-phone's telephone jack at J1 and a working, correctly-wired phone jack. (Defective modular plugs, cords, or jacks are a major cause of telephone malfunction.) If you have not yet connected your external speaker, then connect a speaker to the circuit board. For decent audio quality, use a speaker at least 3″ in diameter. A properly enclosed speaker will sound better and louder than a speaker lying naked on a workbench.

Now press S1 to the "in" (On) position. You should hear a normal dial tone. Adjust volume control as desired. If you do not get a dial tone, recheck all your work, starting with the phone line and speaker wires. With S1 still on, tap the microphone and you should hear the sound in the speaker. Speak in a normal tone of voice near the microphone, and the dial tone should drop out when you speak and return when you are quiet. With S2 (Mute/Hold) pressed IN, you should still hear the dial tone, but the microphone should not respond even to a loud yell.

Finally, turn S1 to the OFF position; this is the equivalent of putting a phone handset back "on the hook." If everything above checks out, you are ready for your first speaker-phone conversation. Ask someone to call you. When you hear the ringing (on any other phone), you can answer the call by pressing S1. You should be able to carry on a normal chat while you are anywhere in an average-sized room. Be aware that since the Speaker-phone is switching between "receive" and "transmit" there will be a short delay or "pumping" effect as you move from talking to listening. Do not forget to "hang up" when you are finished.

Initiating speaker-phone calls

While the MC34118 IC makes provision for adding a DTMF tone dialer, for the simplest and least expensive approach, rate an ordinary telephone set as part of our Speaker-phone installation. The average electronics buff is well aware that suitable phones are abundant at negligible cost, often with memory and other handy features. Simply connect the auxiliary phone and your speaker-phone to the phone line with a commonly available dual-modular adapter, or use a dual wall jack, etc. Use the auxiliary phone for dialing, incoming rings and situations where you prefer a handset. Enjoy the Speaker-phone for efficient, hands-free communication. Be sure to hang up the auxiliary phone after turning on your Speaker-phone.

Optimizing your speaker-phone

There are three variables to be considered in any line-powered Speaker-phone installation: The first concern is the positioning of the speaker in relation to the microphone. Next, you must consider your distance from the telephone switching complex which supplies line voltage. (If the voltage is marginal because you are on a distant point of the phone company "loop," supplementary battery power may be needed for this or any other Speaker-phone.) You must also consider the basic "talk power" of the party on the other end. Let's consider the three details. First, if your Speaker-phone works perfectly from the moment you set it up, then you are in great shape. Otherwise, consider the following: In general, some degree of acoustical separation must be provided between microphone and speaker. Put simply, they should not "point" at each other, or there will be some form of oscillation or erratic operation. Some experimenting may be needed.

Usually, telephone line voltage is sufficient for good Speaker-phone performance. However, the available voltage is determined by the distance of your location from the central office and local factors. The easiest way to determine if a supplementary battery is needed is just to try it out. Please study the section Installing Optional Battery before making any type of battery connection. If the battery improves performance, then continue using it for that particular location. It is switched in and out of the circuit by the "B" section of S1.

Finally, you may encounter situations where the person on the other end is just not speaking loud enough for positive SP1 switching between receive and transmit. This can be caused either by the other party's use of a poor-quality phone or just because they are speaking too softly. You may need to ask them to try a different phone, or just to speak up, or to speak more directly into their handset microphone; or redial for a better connection.

Notes on speaker placement

In general, the Speaker-phone's speaker audio will sound best if the speaker is of reasonably good size and quality and is properly enclosed or "baffled." It should face the same direction as the microphone or away from it, not directly at it. Remember that it is a basic function of the Speaker-phone to detect incoming audio and quickly attenuate the microphone circuit. Therefore, audio feedback of the familiar "squealing" kind is not a common occurrence. The feedback resulting from poor

speaker positioning is more subtle; greatly reduced speaker volume, erratic switching action, possibly with a slight ringing sound.

Intercom operation

Two Speaker-phone units may be used as a hands-free half-duplex wired intercom system. This offers many handy problem-solving applications for home and business alike. If you have the practical application, we see no reason why additional units could not be put "on line." Phone-line Speaker-phone operation need not be sacrificed in such local intercom installations. If you successfully built one or more Speaker-phones, you probably already have some handy switching ideas. We will leave those to your ingenuity.

To set up two Speaker-phone circuits for hands-free intercom service, just do the following: First, solder a jumper from A to B. Interconnect both units with a modular phone extension cord, or any other 2-wire pair practical for your setup. Interconnection of further units is up to user testing and experimenting. In general, keep all "tip" and "ring" lines common to each other.

The Speaker-phone is designed using the latest Motorola chip-set and should serve you and your family for many years into the future.

Chapter 24

Telephone Scrambler

Parts list

Parts Bin

Telephone Scrambler

R1   2.2 megohms, $\frac{1}{4}$ W,
   5% resistors

R2, R3   470 ohms, $\frac{1}{4}$ W,
   5% resistors

R4, R5   100 ohms, $\frac{1}{4}$ W,
   5% resistors

R6, R7   22 K ohms, $\frac{1}{4}$ W,
   5% resistors

R8, R10, R14, R15, R17
   1 K ohms, $\frac{1}{4}$ W,
   5% resistors

R12   8.2 K ohms, $\frac{1}{4}$ W,
   5% resistors

R13   1 K ohm
   potentiometer

R16   10 K potentiometer

R18, R19, R20   10 K ohms,
   $\frac{1}{4}$ W, 5% resistors

R21   33 K ohms, $\frac{1}{4}$ W,
   5% resistors

C1   22 pF, 35 V,
   NPO capacitor

C2, C3   82 pF, 35 V,
   NPO capacitor

C4, C5, C11, C14,
   C16   .01 µF, 35 V disc

C6, C9, C10   1 µF, 35 V,
   electrolytic capacitor

C7, C8   10 µF, 35 V
   electrolytic capacitor

C12   470 pF, 35 V, disc

C13, C15   470 µF, 35 V,
   electrolytic capacitor

U1   74HC86 - quad
   2-input exclusive
   OR gate

U2, U7   74HC161 -
   synchronous 4-bit
   binary counter

U5, U6   TP3054N codec
   (National)

U8   LM7905; −5 V
   regulator

U9   LM7805; +5 V
   regulator

Q1   2N3565 or 2N3904 NPN
   transistor

S1, S2   DPDT switch

J1, J2   RJ-11 -
   4-conductor modular
   phone jack

J3 $\frac{1}{8}$ inch coaxial
   power jack

XTL   3 MHz crystal
   (2.5-4 MHz- usuable)

Misc   PC board, wire,
   6-14 V, 100 ma power
   supply, IC sockets,
   enclosure, etc

There are many situations when you wish you had a Telephone Scrambler for privacy or security. The military has long been known for using sophisticated voice-scrambling systems, but there is a definite need for less costly ones in everyday life, in order to discourage the casual eavesdropper. The voice-encryption systems, or the voice scrambler described here inverts the frequency spectrum of the speech, either side of a

Figure 24-1 *Telephone scrambler system. Courtesy of Poptronix*

reference frequency to scramble the audio, and reinverts it to descramble the speech, see Figure 24-1.

Although the system is intended primarily to scramble telephone conversations, it is not limited to that. The device can also scramble tape recordings, which

will be made intelligible only with the correct descrambler.

This method of speech scrambling is accomplished by mixing the audio input to be scrambled with a carrier tone, as shown in Figure 24-2.

The mixing process is carried out with a balanced modulator, which results in a double-sideband suppressed-carrier signal. The two resulting sidebands are the lower sideband audio frequencies in the voice range (about 150–3,000 Hz) and upper sideband frequencies (about 3,000–7,000 Hz).

Since most voice circuits are designed for frequencies in the lower sideband range, the upper sideband is filtered out. The lower sideband contains frequencies that are similar to the original voice frequencies, but it has an

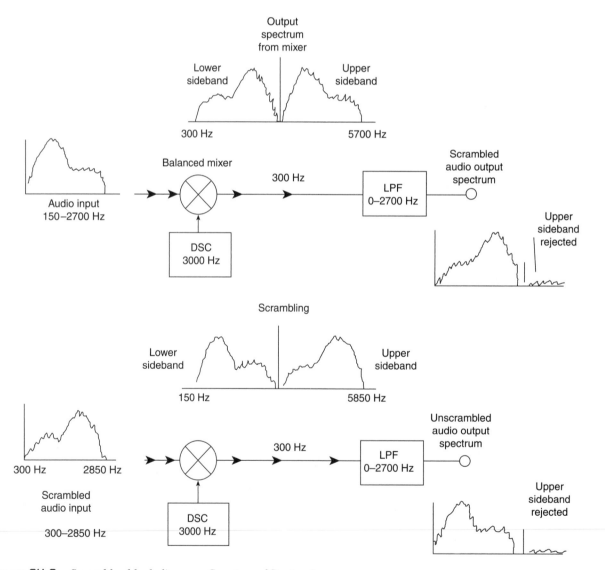

Figure 24-2 *Scrambler block diagram. Courtesy of Poptronix*

inverted spectrum. Assuming a 3,000-Hz carrier signal, an input signal of 500 Hz will produce a 2,500-Hz output, and a 1-KHz signal will produce a 2-KHz output. The spectral energy of a human voice is more concentrated at the ends of the voice spectrum, mainly the 300–1,000 Hz range, and somewhat less in the 2,000–2,500 Hz range. The resulting output will therefore have a very high pitched sound, and be unintelligible. It can, however, be carried over normal telephone lines without being understood by eavesdroppers.

A digital voice-scrambling method is used in the circuit because it requires fewer parts than an analog system, needs no adjustment, and requires no switching. Because the descrambling process is the inverse of the scrambling process, the same circuit can be used for both functions. The encryption system has two channels for full-duplex operation, which allows easy two-way communication. Note that two complete systems at each end of a phone line are required for two people to carry on a scrambled conversation.

An audio input signal is first filtered with an active switched-capacitor bandpass filter to limit the frequency range to between 150 and 2,700 Hz. The signal is then digitized with a sampling rate of 5.86 KHz, which is more than double the highest audio frequency (2,700 Hz). Every second eight-bit digitized audio sample has its sign bit inverted while being fed to a digital-to-analog converter. That has the effect of inverting the spectrum of the analog output signal after conversion from the digitized audio. The signal is then fed to a bandpass filter to remove switching components, leaving the final audio signal as one that corresponds to the input signal, except that its spectrum is folded around one fourth of the sampling frequency or 1,465 Hz.

In Figure 24-3 it can be seen that a 1-KHz sine-wave sampled as shown, with even- numbered samples inverted, results in a lower-frequency sine-wave (b). The process also works in reverse; if the lower-frequency waveform (2-b) is sampled at the same points, and alternate samples inverted, the original waveform can be regenerated.

Circuit description

The voice scrambler/descrambler schematic is illustrated in Figure 24-4. Two National Semiconductor TP3054 coder/decoder, or codec chips,—form the heart of the telephone scrambler circuit.

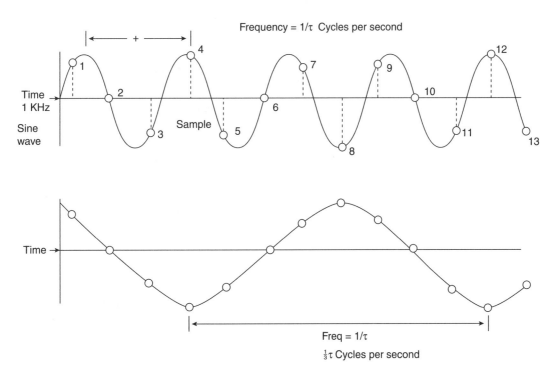

Figure 24-3 *Sampled 1-KHZ sine wave. Courtesy of Poptronix*

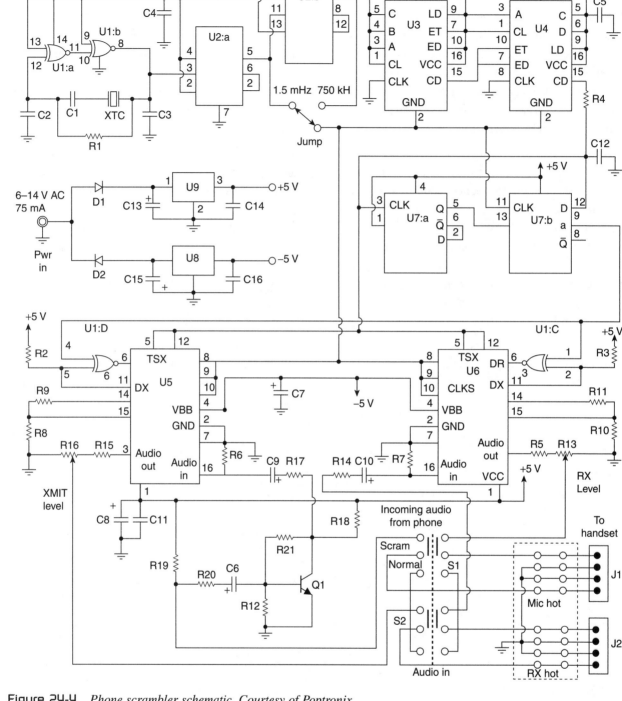

Figure 24-4 *Phone scrambler schematic. Courtesy of Poptronix*

The two integrated circuits, U5 and U6, contain all of the necessary A/D and D/A converters, switched-capacitor filters, and associated tuning and control circuitry.

The scrambler's "ground" must be isolated with respect to true earth ground. Therefore the PC board of the scrambler should be mounted on insulated standoffs

and fed about 75 mA of isolated, low voltage AC from a wall-mounted transformer Do not connect the unit to the AC line without such a suitable isolating transformer

Most of the rest of the circuit is clock and control circuitry that supports the two codecs. The clock signal is generated by an oscillator made up of 3-MHz crystal XTL and U1:a and U1:b. The 3-MHz signal is divided in half by U2:a to produce the main 1.5-MHz clock signal, and U2:b again divides by 2 to produce an optional clock frequency of 750 KHz. That signal is further divided down by U3 and U4 to produce 5.86- and 2.93-KHz signals. The D-type flip-flop at U4 produces a 2.93-KHz pulse train that is used for bit-sign inversion.

The 5.86-KHz pulse shifts a serial data stream, eight clock pulses wide, from the codec's A/D converter to the D/A converter. Data from an A/D converter (pin 11 of U5 or U6) is fed to U1:d or U1:c, respectively. These Exclusive OR gates act as inputs, and are held high, or as straight-through non-inverting buffers if the opposite input is held low. By applying a 2.93 KHz pulse on one input, alternate data-stream sign bits (which occur at a 5.86 KHz rate) are inverted. Therefore, the data from pin 11 of U5 (or U6) that is fed back to the D/A converter section (pin 6) has every other sample reversed in sign. That has the aforementioned effect of inverting the frequency spectrum of the reconstructed analog signal.

The circuitry required to interface the voice-encryption system to a telephone to the scrambler circuitry is shown on the bottom right of the schematic. Resistor R17 couples audio from amplifier Q1 to U5. Transistor Q1 provides about 10 dB voltage gain.

Modular jacks J1 and J2 connect S1 and S2 via jumpers that are configured for your particular telephone set. Because direct insertion of the device in a telephone line would not be feasible without a lot of switching due to ringing and signaling considerations, it is necessary to install this device in the handset line. This way, only the microphone and earphone have to be considered. The TP3054 transistor can drive a 600 ohm load (the impedance of a telephone line) directly if telephones are not being used; simply use the input and output pins of each codec directly. To have the chip drive a loudspeaker a small audio amplifier such as that shown in Figure 24-5 is required.

Note that when using a microphone to input audio to the codec, some microphones have internal audio amplifiers and can produce well over one volt of audio. Those microphone outputs can be input directly to the codec. Low-output microphones require amplification. A switching network (S1 and S2) is added on the PC board to switch the scrambler in or out of the

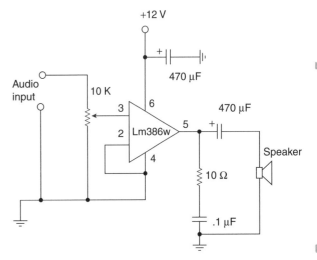

Figure 24-5 *Audio amplifier*

telephone circuit. Resistor R13 sets the sound level at the telephone receiver and R1 is set for optimum reception at the other end of the telephone line.

Figure 24-6 shows two more possible applications using the Telephone Scrambler circuit; 6-a shows how the system can be used to make scrambled audio recordings, and 6-b shows how a radio transceiver can be fitted with this device.

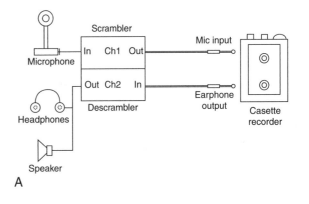

A

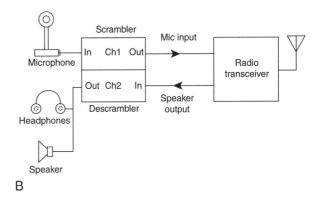

B

Figure 24-6 *Scrambler recording. Courtesy of Poptronix*

(Bear in mind that in services such as amateur radio, it is illegal to use encryption. so check FCC rules to verify the legality for any intended application.)

Circuit assembly

Before we begin constructing the Telephone Scrambler, you will need to locate a clean, well-lit worktable or workbench. Next we will gather a small 25–30 W pencil tipped soldering iron, a length of 60/40 tin–lead solder, and a small canister of Tip Tinner, available at your local RadioShack store; Tip Tinner is used to condition the soldering tip between solder joints. You drive the soldering tip into the compound and it cleans and prepares the soldering tip. Next grab a few small hand tools, and try to locate small flat blade and Phillips screwdrivers. A pair of needle-nose pliers, a pair of tweezers along with a magnifying glass, and a pair of end-cutters and we will begin constructing the project. Now locate your schematic diagram and parts layout diagram along with all of the components needed to build the project. Once all the components are in front of you, you can check them off against the project parts list to make sure you are ready to start building the project. The Telephone Scrambler can be constructed on a double-sided printed circuit board. Because the PC board does not have plated-through holes, first install through-board jumpers at all locations marked with an "X," and solder them on both sides of the board. All component leads that pass through a hole with copper on both sides must also be soldered on both sides.

Table 24-1 illustrates the resistor code chart and how to read resistors. Note that each resistor will have either three or four color bands, which begin at one end of the resistor's body. The first colored band represents the resistor's first digit, while the second color represents the second digit of the resistor's value. The third colored band represents the resistor's multiplier value, and the fourth color band represents the resistor's tolerance value. If there is no fourth colored band then the resistor has a 20% tolerance value. If the resistor's fourth band is silver then the resistor has a 10% tolerance value and if the fourth band is gold then the resistor's tolerance value will be 5%.

With your PC board in front of you, we can now begin populating the circuit board. We are going to place and

solder the lowest height components first, the resistors and diodes. First, locate our first resistor R1. Resistor R1 has a value of 2.2 megohms and is a ¼ W 5% tolerance type. Now take a look for a resistor whose first color band is red, red, green, and gold. From the chart you will notice that red is represented by the digit (2), and again the second band is red, which is represented by (2). Notice that the third color is green and the multiplier is (100,000), so (2) (2) × 100,000 = 2,200,000 or 2.2 megohm. Place R1 on the PC board in its respective location and solder it in place. Then take a pair of end-cutters and trim the excess resistor leads flush to the edge of the circuit board. Now locate the remaining resistors and install them in their proper locations on the PC board. Solder them all in place and then remember to trim the excess leads with your end-cutters.

Note that capacitors generally come from two major groups or types. Non-polar types, such as the ones in this project, have no polarity and can be placed in either direction on the circuit board. The other type of capacitors are called polarized types, these capacitors have polarity and have to be inserted on the board with respect to their polarity marking in order for the circuit to operate correctly. On these types of capacitors, you will observe that the components actually have a plus or minus marking or a black or white band with a plus or minus marking on the body of the capacitors.

Now locate the capacitor chart shown in Table 24-2; this chart will help you identify small capacitors which may be marked with only a three-digit code instead of an actual value.

Locate the capacitors from the component pile and try and locate capacitor C1, which should be marked 22 p or 22 pF, or from the capacitor chart you may see 220 marked on the capacitor body. Now take your end-cutters and cut the excess component leads flush to the edge of the circuit board. Once you have installed the first capacitor, you can move on to installing the remaining capacitors onto the circuit board; solder them in place and then remember to trim the excess leads.

Next locate and install the remaining capacitors, first installing the low profile capacitors followed by the larger electrolytic capacitors. Use good quality electrolytic or tantalum capacitors. Remember the larger electrolytic capacitors have polarity which must be observed if the circuit is to work properly. Failure to

Table 24-1
Resistor color code chart

Color Band	1st Digit	2nd Digit	Multiplier	Tolerance
Black	0	0	1	
Brown	1	1	10	1%
Red	2	2	100	2%
Orange	3	3	1,000 (K)	3%
Yellow	4	4	10,000	4%
Green	5	5	100,000	
Blue	6	6	1,000,000 (M)	
Violet	7	7	10,000,000	
Gray	8	8	100,000,000	
White	9	9	1,000,000,000	
Gold			0.1	5%
Silver			0.01	10%
No color				20%

Table 24-2
Three-digit capacitor codes

pF	nF	μF	Code	pF	nF	μF	Code
1.0			**1R0**	3,900	3.9	.0039	**392**
1.2			**1R2**	4,700	4.7	.0047	**472**
1.5			**1R5**	5,600	5.6	.0056	**562**
1.8			**1R8**	6,800	6.8	.0068	**682**
2.2			**2R2**	8,200	8.2	.0082	**822**
2.7			**2R7**	10,000	10	.01	**103**
3.3			**3R3**	12,000	12	.012	**123**
3.9			**3R9**	15,000	15	.015	**153**
4.7			**4R7**	18,000	18	.018	**183**
5.6			**5R6**	22,000	22	.022	**223**
6.8			**6R8**	27,000	27	.027	**273**
8.2			**8R2**	33,000	33	.033	**333**
10			**100**	39,000	39	.039	**393**
12			**120**	47,000	47	.047	**473**
15			**150**	56,000	56	.056	**563**
18			**180**	68,000	68	.068	**683**
22			**220**	82,000	82	.082	**823**
27			**270**	100,000	100	.1	**104**
33			**330**	120,000	120	.12	**124**
39			**390**	150,000	150	.15	**154**
47			**470**	180,000	180	.18	**184**

(Continued)

Table 24-2—cont'd

Three-digit capacitor codes

pF	nF	μF	Code	pF	nF	μF	Code
56			**560**	220,000	220	.22	**224**
68			**680**	270,000	270	.27	**274**
82			**820**	330,000	330	.33	**334**
100		.0001	**101**	390,000	390	.39	**394**
120		.00012	**121**	470,000	470	.47	**474**
150		.00015	**151**	560,000	560	.56	**564**
180		.00018	**181**	680,000	680	.68	**684**
220		.00022	**221**	820,000	820	.82	**824**
270		.00027	**271**		1,000	1	**105**
330		.00033	**331**		1,500	1.5	**155**
390		.00039	**391**		2,200	2.2	**225**
470		.00047	**471**		2,700	2.7	**275**
560		.00056	**561**		3,300	3.3	**335**
680		.00068	**681**		4,700	4.7	**475**
820		.00082	**821**			6.8	**685**
1,000	1.0	.001	**102**			10	**106**
1,200	1.2	.0012	**122**			22	**226**
1,500	1.5	.0015	**152**			33	**336**
1,800	1.8	.0018	**182**			47	**476**
2,200	2.2	.0022	**222**			68	**686**
2,700	2.7	.0027	**272**			100	**107**
3,300	3.3	.0033	**332**			157	**157**

Code = 2 significant digits of the capacitor value + number of zeros to follow

For values below 10 pF, use "R" in place of decimal e.g. 8.2 pF = 8R2

10 pF = 100
100 pF = 101
1,000 pF = 102
22,000 pF = 223
330,000 pF = 334
1 μF = 105

install the electrolytic capacitors correctly can damage the component or other components when power is first applied to the circuit upon power-up. Install the capacitors in their correct positions on the PC board, and then solder them in place. Next trim the excess component leads flush to the edge of the circuit board using your end-cutters.

The Phone Scrambler project contains two silicon diodes. These diodes also have polarity which must be observed if the circuit is going to work correctly.

You will notice that each diode will have a black or white colored band at one edge of the diode's body. The colored band represents the diode's cathode lead. Check your schematic and parts layout diagrams when installing these diodes on the circuit board. Place all the diodes on the circuit board and solder them in place. Then cut the excess component leads with your end-cutters.

The telephone Scrambler circuit utilizes single NPN transistors at Q1. Transistors generally have three leads, a Base lead, a Collector lead and an Emitter lead.

Refer to the schematic and you will notice that there will be two diagonal leads pointing to a vertical line. The vertical line is the transistor's base lead and the two diagonal lines represent the collector and the emitter. The diagonal lead with the arrow on it is the emitter lead, and this should help you install the transistor on to the circuit board correctly. Go ahead and install transistor Q1 on the PC board and solder it in place. Now follow-up by cutting the excess component leads from Q1 and Q2, with your end-cutters.

The Scrambler utilizes a number of ICs. ICs must be handled carefully and installed properly in order for the circuit to work perfectly. Use anti-static techniques when handling the ICs to avoid damage from static electricity when moving about and installing it. It is wise to install IC sockets as a cheap form of insurance against a possible circuit failure at some later point in time. Install the IC sockets on the PC board and solder them in place. Each integrated circuit has a locator either in the form of a small indented circle, or a notch, or cut-out at the top end of the IC package. Pin 1 of the IC will be just to the left of the locator. Go ahead and install the ICs into their respective IC sockets.

Now is a good time to take a short break. After the break we will inspect the circuit board for any possible "cold" solder joints or "short" circuits. Take a careful look at each of the PC solder joints, they should all look clean and shiny. If any of the solder joints look dull, dark, or blobby, then you unsolder the joint, remove the solder and then resolder the joint all over again. Next, examine the PC board for any short circuits. Short circuits can be caused from two circuit traces touching each other due to a stray component lead that stuck to the PC board from solder residue or from solder blobs bridging the solder traces.

Before the ICs are inserted in their sockets, apply 6 to 12 V AC to the junction of D1 and D2 and ground. Check for 5 V at pin 4 of U5 and U6, and pin 16 of U2, U3, and U4, and pin 14 of U1. Next, verify −5 V at pin 1 of U5 and U6. If these voltages check out, insert the ICs into their sockets.

Testing

Verify that a 5-V peak-to-peak 1.5-MHz signal exists at pin 2 of U3. Check for a 5.8 KHz pulse train at pin 15 of

U4, pins 5 and 12 of U5, and U6. Check for 2.93-KHz pulse train at pin 1 U1:c and pin 4 of U1:d. Due to the short pulse width (250 nanoseconds), it might be difficult to see these pulses with an economy model oscilloscope.

If all checks out, apply a 0.5-V peak-to-peak, 1-KHz tone to the junction of R17 and C9:a; a 2 KHz tone should be produced by U5. Now temporarily connect pin 3 of U5 to the junction of C10 and R14 using a 100 K resistor. Pin 3 of U6 should produce the original 1-KHz tone.

Next apply an audio signal from a tape deck or radio with about 2 V peak-to-peak to the junction of R17 and C9. Listen to the output at pin 3 of U5; it should sound "scrambled." Now listen to the output of channel 2. It should be normal, but note that the high and low frequencies might sound somewhat attenuated due to the narrow band-width of the system.

Adapting a phone

Note that this unit cannot be connected directly to the phone lines. It will handle speech audio, but not pass ringing signals or rotary dialing pulses. It will also distort dialing tones. You must use only a phone whose handset has accessible microphone and receiver connections. You cannot use a unified telephone (where the dial or pad is built into the handset). The handset should preferably have an elecret microphone. However, carbon microphones (found in older phones) can be used, if necessary, but R19 should be changed to about 1 K, and R20 may have to be increased if excessive audio from a carbon microphone over-drives Q1, causing distortion.

Depending on the phone you have, you must make the proper jumper connections on the PC board near J1 and J2. A Radio Shack telephone model No. ET-171 (cat No. 43-374) was used in the prototype. You must identify the following things on your phone(s).

1. The handset microphone and earpiece connections

2. The type of microphone (electret, dynamic, or carbon)

3. Microphone polarity if it is the electret type

4. The base connections.

There are usually four wires that connect a telephone handset to its base. If you cannot visually identify the wires after disassembling the handset, try connecting a 1.5-V battery to alternate pairs of wires on the handset until you hear a click in the earpiece. Mark these as the receiver leads; there should be between 50 and 1,000 ohms between them. The other two leads are for the microphone.

Check for short circuits between both of the receiver leads and the microphone leads with an ohmmeter on a high resistance range. A low resistance or a short between any two leads indicates that they are the ground leads for the microphone and receiver. If the microphone works, all is well; if not, reverse the microphone connections. This will identify the microphone's hot and ground leads. Once you have identified all of the handset leads, note their positions on the modular connector.

The telephone base connections can to be determined from the positions of the handset leads at the modular connector. When you have all of the telephone connections identified, install the jumpers on the PC board near jacks J1 and J2. Once you know the signal positions at jacks J1 and J2 for your phone, install four jumpers per jack to properly route the signals.

The finished board can be mounted in a suitable case. The case pictured allows the telephone to be placed on top of the scrambler without taking up any extra space. With a pair of scrambler phones in hand, you are ready to start talking. All you need now is someone to talk to, and a confidential topic to discuss.

Chapter 25

Telephone Tattletale

Parts list

Telephone Tattletale

R1, R2, R3 10 K ohms,
$\frac{1}{4}$ w, 5% resistor

R4 1.5 megohms, $\frac{1}{4}$ W,
5% resistor

R5, R8 1 megohm, $\frac{1}{4}$ W,
5% resistor

R6 390 K ohms, $\frac{1}{4}$ W,
5% resistor

R7 22 ohms, $\frac{1}{4}$ W,
5% resistor

R9, R10, R11 100 K ohms,
$\frac{1}{4}$ W, 5% resistor

R13 2.2 K ohms, $\frac{1}{4}$ W,
5% resistor

R14 150 ohms, $\frac{1}{4}$ W,
5% resistor

R15 39 K ohms, $\frac{1}{4}$ W,
5% resistor

R16 15 K ohms, $\frac{1}{4}$ W,
5% resistor

R17 680 ohms, $\frac{1}{4}$ W,
5% resistor

C1 1,000 µF, 15 V
electrolytic capacitor

C2 .47 µF, 15 V
ceramic capacitor

C3 .22 µF, 15 V
ceramic capacitor

C4, C5, C6 .1 µF,
200 V ceramic or mylar

C7 47 µF, 15 V
electrolytic capacitor

C8, C9, C10 .01 µF,
15 V ceramic capacitor

D1, D2, D3, D4 1N2069
silicon diode

D5 1N2070 silicon diode

D6, D7, D8 1N4148
silicon diode

U1 CD4002B dual 4-input
NOR gate IC

U2, U3 CD4081B quad
2-input AND gate IC

U4, U7, U8, U9 LM555
timer IC

U5 CD4040B 12 stage
binary counter IC

U6 CD4011B quad 2-input
NAND gate IC

Q1 MPSA42 300 volt NPN
silicon transistor

Q2 2N3904 NPN silicon
transistor

LED1 red LED

T1 transformer – 115 V
primary/6.3 V @
300 mA secondary

Z1 Varistor

F1 fuse 1-ampere

S1 SPST toggle or slide
switch

S2, S3, S4 normally
open sensors

Misc PC board, wire,
IC sockets, terminal
strips, enclosure, etc.

Do you ever wonder what's happening at your home or business while you are away? Does the thought of break-in, fire, or other calamity worry you? If so, Tattletale, shown in Figure 25-1, is just what you need.

This remarkable device is able to report to you, no matter how far away you are, on the condition of your home or business. All it takes is one phone call. With Tattletale at your service, you will be able to monitor up to three possible emergency situations and have your telephone report back to you when you call. Tattletale is completely automatic and easily activated each time you leave your home or business. A convenient LED indicator provides visual indication that Tattletale is armed and ready to answer that inquiring call.

Emergency situations such as fire, break-in, heating system failure, and flood can be easily monitored by connecting three independent normally open switches or thermostats to Tattletale's three sensing terminals. Normally closed switches can also be used by making a simple circuit modification. The type of emergency which Tattletale can monitor is limited only by your imagination.

Operation of this unit is very simple. It is activated by throwing the power switch on before you leave the premises. When you wish to check on the condition of the sensing switches you call your number, and Tattletale answers immediately. You will hear a series of tones or beeps. One beep every couple of seconds indicates that all is well; two, three, or four beeps tells

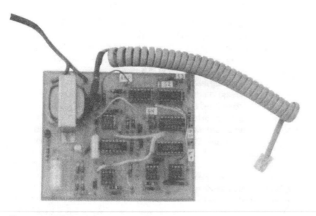

Figure 25-1 *Tattletale circuit board. Courtesy of Cengage Learning*

you that one of the switches has been tripped. The number of beeps indicates which switch, and you have the option of taking action or not as you see fit. Tattletale will continue to report to you for about a minute. Then it will hang up, ready to answer when you call again. You may check Tattletale's status as often as you like.

Circuit details

The Tattletale phone monitor circuit is illustrated in Figure 25-2. The Tattletale circuit contains two sections. The first part is the logic section composed of U1 through U8. The remainder of the circuitry is used to answer the telephone call and place the tones on the telephone line. Each of these sections is discussed in detail. The logic section of the circuit starts with a NOR gate U1:a, which is used as a detector to determine if any of the sensing switches has been tripped. If all sensing switches are normal, U3:a is enabled and the second output of U5 (pin 7) is coupled through U3:a and U1:b to trigger one-shot multi-vibrator U7. Should any of the sensing switches be activated, U3:a is disabled and one of the succeeding outputs of U5 will trigger U7 in a similar manner.

Integrated circuit U8 is an astable multi-vibrator operating at a frequency of about 3 Hz. The output of U8 clocks U5, a binary counter. The first output of U5 (pin 9) is one half the clock frequency and each succeeding output has a period which is twice as long as the preceding stage. The start of the reset pulse generated by U7 is determined by the output signal from U5 that is permitted to pass through AND gates U2 and U3, in accordance with the status of the sensing switches. In this way the count length of U5 is controlled by the switches, and the signal at pin 9 of U5 will contain 1, 2, 3, or 4 pulses. These pulses are used to enable U9, a tone generator operating at a frequency of about 2,000 Hz. Since U7 holds U8 and U5 in reset condition for one or two seconds each time it is triggered, the tones generated by U9 are perceived by you to be in a group of 1, 2, 3, or 4 bursts.

In order to answer a telephone call, the circuit has been designed to respond to the 90-volt, 20-Hz ringing signal. U6:a and U6:b are connected in a latch circuit or flip-flop configuration, which has two stable states.

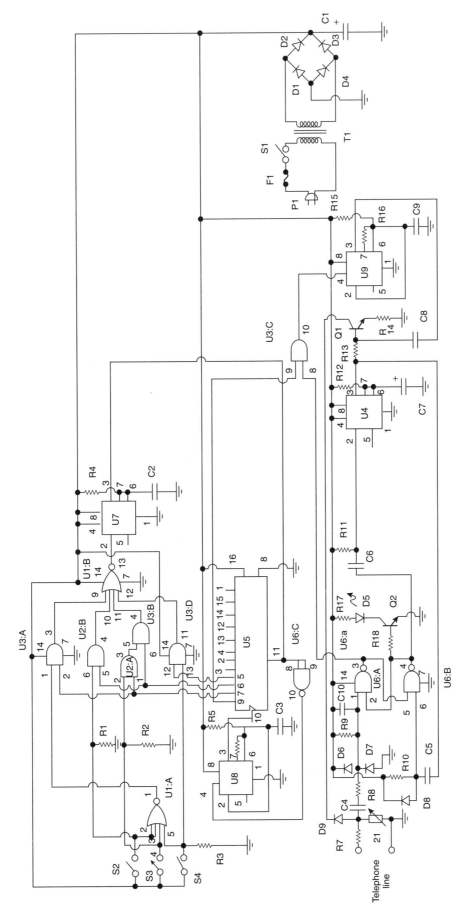

Figure 25-2 *Tattletale circuit. Courtesy of Cengage Learning*

When power is first applied to the circuit the logic level at the output (pin 4) of U6:b will be high. This applies forward bias to Q2, switching it on and illuminating LED1 to indicate that the circuit is armed and ready. The ringing signal which appears across the telephone fine is applied to pin 1 of U6:a through C4 and R8. This signal causes the latch circuit to change state. The negative-going output at pin 4 of U6:b then triggers one shot multi-vibrator U4. The output at pin 3 of U4 then goes positive. The forward bias of Q1 causes this transistor to conduct, connecting R14 across the telephone line and answers the call. Since U4 has a time period of about a minute, the circuit will hold the telephone line for that time interval and then "hang up." The negative-going waveform at pin 3 of U4 is used to reset the latch circuit back to the armed mode so that it is ready to answer another telephone call. The LED is also illuminated again. During the time that Q1 is conducting, the tone bursts from U9 are fed to the base of Q1. Thus, these tone bursts, which contain the audio information indicating the status of the sensing switches, are impressed upon the telephone line.

Construction

The Tattletale project was constructed on a single-sided printed circuit board as illustrated. The prototype circuit board used in this project measured about $3\text{-}\frac{1}{4} \times 4\text{-}\frac{1}{2}''$. You may also construct this circuit on a perfboard, hard wiring the connections. Layout is not critical.

Before we begin building the Tatttletale project, you will need to locate a clean, well-lit worktable or workbench. Next we will gather a small 25–30-W pencil tipped soldering iron, a length of 60/40 tin–lead solder, and a small canister of Tip Tinner, available at your local RadioShack store; Tip Tinner is used to condition the soldering tip between solder joints. You drive the soldering tip into the compound and it cleans and prepares the soldering tip. Next grab a few small hand tools, and try to locate small flat blade and Phillips screwdrivers. A pair of needle-nose pliers, a pair of tweezers along with a magnifying glass, and a pair of end-cutters and we will begin constructing the project. Now locate your schematic diagram and parts layout diagram along with all of the components needed to build the project. Once all the components are in front of you, you can check them off against the project parts list to make sure you are ready to start building the project. Now, locate Tables 25-1 and 25-2, and place them in front of you.

Table 25-1 illustrates the resistor code chart and how to read resistors, while Table 25-2 illustrates the capacitor code chart which will aid you in constructing the project.

Refer now to Table 25-1 which illustrates the resistor color code chart which will help you identify the resistor values. Each resistor will have either three or four color bands, which begin at one end of the resistor's body. The first colored band represents the resistor's first digit, while the second color represents the second digit of the resistor's value. The third colored band represents the resistor's multiplier value, and the fourth color band represents the resistor's tolerance value. If there is no fourth colored band then the resistor has a 20% tolerance value. If the resistor's fourth band is silver then the resistor has a 10% tolerance value and if the fourth band is gold then the resistor's tolerance value will be 5%.

So let's get started, now we are ready to begin assembling the project. The prototype project was constructed on a single-sided printed circuit board or PCB. With your PC board in front of you, we can now begin populating the circuit board. We are going to place and solder the lowest height components first, the resistors and the diodes. First locate our first resistor R1. Resistor R1 has a value of 10 K ohms and is a 5% tolerance type. Now take a look for a resistor whose first color band is brown, black, orange, and gold. From the chart you will notice that brown is represented by the digit (1), and again the second band is black, which is represented by (0). Notice that the third color is orange and the multiplier is (100), so (1) (0) × 100 = 10,000 ohms.

Now, place resistor R1 on the circuit board in its respective location, then solder it in place. Follow up by using a pair of end-cutters to trim the excess component leads flush to the edge of the circuit board. Next locate the remaining resistors and place them on the circuit board at their correct locations and solder them in place. Take your end-cutters and cut the excess component leads, flush to the edge of the circuit board.

Table 25-1

Resistor color code chart

Color Band	1st Digit	2nd Digit	Multiplier	Tolerance
Black	0	0	1	
Brown	1	1	10	1%
Red	2	2	100	2%
Orange	3	3	1,000 (K)	3%
Yellow	4	4	10,000	4%
Green	5	5	100,000	
Blue	6	6	1,000,000 (M)	
Violet	7	7	10,000,000	
Gray	8	8	100,000,000	
White	9	9	1,000,000,000	
Gold			0.1	5%
Silver			0.01	10%
No color				20%

Table 25-2

Three-digit capacitor codes

pF	nF	μF	Code	pF	nF	μF	Code
1.0			1R0	3,900	3.9	.0039	392
1.2			1R2	4,700	4.7	.0047	472
1.5			1R5	5,600	5.6	.0056	562
1.8			1R8	6,800	6.8	.0068	682
2.2			2R2	8,200	8.2	.0082	822
2.7			2R7	10,000	10	.01	103
3.3			3R3	12,000	12	.012	123
3.9			3R9	15,000	15	.015	153
4.7			4R7	18,000	18	.018	183
5.6			5R6	22,000	22	.022	223
6.8			6R8	27,000	27	.027	273
8.2			8R2	33,000	33	.033	333
10			100	39,000	39	.039	393
12			120	47,000	47	.047	473
15			150	56,000	56	.056	563
18			180	68,000	68	.068	683
22			220	82,000	82	.082	823
27			270	100,000	100	.1	104
33			330	120,000	120	.12	124
39			390	150,000	150	.15	154
47			470	180,000	180	.18	184

(Continued)

Table 25-2—cont'd
Three-digit capacitor codes

pF	nF	µF	Code	pF	nF	µF	Code
56			**560**	220,000	220	.22	**224**
68			**680**	270,000	270	.27	**274**
82			**820**	330,000	330	.33	**334**
100		.0001	**101**	390,000	390	.39	**394**
120		.00012	**121**	470,000	470	.47	**474**
150		.00015	**151**	560,000	560	.56	**564**
180		.00018	**181**	680,000	680	.68	**684**
220		.00022	**221**	820,000	820	.82	**824**
270		.00027	**271**		1,000	1	**105**
330		.00033	**331**		1,500	1.5	**155**
390		.00039	**391**		2,200	2.2	**225**
470		.00047	**471**		2,700	2.7	**275**
560		.00056	**561**		3,300	3.3	**335**
680		.00068	**681**		4,700	4.7	**475**
820		.00082	**821**			6.8	**685**
1,000	1.0	.001	**102**			10	**106**
1,200	1.2	.0012	**122**			22	**226**
1,500	1.5	.0015	**152**			33	**336**
1,800	1.8	.0018	**182**			47	**476**
2,200	2.2	.0022	**222**			68	**686**
2,700	2.7	.0027	**272**			100	**107**
3,300	3.3	.0033	**332**			157	**157**

Code = 2 significant digits of the capacitor value + number of zeros to follow

For values below 10 pF, use "R" in place of decimal e.g. 8.2 pF = 8R2

10 pF = 100
100 pF = 101
1,000 pF = 102
22,000 pF = 223
330,000 pF = 334
1 µF = 105

Next we will identify and install the capacitors for the project. Note that capacitors generally come from two major groups or types. Non-polar types, such as the ones in this project, have no polarity and can be placed in either direction on the circuit board. The other type of capacitors are called polarized types, these capacitors have polarity and have to be inserted on the board with respect to their polarity marking in order for the circuit to operate correctly. On these types of capacitors, you will observe that the components actually have a plus or minus marking or a black or white band with a plus or minus marking on the body of the capacitors.

Many capacitors are quite small in physical size, so manufacturers have devised a three-digit code which can be placed on the capacitor instead of the actual value. The code occupies a much smaller space on the capacitor and can usually fit nicely on the capacitor body, but without the chart the builder would have a difficult time determining the capacitor's value.

Next, locate the capacitors from the component pile. It is easier to first install the lowest profile components before installing the larger components. With that in mind we are going to install the smaller profile capacitors first—capacitors C2 and C3 and so forth. Capacitor C2 is marked .47 µF, while C3 is labeled .22 µF. Locate these two capacitors first, and refer to Table 25-2, which will help identify the smaller capacitors, which may have the three-digit codes on them. Use your schematic and parts layout diagrams to find where C2 and C3 are located. After installing these two capacitors, you can move and solder them in place, remembering to trim the excess component leads. Next find the remaining small physical sized capacitors and locate them on the circuit board, then solder them in place. Trim the extra leads as necessary. Finally locate the larger electrolytic capacitors, such as C1 and C7. These capacitors are polarity sensitive and must be oriented correctly for the circuit to work properly. Failure to install them correctly could damage the components and could potentially damage the entire circuit when first powered up. Install the electrolytic capacitors, solder them in place, then trim the excess leads.

This project contains eight silicon diodes at D1 through D8. These diodes also have polarity which must be observed if the circuit is going to work correctly. You will notice that each diode will have a black or white colored band at one edge of the diode's body. The colored band represents the diode's cathode lead. Check your schematic and parts layout diagrams when installing these diodes on the circuit board. Note that diodes D1 through D4 are 1N4001 silicon diodes in the power supply bridge circuit, while diode D9 is a 1N4004, which must also have a 300-V rating since it is across the telephone line input. The remaining diodes D6, D7, and D8 are all 1N4148 small signal diodes. Place all the diodes on the circuit board and solder them in place. Then trim the excess component leads with your end-cutters.

The Tattletale circuit uses two LEDs, which also must be oriented correctly if the circuit is to function properly. Remember that LEDs have two leads, a cathode and an anode lead. The anode lead will usually be the longer of the two leads and the cathode lead will usually be the shorter lead, just under the flat side edge of the LED; this should help orient the LED on the

PC board. Solder the LEDs to the PC board and then trim the excess leads.

The Tattletale circuit also utilizes two transistors at Q1 and Q2. Transistors generally have three leads, a Base lead, a Collector lead and an Emitter lead. Refer to the schematic and you will notice that there will be two diagonal leads pointing to a vertical line. The vertical line is the transistor's base lead and the two diagonal lines represent the collector and the emitter. The diagonal lead with the arrow on it is the emitter lead, and this should help you install the transistor onto the circuit board correctly. Note that the collector-to-emitter voltage rating of Q1 must be at least 300 volts to withstand the 90-volt rms ringing signal from the telephone line. Refer to the manufacturer's specification sheets as well as the schematic and parts layout diagrams when locating and mounting the transistors. Go ahead and install the transistors on the PC board and solder them in place, then follow-up by cutting the excess component leads.

The Tattletale project requires a number of ICs. ICs must be installed correctly in order for the circuit to function properly. It is recommended that you construct the circuit using integrated circuit sockets as a low cost form of insurance in the event of a circuit failure. It is much easier to simply unplug an IC rather than trying to successfully unsolder a 16-pin IC and install a new one. Install the IC sockets but do not install the integrated circuits at this time.

Sensor switches

The Tattletale circuit was designed for normally open type sensors or sensor switches, however, if one or more of your sensing switches needs to be a normally closed type, you may use it by adding an inverter to the circuit to reverse the logic. Note that U6 has one spare section which is not used. This section may be utilized for implementation of a normally closed sensing switch. Figure 25-3 illustrates the circuit modification using a normally closed switch (S5) and U6(d) in place of S2.

Notice that R1 which is normally connected to pins 2 and 3 of U1 is now placed at the input of U6:d (pins 12 and 13). S5, a normally closed switch, is used in place of S2. Due to the inverting action of U6(d), pins 2 and 3 of U1 will be at zero logic level when no emergency exists. If your requirement for Tattletale is to

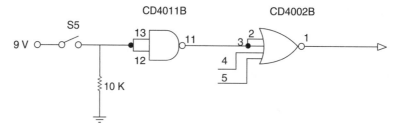

Figure 25-3 *Normally closed sensor option*

use more than one normally closed switch, another chip such as CD40118 or CD4069B may be added to the circuit. Remember, when you use one section of a CD40118 as an inverter, you must connect both inputs together. CMOS digital chips should never have input terminals which are left floating.

Tattletale's logic circuit has not been designed to differentiate between two or more sensing switches, when they are simultaneously activated. In this circuit S2 has priority over S3 and S4, and S3 has priority over S4. For this reason you should use S2 for your most urgent emergency situation, followed by S3 and then S4.

To make the connection to the telephone line it is suggested that you obtain a standard modular plug and cord which is available from any outlet selling telephone accessories. The use of a quick disconnect plug is required by FCC regulations and makes it easy to connect and disconnect Tattletale from your phone line at any time.

It is important that the polarity of your telephone connection is correct as indicated on the schematic. If you use the red and green wires of your modular cord set, the polarity should be correct. If you have any doubt about the polarity of your telephone line, verify it by checking it with any DC voltmeter capable of measuring 50 V. The black and yellow wires of the telephone cord are not used. For a professional touch you may wish to mount the circuit board in a suitable enclosure. Place the LED and S1 where they may easily be seen. To make the connections to the sensing switches use a terminal strip containing four screw connections. Once you have installed Tattletale you may leave it connected to the telephone line and use power switch S1 to turn it on and off. The LED will indicate when Tattletale is activated.

Once the circuit board has been completed, take a short break and then we will inspect the PC board for possible "cold" solder joints and "short" circuits. Pick

up the PC board with the foil side facing upwards toward you. Carefully inspect the solder joints, these should look smooth, clean, and bright. If any of the solder joints look dull, dark, or blobby, then you should remove the solder with a solder sucker or a solder wick and then resolder the joint all over again.

Now we will inspect the PC board for possible short circuits. Short circuits can be caused from small solder blobs bridging the circuit traces or from cut component leads which can often stick to the circuit board once they have been cut. A sticky residue from solder will often trap component leads across the circuit traces. Look carefully for any bridging wires across circuit traces. It is much easier to correct a problem at this stage rather than trying to troubleshoot a non-working circuit later. Once this task is complete, we can move on to installing the circuit board into some sort of plastic enclosure.

Locate a suitable metal chassis box to house the Tattletale project. The prototype circuit board contained all the components including the transformer. Center the circuit board inside the chassis box and use four $\frac{1}{4}$ inch standoffs to lift and mount the PC board to the bottom of the chassis box. You will want to mount the power switch and LED on the front panel of the chassis box. The power line cord should exit the rear panel of the chassis through a grommet. You should install a panel mounted fuse holder at the rear panel of the box. Depending upon your particular needs you could elect to mount a 6-position terminal strip on the front panel to accept connections to the input sensors. You will also need to supply a 2-conductor phone line cord in order to connect the Tattletale to the phone line. One method of solving this problem would be to install cord with a modular telephone plug attached to one end and hard-wired to the circuit at the other end. You could install a modular jack on the chassis box and then simply use a telephone jumper cable with a modular plug at both ends.

Tattletale testing

Before placing any of the ICs in their sockets apply 115 V AC power to the primary of T1 and measure the voltage across C1. This voltage should be about 9-volts DC, with polarity as indicated in the schematic. With the negative lead of the voltmeter connected to the negative side of C1, check each IC socket for the presence of +9 volts at the power supply input terminals. These terminals are pin 8, 14, and 16 for 8, 14-, and 16-pin ICs respectively. If you read the correct voltages, disconnect the AC power from the unit and allow time for C1 to discharge. If incorrect voltages are obtained, troubleshoot the circuit until you obtain the correct voltage readings.

You may now insert the integrated circuit chips into their sockets. Be careful to orient them properly. All integrated circuits have some sort of markings which indicate the positioning information. Most ICs will have either a small indented circle, a cutout, or notch at the top end of the IC package. Pin 1 will be just to the left of the notch, or cutout; this information will help you to position the component correctly in the circuit. Use the IC layout information diagram as well as the schematic to help orient the IC correctly. If you have any doubts in your ability to install the integrated circuits, have a knowledgeable electronics enthusiast help you. It is recommended that the circuit builder use IC sockets for this project as a low cost insurance policy against future circuit failure. It is much easier to simply unplug a defective IC, rather than trying to unsolder 14 or 16 pins successfully without damaging the circuit board.

To check the circuit waveforms it is best if you use an oscilloscope. If an oscilloscope is not available, you may use a voltmeter with an input resistance of at least 1 megohm to check DC levels and slow-moving waveforms.

The first part of the checkout is the call answering circuit. Turn AC Power on; the LED should light. Take a piece of wire and momentarily short pin 1 of U6 to ground. The LED should extinguish and remain so for about a minute, then it should automatically light again. If you do not get this response check the flip-flop circuit of U6:a and U6:b by examining pin 4 of U6 as you momentarily short to ground first pin 1 then pin 6 of U6. This action should cause the logic level at pin 4 to first go high, then low.

You can check the one-shot action of U4 by momentarily shorting pin 2 of U4 to ground while examining the output at pin 3. Pin 3 should read about 9 volts when you trigger U4 at pin 2, remain so for about a minute, and then return to zero volts. When you are satisfied that the U6 and U4 circuitry are operating properly, proceed with the next check.

Set sensing switch S4 to the closed position; S2 and S3 should be left in the open position. Apply AC-line power to the circuit and check pin 9 of U5 for four discrete pulses, repeating every second or two. Next, close S3 and check for three pulses at pin 9 of U5. Then close S2; the number of pulses should now be two. With all sensing switches open, there should be just one pulse every second or two.

If you do not get the specified response, you can check the counter chain by temporarily removing U7 from its socket and shorting pin 11 of U5 to ground. This will enable U5 to count continuously since it will now be clocked by U8. You can then follow the waveforms produced by U5 through the AND gates of U2 and U3 to the trigger input terminal (pin 2) of U7.

U7 can be checked manually for its one-shot action by momentarily shorting pin 2 to ground and examining the one second pulse at pin 3. Pin 3 should go to about 9 volts for about a second and then return to zero. U9 can be checked for 2,000-Hz astable operation by temporarily removing U3 and connecting pins 4 and 8 of U9 together. You should obtain a 2,000-Hz square wave at pin 3 of U9. If the unit passes these tests remove any temporary connections and insert the ICs into their sockets.

With AC power off connect Tattletale to the telephone line using the modular connector. Turn the AC power on; the LED should be lighted. Make an operational check. Call your number from another telephone which is not connected to the same line. You should get an immediate answer in the form of one beep every second. Stay on the line. Tattletale should disconnect after about a minute. You can further check the operation of the unit by closing one of the sensing switches and verifying that the proper number of pulses is produced when you call. This completes the checkout of your unit and you may place it in service.

For protection against break-in you may use any of the readily available magnetically or mechanically operated door and window switches. It is also possible to use a continuous closed circuit composed of foil placed around the periphery of glass panes. Remember, a continuous loop of foil such as this constitutes a normally closed circuit and you will need to implement the circuit modification of Figure 25-3.

To detect fire, heating system, refrigeration, and/or air conditioning failure it is easiest to install a suitable thermostat in the affected area. For heating system failure, use a thermostat whose contacts close on a fall in temperature. For fire, or refrigeration, and air conditioning failure use a thermostat with contacts that close on a rise in temperature.

To detect a flood you can purchase or construct a simple float switch which is activated when water enters the protected area. For an extra early warning it would be prudent to locate the float switch in a well or depression such as the bottom of the sump where a pump is located. Be sure that water is not permitted to touch any of the electrical connections of your switch. Now that Tattletale is ready to operate, do not forget to turn it on when you leave the premises. The LED indicator will provide assurance that your unit is armed and ready to answer your call. Be sure to turn the unit off again when you return so that you do not miss your incoming calls.

Chapter 26

DTMF Telephone Controller Project

Parts list

Parts Bin

DTMF Telephone Controller

R1   10 K ohm, $\frac{1}{4}$ W,
  5% resistor

R2   150 K ohm, $\frac{1}{4}$ W,
  5% resistor

R3   82 K ohm, $\frac{1}{4}$ W,
  5% resistor

R4 R_A - timing
  component U5 (see
  text) set for 60 s

R5 R_B - timing
  component U6 (see
  text) set for 5 s

R6, R9   4.7 K, $\frac{1}{4}$ W,
  5% resistor

R7 R_D - timing
  component U7(see text)
  set for 5 s

R8 R_C - timing
  component U7(see text)
  set for 5 s

R10   390 K, $\frac{1}{4}$ W,
  5% resistor

R11, R12, R13, R14
  330 ohm, $\frac{1}{4}$ W,
  5% resistor

R15, R16, R17   330 ohm,
  $\frac{1}{4}$ W, 5% resistor

R18, 19   4.7 K, $\frac{1}{4}$ W,
  5% resistor

R20, R21, R22, R23
  4.7 K, $\frac{1}{4}$ W,
  5% resistor

C1   .22 µF, 200 V
  capacitor

C2   10 µF, 200 V
  capacitor

C3 C_A - timing
  component U5(see text)
  set for 60 sec

C4, C6, C8, C9   1 µF,
  35 V capacitor

C5 C_B - timing
  component U6(see text)
  set for 5 sec

C7 C_C - timing
  component U7(see text)
  set for 5 sec

C10   .1 µF, 35 V disc
  capacitor

Q1, Q2, Q3, Q4   BC547
  NPN transistor

Q5, Q6, Q7, Q8   BC547
  NPN transistor

U1   MT8870 Touch-Tone
  decoder

U2   7447 IC-BCD
  7-segment decoder

U3   74154 IC - 4 to 16
  line decoder

U4   MCT2E opto-coupler

U5, U6, U7   NE555 Timer
  IC

U8, U12   7408 - Quad
  2-input AND gate IC

U9, U10   7474 -
  D-Flip-Flop IC

```
U11  74126 - Quad
     3-state buffer IC
U13  7404 - Hex
     inverter IC
U14  LM7805 regulator IC
XTL  3.58 MHZ crystal
DSP  7-segment display
     (see text)
RLY1, RLY2  5-V reed
     relay
RLY3, RLY4, RLY5  5-V
     reed relay
```

The DTMF Telephone Controller is ideal for remote control applications around your home, office, shop, or barn. You can remotely control almost any device with this controller project. The DTMF Telephone Controller unit continually monitors your phone line and looks for the proper user touch-tone sequences to remotely control devices around your home or office (see Figure 26-1).

Once it is heard, the user may select which of the trigger relays to turn on or off. Now you can call home and, when your answering machine picks up, control lights, security systems, or just about any number of other custom applications! The DTMF controller offers full control from any telephone anywhere in the world!

The Telephone remote controller system board utilizes the Dual Tone Multi Frequency Tone (DTMF) encoding or touch-tone dialing technology, which was invented by the Bell System many years ago. When you press a numeric button in the telephone keypad, the keypad starts up two oscillators and generates two

tones at the same time. These two tones are taken from a row frequency and a column frequency. The resultant frequency signal is called "Dual Tone Multiple Frequency." A DTMF signal is the algebraic sum of two different audio frequencies. Each of the low and high frequency groups comprise four frequencies from the various keys present on the telephone keypad; two different frequencies, one from the high frequency group and another from the low frequency group are used to produce a DTMF signal to represent the pressed key.

The frequencies were originally chosen so that they were not harmonically related to each other. The frequencies associated with touch-tone keys on the telephone encoder keypad are shown in Table 26-1.

When you send these DTMF signals to the telephone exchange, the telephone exchange identifies these signals and makes the connection through computers and switching circuits to the person you are calling.

When you press the digit 5 on a telephone keypad it generates a resultant tone signal which is made up of frequencies 770 and 1336 Hz. Pressing digit 8 will produce the tone taken from tones 852 Hz and 1336 Hz. In both the cases, the column frequency 1336 Hz is the same. These signals are digital signals which are symmetrical with the sinusoidal wave.

Circuit description

The Telephone Controller consists of eight sub-circuit blocks, beginning with the ring detector, a DTMF decoder, 4-16 line decoder/demultiplexer, some D-flip-flops, display driver, relay driver circuits, and power supply, etc. The overall DTMF Telephone Controller schematic is shown in Figure 26-2.

Figure 26-1 *DTMF controller*

Table 26-1
Touch tone frequency chart

		High Frequency Group		
Low		1,209 Hz	1,330 Hz	1,447 Hz
Frequency	697 Hz	1	2	3
Group	770 Hz	4	5	6
	852 Hz	7	8	9
	941 Hz	0	0	#

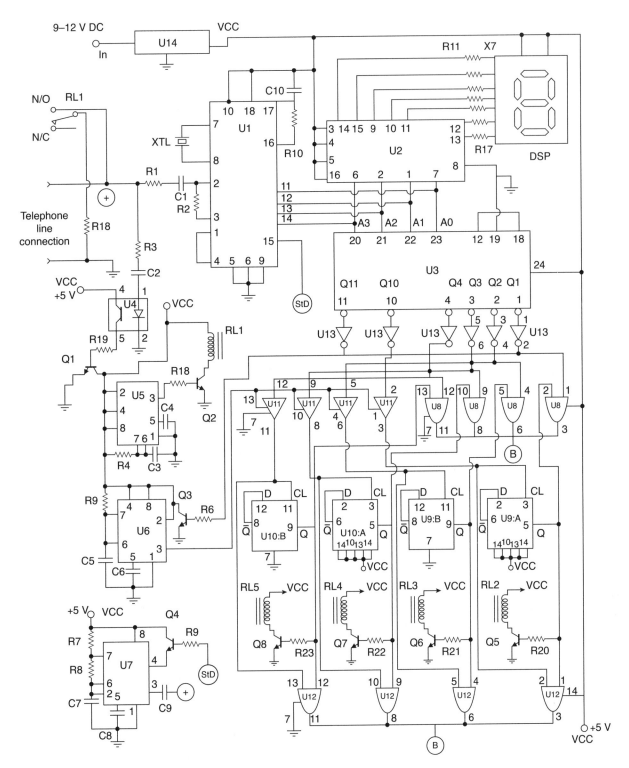

Figure 26-2 *Touch tone controller*

Our discussion will begin with the ring detection circuit and then progress through the entire circuit, describing the circuit in detail.

Ring detector circuit

The ring detector circuit essentially "picks up the phone" automatically after a set period of time. This circuit identifies the ringing signal sent from the telephone exchange. On getting the ringing signal this circuit connects the controller unit to the telephone line. The incoming AC ringing signal is bypassed by resister R3 and capacitor C2 and is then applied to the opto-coupler MCT2E at U4. Note that this opto-coupler is a 6 pin IC. The opto-coupler is made up of an internal LED and a transistor. When the internal LED glows, the light falls on the emitter-collector junction of the transistor. This transistor becomes forward biased and the output is obtained at the emitter of the transistor. On applying the signal to the anode of the opto-coupler, grounding the cathode, on the positive cycle of the signal the LED glows; as a result +5 V output is obtained at the emitter of the opto-coupler at pin number 4 of U4.

The ring detector circuit is built around a monostable multi-vibrater constructed around timer IC 555, at U5. When a negative going pulse is applied to its triggering input at pin 2, the output of the IC goes high. This output is available at pin 3 of this IC. This will remain high for the time period designed by the RC combination, depending on values of resister R_A and capacitor C_A. A high signal on pin 3 of this timer IC biases the transistor Q2 in the relay driver circuit which in turn switches ON the relay. This relay puts a resistance loop of about 220 ohms across the telephone line. This resistance loop coming on-line causes the telephone line voltage to drop from 50 to 12 V. This is the same as lifting the receiver of the telephone handset (hook-off state). This timing circuit is designed for a period of 60 seconds, calculated by the formula: $Td = 1.1\ R_A\ C_A$.

After the timing period the output of this IC goes low which in turn switches OFF the transistor T2. By varying the values of the R_A and C_A the ON period of the monostable multi-vibrator is changed according to the formula given above. In the relay driver circuit a resister is used to provide the necessary base current to the transistor so that it can bias properly. Note that the

other NE555 timer chips at U6 and U7 also have their timing components listed as R_b, R_c .... and C_b, C_c .... , you must also compute the optimum timing components for these two ICs.

Signal decoding section

After passing through the ring detector, the circuit is ready to receive a DTMF signal from the remote telephone or touch-tone encoder. The MT8870 DTMF decoder chip is the heart of the Telephone Controller. The decoder chip takes DTMF signal coming via the telephone line and converts that signal into a respective 4-bit BCD number output obtained from pins 11 to 14; see the block diagram in Figure 26-3 and Truth Table 26-2.

The oscillator frequency of the decoder is controlled by a 3.85 MHz crystal. The MT-8870 is a full DTMF receiver that integrates both band split filter and decoder functions into a single 18-pin DIP. Its filter section uses switched capacitor technology for both the high and low group filters and for dial tone rejection. Its decoder uses digital counting techniques to detect and decode all 16 DTMF tone pairs into a 4-bit code. External component count is minimized by provision of an on-chip differential input amplifier, clock generator, and latched tri-state interface bus. Minimal external components required include a low-cost 3.579545 MHz crystal, a timing resistor, and a timing capacitor. The MT-8870-02 can also inhibit the decoding of fourth column digits.

The MT-8870 operating functions include a band split filter that separates the high and low tones of the received pair, and a digital decoder that verifies both the frequency and duration of the received tones before passing the resulting 4-bit code to the output bus. The low and high group tones are separated by applying the dual-tone signal to the inputs of two 6th order switched capacitor band pass filters with bandwidths that correspond to the bands enclosing the low and high group tones.

The filter section of the DTMF controller incorporates notches at 350 and 440 Hz, providing excellent dial tone rejection. Each filter output is followed by a single-order switched capacitor section that smooths the signals prior to limiting. Signal limiting is performed by high gain comparators that provide hysteresis to prevent detection of unwanted low-level signals and noise. The MT-8870 decoder uses

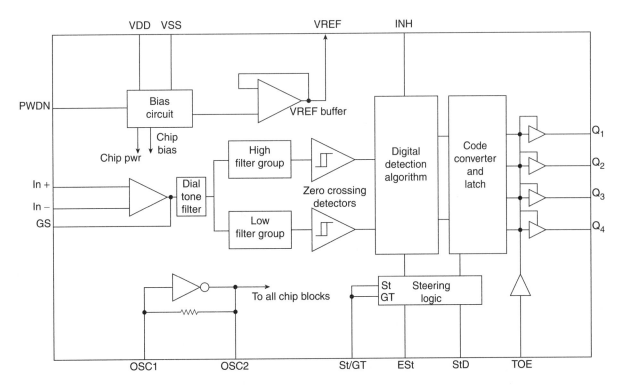

Figure 26-3 *Touch tone decoder block diagram*

a digital counting technique to determine the frequencies of the limited tones and to verify that they correspond to standard DTMF frequencies. When the detector recognizes the simultaneous presence of two valid tones (known as signal condition), it raises the Early Steering flag (ESt). Any loss of signal condition will cause ESt to fall. Before a decoded tone pair is registered, the receiver checks for valid signal duration (referred to as character-recognition-condition). This check is performed by an external RC time constant driven by ESt. A short delay to allow the output latch to settle, and the delayed steering output flag (StD) goes high, signaling that a received tone pair has been registered. The contents of the output latch are made

Table 26-2
MT8870 Truth table

| | BCD Outputs Q3–Q0 | | | | |
Keys	Q3	Q2	Q1	Q0	Device
1	0	0	0	1	1
2	0	0	1	0	2
3	0	0	1	1	3
4	0	1	0	0	4
5	0	1	0	1	5
6	0	1	1	0	6
7	0	1	1	1	7
8	1	0	0	0	8
9	1	0	0	1	9
0	1	0	1	0	10

available on the 4-bit output bus by raising the three state control input (OE) to logic high. The inhibit mode is enabled by a logic high input to pin 5 (INH). It inhibits the detection of 1633 Hz.

The output code will remain the same as the previous detected code. On the M-8870 models, this pin is tied to ground (logic low). The input arrangement of the MT-8870 provides a differential input operational amplifier as well as a bias source (VREF) to bias the inputs at mid-rail. Provision is made for connection of a feedback resistor to the op-amp output (GS) for gain adjustment. The internal clock circuit is completed with the addition of a standard 3.579545 MHz crystal. The input arrangement of the MT-8870 IC provides a differential input operational amplifier as well as a bias source (VREF) to bias the inputs at mid-rail. Provision is made for connection of a feedback resistor to the op-amp output (GS) for gain adjustment.

The DTMF Controller utilizes three NE555 timer chips. The NE555 integrated circuit is a timing chip that is capable of producing accurate timing pulses. This IC is used as a multi-vibrator. By using this IC we can construct two types of multi-vibrator, monostable and astable. The monostable multi-vibrator produces a single pulse when a triggering pulse is applied to its triggering input. The astable multi-vibrator produces a train of pulses depending on the resister-capacitor combination wired around it.

With a monostable operation, the time delay is controlled by one external resistor and one capacitor connected between Vcc-Discharge (R), and Threshold-Ground (C). With an astable operation, the frequency and pulse width are produced by two external resistors and one capacitor connected between Vcc-Discharge (R), Discharge-Threshold (R), and Threshold-Ground (C). In integrated circuit U5, the first NE555 is driven by the ring detector and control the phone line connection. The second NE555, at U6 drives the 74126 chips and the third NE555 chip is driven by the touch-tone decoder chip's StD line.

As previously mentioned, the output of the Touch-Tone decoder is then fed to the 4-16 line decoder IC74154. This IC takes the BCD number and decodes. According to that BCD number it selects the active low output line from 1 to 16 which is decimal equivalent of the BCD number present at its input pins. The output of

the 74154 the needs to get inverted to get a logical high output. This inversion is carried out by TTL hex inverter IC 7404. This IC inverts the data on its input terminal and gives inverted output.

The SM74154 integrated circuit is a 4-16 line decoder; it takes the 4 line BCD input and selects respective output, one among the 16 output lines. It is active low output IC so when any output line is selected it is indicated by an active low signal; the rest of the output lines will remain active high. This 4-line-to-16-line decoder utilizes TTL circuitry to decode four binary-coded inputs into one of sixteen mutually exclusive outputs when both the strobe inputs, G1 and G2, are low. The de-multiplexing function is performed by using the 4 input lines to address the output line, passing data from one of the strobe inputs with the other strobe input low. When either strobe input is high, all outputs are high. These de-multiplexers are ideally suited for implementing high-performance memory decoders. All inputs are buffered and input clamping diodes are provided to minimize transmission-line effects and thereby simplify system design; see the multiplexer/demultiplexer diagram in Figure 26-4 and truth Table 26-3.

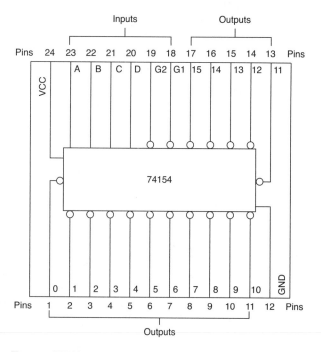

Figure 26-4 *74154 Decoder chip*

Table 26-3
CSN74154 Decoder truth table

G1	G2	D	C	B	A	0	1	2	3	4	5	6	7	8	9	10	11	12	13	14	15
L	L	L	L	L	L	L	H	H	H	H	H	H	H	H	H	H	H	H	H	H	H
L	L	L	L	L	H	H	L	H	H	H	H	H	H	H	H	H	H	H	H	H	H
L	L	L	L	H	L	H	H	L	H	H	H	H	H	H	H	H	H	H	H	H	H
L	L	L	L	H	H	H	H	H	L	H	H	H	H	H	H	H	H	H	H	H	H
L	L	L	H	L	L	H	H	H	H	L	H	H	H	H	H	H	H	H	H	H	H
L	L	L	H	L	H	H	H	H	H	H	L	H	H	H	H	H	H	H	H	H	H
L	L	L	H	H	L	H	H	H	H	H	H	L	H	H	H	H	H	H	H	H	H
L	L	L	H	H	H	H	H	H	H	H	H	H	L	H	H	H	H	H	H	H	H
L	L	H	L	L	L	H	H	H	H	H	H	H	H	L	H	H	H	H	H	H	H
L	L	H	L	L	H	H	H	H	H	H	H	H	H	H	L	H	H	H	H	H	H
L	L	H	L	H	L	H	H	H	H	H	H	H	H	H	H	L	H	H	H	H	H
L	L	H	L	H	H	H	H	H	H	H	H	H	H	H	H	H	L	H	H	H	H
L	L	H	H	L	L	H	H	H	H	H	H	H	H	H	H	H	H	L	H	H	H
L	L	H	H	L	H	H	H	H	H	H	H	H	H	H	H	H	H	H	L	H	H
L	L	H	H	H	L	H	H	H	H	H	H	H	H	H	H	H	H	H	H	L	H
L	L	H	H	H	H	H	H	H	H	H	H	H	H	H	H	H	H	H	H	H	L
L	H	X	X	X	X	H	H	H	H	H	H	H	H	H	H	H	H	H	H	H	H
H	L	X	X	X	X	H	H	H	H	H	H	H	H	H	H	H	H	H	H	H	H
H	H	X	X	X	X	H	H	H	H	H	H	H	H	H	H	H	H	H	H	H	H

Number display section

This section of the circuit displays the received device code from the telephone line dialed from remote section. This section consists of a BCD to seven segment decoder, the SM74LS47 and a seven segment common anode display, at DSP. The DM74LS47 chip accepts four lines of BCD (8421) input data from the decoder chip and generates their complements internally and decodes the data with seven AND/OR gates having open-collector outputs to drive indicator segments directly. Each segment output is guaranteed to sink 24 mA in the ON (LOW) state and withstand 15 V in the OFF (HIGH) state with a maximum leakage current of 250 mA. Auxiliary inputs provided blanking, lamp test and cascadable zero-suppression functions. The display driver IC has 10 pins, two of these pins are common for all LEDs and remaining are another polarity terminals of the LED. When common anode seven segment display is used, two common terminal pins are connected to +5 V or logic high state and another terminal are kept at logic low state, allowing the respective LED to glow.

Device control section

The Device Control Section consists of device status check unit, device switching unit, device status feedback unit, relay driver circuit and beep tone generator unit. Before switching On/Off any device, there can be confusion about its present status of the outputs. This section handles the device status. The inverted output of 4-16 line decoder and the output of respective flip-flop are fed to the independent block of AND gates at U8 and U12 using 7408. If the device is already in the ON state, then we will hear a beep sound. The outputs of each AND gate are connected to the beep tone generator unit by using a transistor. This beep generator unit produces a short duration beep indicating than the device is already in switched ON state. If device is in OFF state then no beep will be heard.

Device switching section

This Device Switching section consists of a tri-state buffer at U11 and two flip-flops at U9 and U10. The tri-state

buffer, an LM74126, contains four independent gates each of which performs a non-inverting buffer function. The outputs have the 3-STATE feature. When control signal is at high state, the outputs are nothing but the data present at its input terminals. When control signal is at low state, the outputs are held at high impedance state. So no output will be available at the output terminal. The LM7474 is a conventional D-flip-flop IC. This IC consists of two D flip-flops. These flip-flops are used to latch the data that present at its input terminal. Each flip-flop has one data, one clock, one clear, one preset input terminals.

After making confirmation of current status of the device to alter the status of that device, you have to change the mode of the tri-state buffer by making the control input high. This is done by pressing the "#" key. When this key is pressed the output of the 4-16 line decoder goes low. This gives a triggering pulse to the monostable multi-vibrator which is built around the U6. This will keep the output high for about 5 seconds. Working of the monostable multi-vibrator already discussed. In this time interval the output of the tri state buffer will be the signal at its input terminal.

Now the device code of the chosen relay whose status is to be altered is again pressed. The output of the tri-state buffer is latched by using a D flip-flop. This D flip flop is used in the toggle mode. For each positive going edge of the clock a pulse will trigger the flip flop. After a period of 5 s the output of the U6 goes low and puts the tri state buffer in the high impedance state. Therefore to change the status of any other device to be done after the output of U6 goes low, again "#" key is pressed to make the tri-state buffer act as input–output state and the respective code of the device is pressed.

Device feedback section

After changing the present status of the device, you can confirm the operation that you performed. The DTMF Controller unit will now give the feedback tone after switching ON any device. This status feedback circuitry, utilizes a dual input AND gate, the output of the flip flop and a tri state buffer to provide the input for the feedback. When the both inputs are high that indicates that the device is switched ON, then the output of the AND gate goes logic high state. This output is fed to the beep generator unit through switching a transistor. Until you press the key the feedback tone is heard. This feedback tone is heard only when the device is switched ON. After switching OFF the device, this tone is not heard.

Beep tone generator section

The beep tone generator unit produces a beep tone of audible frequency. This unit is constructed using a 555 timer chip. Here it is wired as an astable multi-vibrator with a few external components like resister and capacitor, required along with the timer 555 chip set. This frequency comes in the audible range between 40 to 650 Hz. It should be less than 650 Hz otherwise it will mix up with the DTMF tone. When it is less than 650 Hz the frequency which causes the false triggering is filtered-off by the external structure of DTMF decoder IC at U1.

Relay driver circuit

In order to switch the appliances On or Off, small PC board type relays were used. Since the output of the D flip flop is normally +5 V or it is the voltage of logic high state, we cannot use this output to run the device or appliances. Therefore here we use relays which can handle a high voltage of 230 V or more, and a high current in the rate of 10 A to energize the electromagnetic coil of the relays +5 V is sufficient. Here we use the transistors to energize the relay coil. The output of the D flip-flop is applied to the base of the transistor Q5–Q8 via a resistor. When the base voltage of the transistor is above 0.7 V the emitter-base (EB) junction of the transistor becomes forward biased. As a result the transistor goes into saturation, thus switching ON the transistor. Switching the transistor ON, thus activates the relay. When the output of D flip-flop goes low the base voltage drops below 0.7 V and as a result the device also switches OFF.

The DTMF controller is a relatively compact unit with four 5-A relay outputs, each with its own screw terminals. The DTMF controller requires 9–12 V DC at 300 mA, which can be obtained from a small "wall cube" or "wall-wart" power supply, available through a local RadioShack store or parts supplier outlet.

Power supply section

In order for the DTMF Controller to operate, the circuit must have a power supply and a permanent back up power supply. This is achieved by using a 5 V regulated power supply from a voltage regulated IC 7805. This 5 V source is connected to all ICs and relays, and the IC gets a backup from a 9 V battery.

Assembling the DTMF controller

Before we begin constructing the DTMF Controller, you will need to locate a clean, well-lit worktable or workbench. Next we will gather a small 25–30-W pencil tipped soldering iron, a length of 60/40 tin–lead solder, and a small canister of Tip Tinner, available at your local RadioShack store; Tip Tinner is used to condition the soldering tip between solder joints. You drive the soldering tip into the compound and it cleans and prepares the soldering tip. Next grab a few small hand tools; try to locate small flat blade and Phillips screwdrivers. A pair of needle-nose pliers, a pair of tweezers along with a magnifying glass, and a pair of end-cutters and we will begin constructing the project. Now locate your schematic diagram and parts layout diagram along with all of the components needed to build the project.

Once all the components are in front of you, you can check them off against the project parts list to make sure you are ready to start building the project. Now, locate Tables 26-4 and 26-5, and place them in front of you. Table 26-4 illustrates the resistor code chart and how to read resistors, while Table 26-5 illustrates the capacitor code chart which will aid you in constructing the project.

Refer now to the resistor color code chart in Table 26-4, which will help you identify the resistor values for the project. Note that each resistor will have either three or four color bands, which begin at one end of the resistor's body. The first colored band represents the resistor's first digit, while the second color represents the second digit of the resistor's value. The third colored band represents the resistor's multiplier value, and the fourth color band represents the resistor's tolerance value. If there is no fourth colored band then the resistor has a 20% tolerance value. If the resistor's fourth band

Table 26-4

Resistor color code chart

Color Band	1st Digit	2nd Digit	Multiplier	Tolerance
Black	0	0	1	
Brown	1	1	10	1%
Red	2	2	100	2%
Orange	3	3	1,000 (K)	3%
Yellow	4	4	10,000	4%
Green	5	5	100,000	
Blue	6	6	1,000,000 (M)	
Violet	7	7	10,000,000	
Gray	8	8	100,000,000	
White	9	9	1,000,000,000	
Gold			0.1	5%
Silver			0.01	10%
No color				20%

Table 26-5

Three-digit capacitor codes

pF	nF	µF	Code	pF	nF	µF	Code
1.0			1R0	3,900	3.9	.0039	392
1.2			1R2	4,700	4.7	.0047	472
1.5			1R5	5,600	5.6	.0056	562
1.8			1R8	6,800	6.8	.0068	682
2.2			2R2	8,200	8.2	.0082	822
2.7			2R7	10,000	10	.01	103
3.3			3R3	12,000	12	.012	123
3.9			3R9	15,000	15	.015	153
4.7			4R7	18,000	18	.018	183
5.6			5R6	22,000	22	.022	223
6.8			6R8	27,000	27	.027	273
8.2			8R2	33,000	33	.033	333
10			100	39,000	39	.039	393
12			120	47,000	47	.047	473
15			150	56,000	56	.056	563
18			180	68,000	68	.068	683
22			220	82,000	82	.082	823
27			270	100,000	100	.1	104
33			330	120,000	120	.12	124
39			390	150,000	150	.15	154
47			470	180,000	180	.18	184

Table 26-5—cont'd

Three-digit capacitor codes

pF	nF	μF	Code	pF	nF	μF	Code
56			**560**	220,000	220	.22	**224**
68			**680**	270,000	270	.27	**274**
82			**820**	330,000	330	.33	**334**
100		.0001	**101**	390,000	390	.39	**394**
120		.00012	**121**	470,000	470	.47	**474**
150		.00015	**151**	560,000	560	.56	**564**
180		.00018	**181**	680,000	680	.68	**684**
220		.00022	**221**	820,000	820	.82	**824**
270		.00027	**271**		1,000	1	**105**
330		.00033	**331**		1,500	1.5	**155**
390		.00039	**391**		2,200	2.2	**225**
470		.00047	**471**		2,700	2.7	**275**
560		.00056	**561**		3,300	3.3	**335**
680		.00068	**681**		4,700	4.7	**475**
820		.00082	**821**			6.8	**685**
1,000	1.0	.001	**102**			10	**106**
1,200	1.2	.0012	**122**			22	**226**
1,500	1.5	.0015	**152**			33	**336**
1,800	1.8	.0018	**182**			47	**476**
2,200	2.2	.0022	**222**			68	**686**
2,700	2.7	.0027	**272**			100	**107**
3,300	3.3	.0033	**332**			157	**157**

Code = 2 significant digits of the capacitor value + number of zeros to follow

For values below 10 pF, use "R" in place of decimal e.g. 8.2 pF = 8R2

10 pF = 100
100 pF = 101
1,000 pF = 102
22,000 pF = 223
330,000 pF = 334
1 μF = 105

is silver then the resistor has a 10% tolerance value and if the fourth band is gold then the resistor's tolerance value will be 5%.

Finally we are ready to begin assembling the project, so let's get started. The prototype project was constructed on a single-sided printed circuit board or PCB. With your PC board in front of you, we can now begin populating the circuit board. We are going to place and solder the lowest height components first, the resistors and the diodes. First, locate our first resistor R1. Resistor R1 has a value of 10 K ohms and is a 5% tolerance type. Now take a look for a resistor whose first color band is brown, black, and orange. From the chart you will notice that brown is represented by the digit (1), the second color band is black, which is represented by (0). Notice that the third color is orange, so the multiplier is (100), so (1) (0) × 100 = 10,000 ohms or 10 K ohms.

Go ahead and locate where resistor R1 goes on the PC and install it. Next solder it in place on the PC board, and then with a pair of end-cutters trim the excess component leads flush to the edge of the circuit board. Next locate the remaining resistors and install them in their respective locations on the main controller PC board. Solder the resistors to the board, and remember to cut the extra lead with your end-cutters.

Next refer to Table 26-5; this chart helps to determine value of small capacitors. Many capacitors are quite small in physical size, so manufacturers have devised a three-digit code which can be placed on the capacitor instead of the actual value. The code occupies a much smaller space on the capacitor and can usually fit nicely on the capacitor body, but without the chart the builder would have a difficult time determining the capacitor's value.

Capacitors must be installed properly with respect to polarity in order for the circuit to work properly. Failure to install electrolytic capacitors correctly could cause damage to the capacitor as well as to other components in the circuit when power is first applied.

Note that capacitors generally come from two major groups or types. Non-polar types, such as the ones in this project, have no polarity and can be placed in either direction on the circuit board. The other type of capacitors are called polarized types, these capacitors have polarity and have to be inserted on the board, with respect to their polarity marking in order for the circuit to operate correctly. On these types of capacitors, you will observe that the components actually have a plus or minus marking or a black or white band with a plus or minus marking on the body of the capacitors.

Now look in the component stack for the capacitors and locate capacitor C1; it is a rather large capacitor labeled .22 µF, 250 v -NP. The NP refers to non-polarized. This large capacitor is placed across the telephone line and will have to be a 250 V type. When installing the remaining capacitors, refer to the Table 26-5 to find the three-digit code on the smallest capacitors, if you do not see the actual value printed on them. Go ahead and place the capacitors on the circuit board and solder them in place. Now take your end-cutters and cut the excess component leads flush

to the edge of the circuit board. Once you have installed these components, you can move on to installing the remaining capacitors.

The DTMF Telephone Controller utilizes eight conventional transistors. Transistors are generally three lead devices, which must be installed correctly if the circuit is to work properly. Careful handling and proper orientation in the circuit is critical. A transistor will usually have a Base lead, a Collector lead, and an Emitter lead. The base lead is generally a vertical line inside a transistor circle symbol. The collector lead is usually a slanted line facing towards the base lead and the emitter lead is also a slanted line facing the base lead, but it will have an arrow pointing towards or away from the base lead. Now, identify the two transistors, referring to the manufacturer's specification sheets as well as the schematic and parts layout diagrams when installing the two transistors. If you have trouble installing the transistors, get the help of an electronics enthusiast with some experience. Install the transistors in their respective locations and then solder them in place. Remember to trim the excess component leads with your end-cutters.

The DTMF Controller circuit is a number of ICs. ICs are often static sensitive, so they must be handled with care. Use a grounded anti-static wrist strap and stay seated in one location when handling the ICs. Take out a cheap insurance policy by installing IC sockets for each of the ICs. In the event of a possible circuit failure, it is much easier to simply unplug a defective IC, than trying to unsolder 14 or 16 pins from a PC board without damaging the board. ICs have to be installed correctly if the circuit is going to work properly. IC packages will have some sort of markings which will help you orient them on the PC board. An IC will have either a small indented circle, a cutout, or notch, at the top end of the IC package. Pin 1 of the IC will be just to the left of the notch, or cutout. Refer to the manufacturer's pin-out diagram, as well as the schematic when installing these parts. If you doubt your ability to correctly orient the ICs, then seek the help of a knowledgeable electronics enthusiast to help you.

Next, you will need to identify the opto-isolator which is installed at U4. It is a 6-pin device which looks like a small DIP IC. It should have a located mark to help you identify pin number one, usually at the top left of the IC. Go ahead and install the opto-isolator on the PC using an 8-pin DIP IC socket.

The DTMF Controller circuit utilizes a single 7-segment common anode LED display at DSP, which also must be oriented correctly if the circuit is to function properly. The 7-segment display has 10 pins. Out of this, two pins are common for all LEDs and remaining are other polarity terminals of the LED. When common anode seven segment display is used, two common terminal pins are connected to +5 V or logic high state and other terminals are kept at logic low state. Then respective LED glows. You may want to lead about ¼″ lead length on the LED, in order to have the LED protrude through the enclosure box. Solder the LED to the PC board and then trim the excess leads.

Once the DTMF Telephone Controller circuit board has been completed, take a short break and then we will inspect the PC board for possible "cold" solder joints and "short" circuits. Pick up the PC board with the foil side facing upwards toward you. Carefully inspect the solder joints, these should look smooth, clean, and bright. If any of the solder joints look dull, dark, or blobby, then you should remove the solder with a solder sucker or a solder wick and then resolder the joint all over again. Now we will inspect the PC board for possible short circuits. Short circuits can be caused from small solder blobs bridging the circuit traces or from cut component leads which can often stick to the circuit board once they have been cut. A sticky residue from solder will often trap component leads across the circuit traces. Look carefully for any bridging wires across circuit traces.

Once this task is complete, we can move on to installing the circuit board into an aluminum enclosure. Locate a suitable enclosure to house the circuit board within the enclosure. Arrange the circuit board on the bottom of the chassis box. Now, locate four ⅛ inch standoffs, and mount the PC board between the standoffs and the bottom of the chassis box.

Mount a coaxial power jack on the rear panel of the chassis box for the small AC "wall-cube" power supply, which is connected to the input of the regulator at U14. The 9–12 V dc power supply could be a low cost

"wall wart" type power supply. You can mount the status 7-segment number display (DSP) on the circuit board or on the front panel if desired. If you mount the DSP display on the circuit board, you will have to cut out a "window" on the chassis, so that when the PC board is mounted up to the panel the display can shine out the small window. Next you will have to install an RJ11 phone jack or a 2-position terminal strip on the front panel of the chassis to bring the phone into the chassis box. You will most likely want to mount a 10-position screw terminal block on the rear panel to allow connections to the control relays. Install a power On-Off switch on the front panel.

Now you are ready to connect the main controller to the phone line, and apply power to the DTMF Telephone Controller board; and begin testing of the circuit.

Testing the DTMF controller

Once the DTMF Controller has been installed in the chassis box and connected to the telephone line and power supply, you will be ready to begin testing the DTMF Controller circuit. Now switch the On-Off switch so that the telephone line is connected to telephone interfacing unit.

Now place a call to your Telephone Controller set using a remote telephone set or mobile phone. The signal goes to the telephone exchange and the exchange sends a ringing signal to your set through the phone line. The ring detector unit detects the ringing signal and makes the output of the U5 to high state so that local control section of then connected to the telephone for a time interval of 60 s.

Now follow the steps below to test the proper sequence of the DTMF Controller. First, press the respective code of the device whose status is to be checked. The dialed number of the device is displayed on seven segment display. If the device is already switched ON then you will hear a short duration beep tone from beep tone generator unit. Now press the "#" button on the keypad and again press the device number which is displayed; now the device is switched OFF and you will not hear the feedback tone indicating that the device is switched OFF.

Finally, repeat the first step once again. In step one you will not hear the beep tone because the device is switched OFF during above step two. Repeat step two. Now you will hear the feedback tone because the device is switched ON. After 60 s the local control unit will be disconnected from the telephone line so that your money is saved.

Applications

The DTMF Telephone Controller is a very useful project with many possible home or office switching applications. Using a standard telephone as a remote encoder and the conventional telephone system, and you can construct useful control projects. Have fun!

Chapter 27

The BASIC STAMP Microcontroller

A handful of telephone related projects in the latter part of this book illustrates how a microcontroller, which is a small microprocessor, can be used to perform the major functions of or control a complete system. The projects in this book revolve around the Parallax BASIC STAMP 2 microcontroller. The Parallax controllers can perform many functions, are quite flexible, and are relatively low in cost. Parallax probably offers the most support of any microcontroller supplier, so you can get started quickly and easily.

The BASIC STAMP microcontroller is a single-board computer that runs the Parallax PBASIC language interpreter in its microcontroller. The developer's code is stored in an EEPROM, which can also be used for data storage. The PBASIC language has easy-to-use commands for basic I/O, like turning devices on or off, interfacing with sensors, etc. More advanced commands allow the BASIC STAMP module to interface with other integrated circuits, communicate with each other, and operate in networks.

The program code entered into the BASIC STAMP is tokenized or compressed and stored within the STAMP's EEPROM memory. This memory is non-volatile, meaning it retains its program even without power. Every time the BASIC STAMP receives power, it starts running its code starting with the first executable line. This EEPROM can be rewritten immediately, without any lengthy erase cycle or procedure, by simply downloading the code from the STAMP editor again. Each location in the EEPROM is guaranteed for 10,000,000 write cycles before it wears out for the BS 2. The BASIC STAMP microcontroller is also available in a few other configurations such as the BS 2e, BS 2sx, BS 2p, and BS 2pe. EEPROM is guaranteed for 100,000 write cycles before it wears out.

Programming the BASIC STAMP 2 series microcontrollers is performed in PBASIC which is a very simple yet powerful language that allows anyone to program a microcontroller. The foundation of the language is designed around standard BASIC and commands terms as such as PULSOUT or SERIN which are hardware-driven commands. The beauty of the BASIC STAMP series microcontrollers is that you have the power to tell it exactly what to do, without going through any extra processes. This means high reliability and no guesswork. Using PBASIC allows you to save valuable time because the learning curve is very short and the sample codes and resources are abundantly available from www.parallax.com. Parallax offers free editors, manuals, and forums for help with programming the BASIC STAMP series as well as the new Propeller series of microcontrollers.

To begin using the BASIC STAMP controller, you would first connect the hardware sensors or switches to the BASIC STAMP pins, then you simply download the Windows STAMP Editor 2.5 or above. After downloading the editor, you install the editor program. Next you open up the editor program and call-up the program of choice or type in the code you wish to program into the BASIC STAMP controller. Connect up a programming cable between your personal computer or laptop and the STAMP controller. Next, you apply power to the BASIC STAMP controller. You will have to configure the editor program to communicate serially to the STAMP so you will have to establish communications between the PC and the STAMP by setting protocols such as COM port and speed and bits. Once that is down you can press RUN from the drop down tabs and the program should be loading in to the STAMP controller.

On the BASIC STAMP 2 modules from within the editor and with your PBASIC program on the screen, CTRL-M (Windows software) to view the memory map screens. In the DOS software, the first screen shows your RAM space, the second screen shows an expanded view of the program (EEPROM) space and the third screen shows a detailed view of the program space. Press the spacebar to alternate between the three screens.

In the Windows software, all EEPROM and RAM are shown on one window.

The BASIC STAMP 1 was the first of the STAMP series, using a mini version of BASIC or BASIC tokens in order to program the microcontroller. The BASIC STAMP 1C features a 256 byte EEPROM for program memory and runs at 4 mhz, performing 2,000 instructions per second. The STAMP 1 is programmed via the parallel port of a personal computer. All of the projects in this STAMP 2 communications project book will use the STAMP 2 microprocessor which sports a 2,048 byte EEPROM and runs at 20 µHz, performing 4,000 instructions per second. The STAMP 2 runs much faster and has a number of extra commands, such as SHIFTIN/SHIFTOUT and XOUT and will hold larger programs than the original STAMP 1. Table 27-1 illustrates the differences between the BASIC STAMP 1 and the BASIC STAMP 2.

The STAMP 2 is a very versatile and popular microprocessor building block which can be used to develop many different electronic systems, as shown in Figure 27-1 and in the block diagram shown in Figure 27-2.

Figure 27-1 *BASIC STAMP 2 microcontroller. Courtesy of Parallax, Inc*

The BASIC STAMP 2 is available in two different types in a DIP package or the SOIC style package. The original BASIC STAMP 2 is a PIC chip with a PBASIC2 interpreter running at 20 MHz. The STAMP 2 has 16 pins for input/output applications and can talk directly to

Table 27-1
STAMP microcontroller comparison chart

Product	BS-1-1C / STAMP 1	BS2-1C / STAMP 2
Package	14-Pin SIP	24-Pin DIP
Package size	1.4″ x .6″ x.1″	1.2″ × .6″ × .4″
Environment	32–158 degrees F	40–185 degrees F
Processor speed	4 mHz	20 mHz
Program execution speed	~2,000 instructions/s	~4,000 instructions/s
RAM size	16 bytes-2-I/O, 14 variable	32 bytes 6-I/O, 26 variable
Scratch pad RAM	N/A	N/A
EEPROM program size	256 bytes, 80 instructions	2 K bytes, 500 instructions
Number of I/O pins	8	16 + 2 dedicated serial
Voltage requirement	5–15 Vdc	5–15 Vdc
Current draw @ 5 V	1 mA-Run/25 µA sleep	3 mA-Run/50 µA sleep
Source/sink current per I/O	20 mA/25 mA	20 mA/25 mA
Source/sink current per unit	40 mA/50 mA	40 mA/50 mA per 8-I/O pins
PBASIC Commands	32	42
PC program Interface	serial w/BS1 serial adaptor	serial port 9600 baud
Windows text Editor	STAMPw.exe	STAMPw.exe

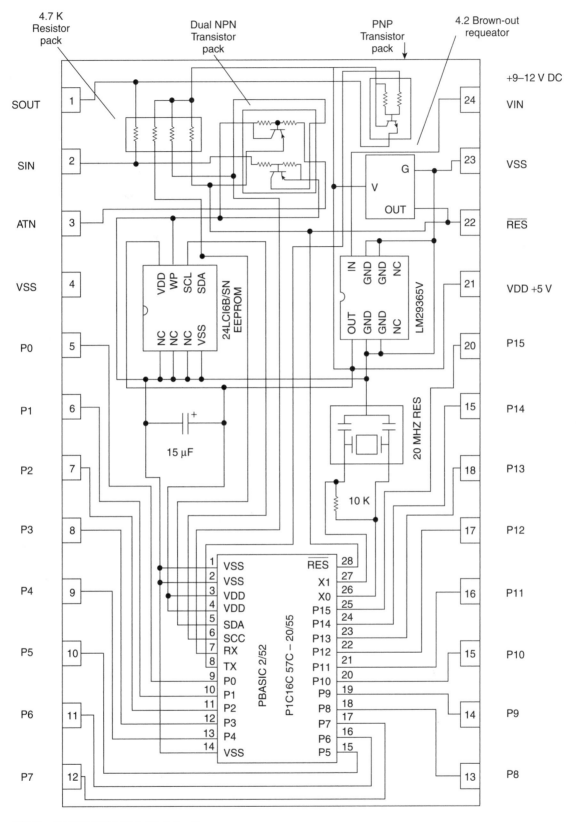

Figure 27-2 *BASIC STAMP 2 (Original)*

a serial port from a personal computer for programming. Parallax which sells the STAMP 2 products offers this in a 24 pin carrier board version which is a self contained STAMP 2 with onboard regulator and brow-out protection. The original STAMP 2 module is offered for about $49. Many first time hobbyists choose the STAMP 2 and a carrier board, the Board of Education or Homework series prototype systems to begin designing and programming their own new projects. The original STAMP 2 is illustrated with its programming cable in Figure 27-3.

The BASIC STAMP 2 is very easy to utilize. Simply connect a 9-V battery to pin 24 and a programming cable to pins 1 through 4 and you are ready to begin. Open up the free STAMP 2 editor, load a program from your computer to the BASIC STAMP 2 and your STAMP 2 will come to life. The original BASIC STAMP 2 utilizes a small 24-pin carrier board which houses the main processor chip, memory, regulators, etc.

Another approach to obtaining the STAMP 2 is to build your own "alternative" STAMP 2; in this configuration you obtain the 28 pin DIP interpreter chip and a few extra support or "glue" components and you can save some money by assembling your own Alternative STAMP 2.

The "alternative" STAMP 2 configuration revolves around the lower cost BASIC STAMP 2 Interpreter

Figure 27-4   *BASIC STAMP 2 Alternative*

chip, a 16C57C-HS/P PIC, and a 28 pin DIP style STAMP 2 as seen in Figure 27-4.

The lower cost 28 pin PIC BASIC interpreter chips, the PBASIC2C/P, requires a few external support chips and occupies a bit more real estate but the cost is significantly lower. The BASIC STAMP 2 costs $49, while "rolling your own" version will cost about half or $11 for the STAMP alternative plus $8 or so for the

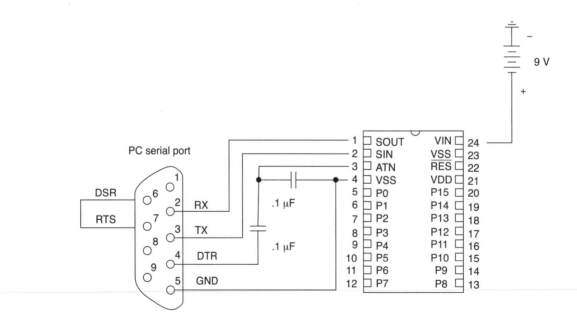

Figure 27-3   *BASIC STAMP 2 (BS2-1C) programming cable*

extra support components. The 28 pin BASIC STAMP 2 interpreter chip requires a few external support components, a 24LC6B memory chip, a MAX232 serial interface chip and a ceramic resonator. The MAX232 requires four 1 uf capacitors, a diode and a resistor, and provides a conventional serial I/O which is used to program the microprocessor. The STAMP 2 "alternative" circuit is depicted in Figure 27-5.

Whereas the original BASIC STAMP 2 contains all the associates microprocessor components on the 24-pin carrier board the alternative STAMP utilizes discrete components wired to the large 28-pin 16C57C-HS/P microprocessor chip.

You can build the alternative STAMP 2 on an experimenter board; with this method you can insert components and the chips, and wire the circuit together using jumper wires. In this method your designs are fluid and can be changed until you complete and debug your project. A second method to building your STAMP 2

projects is to use "perf-board" type construction, using point to point wiring. This type of construction is a little more permanent but can be modified more easily than if you were to construct your own printed circuit board. Printed circuit boards are generally used when design is complete or finalized and debugged.

The design of the alternate STAMP 2 requires you to supply your own 5 volt regulator in order to power the BASIC interpreter, memory, and serial communications chip. You can supply the regulator with 9–12 V DC via a "wall wart" power supply or build your own power supply.

A heat sink on the regulator is advisable, in order to dissipate heat properly. When constructing the alternative STAMP 2, be sure to observe the correct polarity when installing the four 1 µF electrolytic capacitors and the diode at D1. Note that the switch S1 is used to reset the microprocessor if one of the programs stalls. The alternative STAMP 2 has 16 in/outs

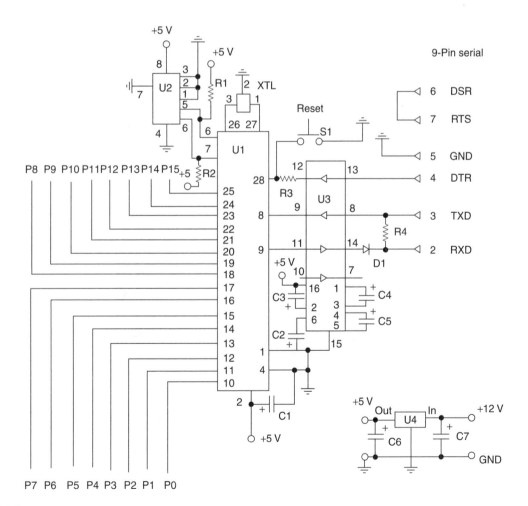

Figure 27-5 *Alternative STAMP 2*

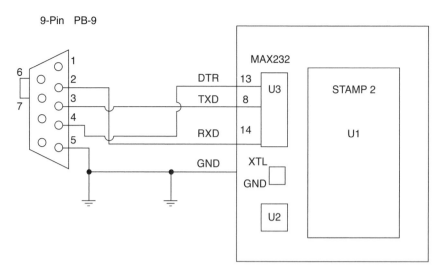

9-Pin PB-9

Figure 27-6   *Alternative STAMP 2 Programming cable*

but the pinouts are different from the original STAMP 2. The alternative STAMP 2 provides P0 through P15 at the interpreter chip pins 10 through 25. A simple trick is to add 10 to the functional designation. The MAX232 chip at U3 allows the PIC processor to communicate to your personal computer, so that you can download your programs via the serial port of your PC (see Figure 27-6).

The BASIC STAMP 2 is also available in the OEM plug-in board version which uses the 16C57C-HS/P chip as shown in Figures 27-7 and 27-8.

The OEM BASIC STAMP 2 is available as a kit or in a pre-wired version for around $38. The OEM version is designed to be a plug-in computer for a commercial product. The entire BASIC STAMP 2 and support components are on the PC board with a 20 pin header at

Figure 27-7   *BASIC STAMP OEM controller. Courtesy of Parallax, Inc*

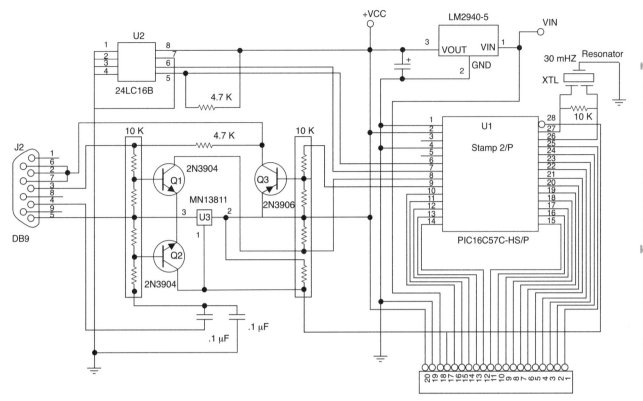

Figure 27-8 *BASIC STAMP 2 OEM/Industrial*

the bottom of the board. The pinouts for the OEM STAMP 2 are again a little different than the original STAMP 2 but lend themselves to all STAMP 2 applications; this is a good alternative to the original BASIC STAMP 2. The OEM STAMP 2 contains the STAMP 2 chip, a regulator, low voltage detector chip, a 20 mHZ resonator, a memory chip, three transistors, two resistor packs, and a few capacitors all on the PC board. You can readily substitute the BASIC OEM module for any of the projects in this book or for any BASIC STAMP 2 project. You must be careful to observe the proper pin-outs for both input and output pins because they are different between the original BASIC STAMP 2, the alternative STAMP 2 and the OEM STAMP 2.

Parallax, the supplier of the STAMP microcontroller products, has been around from the early 1990s and has a very large and dedicated following; and for very good reasons. Parallax supplies a complete line of processors, sensors, support chips as well as good on-line documentation, manuals, example codes, and forums for each of their products; see www.parallax.com

Chapter 28

DTMF Touch-Tone Generator/Decoder Display

Parts list

Parts Bin

Touch-Tone Generator/Dialer

U1 CM8880 Touch-tone encoder/decoder IC

U2 BASIC STAMP 2 9 (original)

U3 LM 7805 5 V regulator IC

U4 LM358 Op-amp

R1 100 K ohm $\frac{1}{4}$ W resistor

R2, R5 10 K ohm $\frac{1}{4}$ W resistor

R3 390 K ohm $\frac{1}{4}$ W resistor

R4 3.3 K ohm $\frac{1}{4}$ W resistor

R6 10 K ohm potentiometer

R7 22 K ohm $\frac{1}{4}$ W resistor

C1 .1 µF 250 V polyester capacitor

C2, 23, C7, .1 µF 35 volt disc capacitor

C4 2.2 µF 35 V electrolytic capacitor

C5, C6 10 µF 35 V electrolytic capacitor

XTL 3.579 color burst crystal

S1 SPST toggle switch

S2 SPST pushbutton switch (normally open) PC type

P1 DB-9, 9 pin female RS-232 connector

Misc circuit board, headers, wire, jack, plugs, etc.

Touch-Tone Decoder/Display

U1 CM8880 Touch-tone encoder/decoder IC

U2 BASIC STAMP 2 (original)

U3 LM 7805 5 V regulator

R1, R2, R5 100 K ohm $\frac{1}{4}$ W resistor

R3 37.5 K ohm $\frac{1}{4}$ W resistor

R4 60 K ohm $\frac{1}{4}$ W resistor

R6 390 K ohm $\frac{1}{4}$ W resistor

R7 10 K ohm $\frac{1}{4}$ W resistor

C1, C2 .1 µF 250 V polyester capacitor

C3, CD4, C5 .1 µF 35 volt disc capacitor

C6 10 µF, 35 V electrolytic capacitor

XTL 3.579 color burst crystal

S1   SPST toggle switch

S2   SPST pushbutton
     switch (normally open)

LX   LCD module Scott
     Edwards ILM-216
     (www.seetron.com)

P1   DB-9, 9 pin female
     RS-232 connector

Misc circuit board,
      headers, wire, etc.

Dual Tone Multifrequency or DTMF tone signaling has been around since the late 1950s and 1960s. The DTMF signaling system originated from the Grand Old Ma Bell. The beauty of this signal system is that two non-resonant tones are transmitted simultaneously and can be sent over land lines or wire or radio systems between and through various repeater systems around the country. The DTMF system is very reliable in that two tones must be decoded at once, so that false outputs are very unlikely. The DTMF system was and still is a very powerful method of signaling and control for the phone system as well as for ham radio operators to control remote repeater systems or remote base transmitters over long distances. Touch-Tone frequencies are illustrated in Table 28-1.

In the past, Touch-Tone frequencies had to be generated and decoded by discrete tone generators and tone decoders. Encoders and decoders were quite complex. Today DTMF encoding and decoding can be accomplished by a single 20 pin integrated circuit chip.

In this chapter, we will take a look at two different projects. The first project is a simple Touch-Tone Generator which can be used to send DTMF signals through a telephone line or radio transceiver. The second project uses the CM8880 to decode DTMF tones

Table 28-1

Touch tone or DTMF frequency tones

	1209	1336	1477	1633
697	1	2	3	A
770	4	5	6	B
852	7	8	9	C
941	0	0	#	D

on a phone line or radio circuit and display the numbers on an LCD panel.

Both Touch-Tone Generator and decoder/display projects use the STAMP 2 microprocessor and the new CM8880 DTMF encoder/decoder chip, and a 20 pin in-line DIP package. This Encoder/Decoder chip can be easily interfaced to the STAMP 2 controller and can be used to both generate and decode Touch-Tone tones. The chip only requires about six external components to form a complete encoder/decoder. The CM8880 chip uses a ubiquitous 3.579 color burst crystal for its oscillator. The chip produces a binary output on pins D0 through D3, which directly connects to the STAMP 2 controller (see Table 28-2). The Est and StGt, i.e. pins 18 and 19 respectively, form a time constant to

Table 28-2

DTMF binary and decimal values

Binary Value	Decimal Value	Keypad Symbol
0000	0	D
0001	1	1
0010	2	2
0011	3	3
0100	4	4
0101	5	5
0110	6	6
0111	7	7
1000	8	8
1001	9	9
1010	10	0
1011	11	0
1100	12	#
1101	13	A
1110	14	B
1111	15	C

Figure 28-1   *DTMF decoder display*

Table 28-3

Control bits CS, RW, and RSO status

CS	RW	RSO	Description
0	0	0	Active: write data (i.e. send DTMF)
0	0	1	Active: write instructions to CM8880
0	1	0	Active: read data (i.e. receive DTMF)
0	1	1	Active: read status from CM8880
1	0	0	inactive
1	0	1	inactive
1	1	0	inactive
1	1	1	inactive

determine a valid signal duration time in which to accept a tone pair.

Three additional bits are used to select the modes of the CM8880. The (CS) or chip select pin, the (RW) or read/write pin and the (RS0) or resister select pin; Table 28-3 depicts all the combinations of the three control pins.

The CM8880 is only active when CS equals 0. The RW bit determines the data direction; 1 = read and 0 = write, and the RS bit determines whether the transaction involves data (DTMF tones) or internal functions, i.e. 1= instructions/status and 0 = data.

Before you can use the CM8880, you have to set it up. The device has two control registers; A and B. In the beginning of the program listing you will notice the set-up of the registers; also refer to Tables 28-4 and 28-5.

Touch-Tone Generator project

Our first project in this chapter is a Touch-Tone Generator, run by the BASIC STAMP 2 microprocessor shown in Figure 28-2.

The Touch-Tone Generator can be used to send Touch-Tone signals through a radio or wire circuit. This circuit can be interfaced directly to a telephone line or to the microphone input of a radio so that you could send DTMF tones over telephone or radio circuits to control things remotely. The Touch-Tone Generator utilizes the STAMP 2 to animate the CM8880 encoder/decoder chip. Note that pins 9, 10, and 11 are used to enable and

Table 28-4

Functions of control register A

Bit	Name	Function
0	Tone out	0 = tone generator disabled
		1 = tone generator enabled
1	Mode control	0 = send and receive DTMF
		1 = send DTMF, receive call progress tone (DTMF bursts lengthened to 104 ms)
2	Interrupt enable	0 = make controller check for DTMF rec'd
		1 = interrupt controller via pin 13 when DTMF rec'd
3	Register select	0 = next instruction write goes to CRA
		1 = next instruction write goes to CRB

mode select the CM8880 chip. Pins 14 through 17 are data output pins which send the binary representation of the number received by the decoder portion of the chip. The Touch-Tone Generator uses pin 8 of the CM8880 to send the analog audio signal to be amplified by U4. The output of the IC amplifier is couple via C6 to a microphone input of a radio transceiver or wire circuit. If you have a walkie talkie or mobile radio you could use this circuit to send tones over radio to a remote receiver with a second CM8880 set up as a decoder to translate the remote tones into control functions (see the Radio Decoder project). In the Touch-Tone Generator program shown below, you will notice the

Table 28-5

Functions of control register B

Bit	Name	Function
0	BURST	0 = output DTMF bursts of 52 or 104 ms
		1 =1 output DTMF as long as enabled
1	Test	0 = normal operating mode
		1 = present test timing bit on pin 13
2	Single/ Dual	0 = output dual (real DTMF) tones
		1 = output separate row or column tones
3	Column/ Row	0 = if above = 1 select row tone
		1 = if above = 1 select column tone

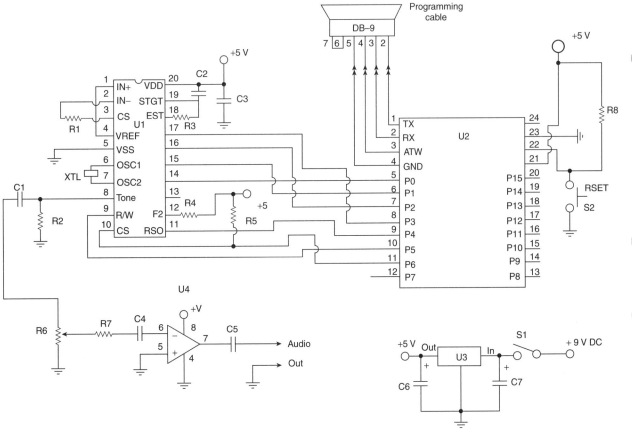

Figure 28-2 *Touch-Tone Generator*

phone number *459-0623*. Simply substitute your own phone number or tone sequence for your particular application.

Touch-Tone decoder/display project

Our second project utilizes the CM8880 DTMF encoder/decoder and the STAMP 2 to form a Touch-Tone Decoder Display project. This decoder will allow you to monitor Touch-Tone signals,decode the phone numbers dialed, and display the number on a serial LCD display. This project can be placed across the phone line to capture phone numbers, or it could be used with a radio transceiver to monitor access tones sent "over the air" and display the captured numbers on the display. The Touch-Tone Decoder project uses the CM8880 in the decoder mode. The tone signals are received via C1 and C2, using the balanced input network shown in Figure 28-3.

The STAMP 2 animates the control pins on pins 9, 10, and 11, and accepts the binary output of the CM8880 via pins 14 through 17. The LCD panel is driven by the BASIC STAMP 2 on Pin P7. The program accepts numbers and displays them sequentially, and if the delay between characters is too long then the display will show a <space> until another stream is decoded. Power switch S1 applies 5 V to the STAMP 2 via the 5 V regulator at U3. Momentary pushbutton S2 is the system reset button, which is used in the event of a system "lock-up."

Circuit assembly

Before we begin constructing the Touch-Tone Generator/Decoder, you will need to locate a clean, well-lit worktable or workbench. Next we will gather a small 25–30 W pencil tipped soldering iron, a length of 60/40 tin–lead solder, and a small canister of Tip Tinner, available at your local RadioShack store; Tip Tinner is used to condition the soldering tip

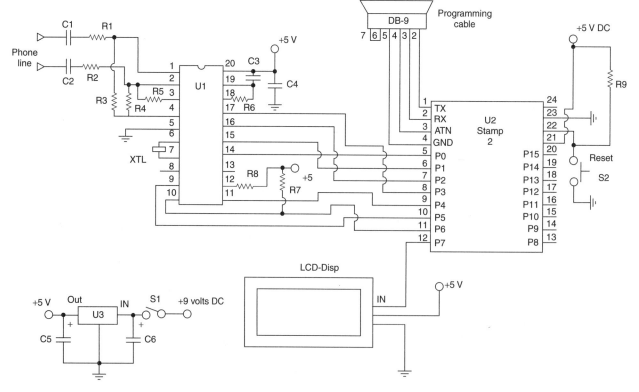

Figure 28-3 *DTMF Decoder/display*

between solder joints. You drive the soldering tip into the compound and it cleans and prepares the soldering tip. Next grab a few small hand tools; try to locate small flat blade and Phillips screwdrivers. A pair of needle-nose pliers, a pair of tweezers along with a magnifying glass, and a pair of end-cutters and we will begin constructing the project. Now locate your schematic diagram and parts layout diagram along with all of the components needed to build the project.

The Touch-Tone Generator/Decoder can be built on a 4″ × 5″ single sided glass epoxy circuit board. There are slight variations in the two circuits. You could build a single Touch-Tone encoder/decoder circuit that will perform both functions if desired, incorporating all the components; note the component designations are a bit different. You could build the Touch-Tone Decoder and simply add the additional components around U4 for Touch-Tone output generation.

Once all the components are in front of you, you can check them off against the project parts list to make sure you are ready to start building the project. Now, locate Tables 28-6 and 28-7, and place them in front of you.

Table 28-6 illustrates the resistor code chart and how to read resistors, while Table 28-7 illustrates the capacitor code chart, which will aid you in constructing the project.

Refer now to the resistor color code chart in Table 28-6, which will help you identify the resistor values. Note that each resistor will have either three or four color bands, which begin at one end of the resistor's body. The first colored band represents the resistor's first digit, while the second color represents the second digit of the resistor's value. The third colored band represents the resistor's multiplier value, and the fourth color band represents the resistor's tolerance value. If there is no fourth colored band then the resistor has a 20% tolerance value. If the resistor's fourth band is silver then the resistor has a 10% tolerance value and if the fourth band is gold then the resistor's tolerance value will be 5%.

Finally, we are ready to begin assembling the project, so let's get started. The prototype project was constructed on a single-sided printed circuit board or PCB. With your Touch-Tone Decoder PC board in front of you, we can now begin populating the circuit board. We are going to place and solder the lowest height

Table 28-6
Resistor color code chart

Color Band	1st Digit	2nd Digit	Multiplier	Tolerance
Black	0	0	1	
Brown	1	1	10	1%
Red	2	2	100	2%
Orange	3	3	1,000 (K)	3%
Yellow	4	4	10,000	4%
Green	5	5	100,000	
Blue	6	6	1,000,000 (M)	
Violet	7	7	10,000,000	
Gray	8	8	100,000,000	
White	9	9	1,000,000,000	
Gold			0.1	5%
Silver			0.01	10%
No color				20%

Table 28-7
Three-digit capacitor codes

pF	nF	μF	Code	pF	nF	μF	Code
1.0			**1R0**	3,900	3.9	.0039	**392**
1.2			**1R2**	4,700	4.7	.0047	**472**
1.5			**1R5**	5,600	5.6	.0056	**562**
1.8			**1R8**	6,800	6.8	.0068	**682**
2.2			**2R2**	8,200	8.2	.0082	**822**
2.7			**2R7**	10,000	10	.01	**103**
3.3			**3R3**	12,000	12	.012	**123**
3.9			**3R9**	15,000	15	.015	**153**
4.7			**4R7**	18,000	18	.018	**183**
5.6			**5R6**	22,000	22	.022	**223**
6.8			**6R8**	27,000	27	.027	**273**
8.2			**8R2**	33,000	33	.033	**333**
10			**100**	39,000	39	.039	**393**
12			**120**	47,000	47	.047	**473**
15			**150**	56,000	56	.056	**563**
18			**180**	68,000	68	.068	**683**
22			**220**	82,000	82	.082	**823**
27			**270**	100,000	100	.1	**104**
33			**330**	120,000	120	.12	**124**
39			**390**	150,000	150	.15	**154**
47			**470**	180,000	180	.18	**184**
56			**560**	220,000	220	.22	**224**

(Continued)

Table 28-7—cont'd
Three-digit capacitor codes

pF	nF	μF	Code	pF	nF	μF	Code
68			**680**	270,000	270	.27	**274**
82			**820**	330,000	330	.33	**334**
100		.0001	**101**	390,000	390	.39	**394**
120		.00012	**121**	470,000	470	.47	**474**
150		.00015	**151**	560,000	560	.56	**564**
180		.00018	**181**	680,000	680	.68	**684**
220		.00022	**221**	820,000	820	.82	**824**
270		.00027	**271**		1,000	1	**105**
330		.00033	**331**		1,500	1.5	**155**
390		.00039	**391**		2,200	2.2	**225**
470		.00047	**471**		2,700	2.7	**275**
560		.00056	**561**		3,300	3.3	**335**
680		.00068	**681**		4,700	4.7	**475**
820		.00082	**821**			6.8	**685**
1,000	1.0	.001	**102**			10	**106**
1,200	1.2	.0012	**122**			22	**226**
1,500	1.5	.0015	**152**			33	**336**
1,800	1.8	.0018	**182**			47	**476**
2,200	2.2	.0022	**222**			68	**686**
2,700	2.7	.0027	**272**			100	**107**
3,300	3.3	.0033	**332**			157	**157**

Code = 2 significant digits of the capacitor value + number of zeros to follow

For values below 10 pF, use "R" in place of decimal e.g. 8.2 pF = 8R2

10 pF = 100

100 pF = 101

1,000 pF = 102

22,000 pF = 223

330,000 pF = 334

1 μF = 105

components first, the resistors and diodes. First, locate our first resistor R1. Resistor R1 has a value of 100,000 ohms and is a 5% tolerance type. Now take a look for a resistor whose first color band is brown, black, yellow, and gold. From the chart you will notice that brown is represented by the digit (1), the second color band is black, which is represented by (0). Notice that the third color is yellow and the multiplier is (10,000), so (1)(0) × 10,000 = 100,000 or 100 K ohms.

Go ahead and locate where resistor R1 goes on the PC and install it; next solder it in place on the PC board, and then with a pair of end-cutters trim the excess component leads flush to the edge of the circuit board. Next locate the remaining resistors and install them in their respective locations on the main controller PC board. Solder the resistors to the board, and remember to cut the extra lead with your end-cutters.

Note that capacitors generally come from two major groups or types. Non-polar types, such as the ones in this project, have no polarity and can be placed in either direction on the circuit board. The other type of capacitors are called polarized types; these capacitors have polarity and have to be inserted on the board with respect to their polarity marking in order for the circuit to operate correctly. On these types of capacitors, you will observe that the components actually have a plus or minus marking or a black or white band with a plus or minus marking on the body of the capacitors. Capacitors must be installed properly with respect to polarity in order for the circuit to work properly. Failure to install electrolytic capacitors correctly could cause damage to the capacitor as well as to other components in the circuit when power is first applied.

Many capacitors are quite small in physical size, so manufacturers have devised a three-digit code which can be placed on the capacitor instead of the actual value. The code occupies a much smaller space on the capacitor and can usually fit nicely on the capacitor body, but without the chart the builder would have a difficult time determining the capacitor's value.

Now look for the capacitors from the component stack and locate capacitors C3 and C4, they are small sized capacitors labeled (104) or .1 µF. Refer to Table 28-7 to find code (104) and you will see that it represents a capacitor value of .1 µF. Capacitors C1 and C2 are also .1 µF but are larger in size and voltage requirements. Go ahead and place these first four capacitors on the circuit board and solder them in place. Now take your end-cutters and cut the excess component leads flush to the edge of the circuit board. Once you have installed these components, you can move on to installing the remaining capacitors at C5, C6, and C7, which are an electrolytic types, so be sure you observe the correct polarity when installing these.

Both the Touch-Tone Generator and the Decoder projects contain ICs. The Touch-Tone encoder utilizes the Touch-Tone encoder/decoder chip as well as the LM358 Op-Amp, while the Touch-Tone decoder uses only the Touch-Tone encoder/decoder chip. ICs are often static sensitive, so they must be handled with care. Use a grounded anti-static wrist strap and stay seated in one location when handling the ICs. Take out a cheap insurance policy by installing IC sockets for each of the ICs. In the event of a possible circuit failure, it is much easier to simply unplug a defective IC than trying to unsolder 20 or 24 pins from a PC board without damaging the board. ICs have to be installed correctly if the circuit is going to work properly. IC packages will have some sort of markings which will help you orient them on the PC board. An IC will have either a small indented circle, a cutout, or notch at the top end of the IC package. Pin 1 of the IC will be just to the left of the notch, or cutout. Refer to the manufacturer's pin-out diagram, as well as the schematic, when installing these parts. If you doubt your ability to correctly orient the ICs, then seek the help of a knowledgeable electronics enthusiast to help you.

When installing the BASIC STAMP 2 microcontroller be sure to handle it very carefully, using anti-static precautions. The BASIC STAMP 2 module has a copper foil dot in the top center of the carrier board; also note that the largest chip on the microcontroller board is at the bottom of the carrier board. This information should help in orienting the BASIC STAMP 2 module. Be slow and careful when installing the BS2, as it represents a significant cost and you do not want to damage it by installing it incorrectly.

Now let's finish up the circuit board, locating the crystal at XTL and installing it in its proper location between pins 6 and 7 of U1. Install a 4-position header for the programming pins 1, 2, 3, and 4 on the BASIC STAMP 2 controller. Install the printed circuit board type momentary pushbutton reset switch at S2. If you are building the Touch-Tone Generator, you will want to install the PC board type potentiometer at R6 and provide two terminals for the audio output from the audio amplifier chip at U4.

An LM7805 5-V regulator was used to step-down the voltage to 5-V for the BASIC STAMP 2, the Touch-Tone decoder chip, and the Op-Amp at U4. Be sure to orient the regulator correctly in order for the circuit to work properly. Check the manufacturer's specification sheet for proper pin-outs of the regulator before installing it.

If you are building the Touch-Tone Decoder circuit you will have to wire up the serial LCD module to the BASIC STAMP 2 circuit. The ground wire of the LCD will go to the system ground, the LCD's power lead will go to the +5 V system power, and the serial signal lead of the LCD will go to pin 12 or P7 on the STAMP 2 module.

Once the Touch-Tone Encoder or Touch-tone Decoder circuit board has been completed, take a short break and then we will inspect the PC board for possible "cold" solder joints and "short" circuits. Pick up the PC board with the foil side facing upwards toward you.

Carefully inspect the solder joints, these should look smooth, clean, and bright. If any of the solder joints look dull, dark, or blobby, then you should remove the solder with a solder sucker or a solder wick and then resolder the joint all over again. Now we will inspect the PC board for possible short circuits. Short circuits can be caused from small solder blobs bridging the circuit traces or from cut component leads which can often stick to the circuit board once they have been cut. A sticky residue from solder will often trap component leads across the circuit traces. Look carefully for any bridging wires across circuit traces.

Next, you will need to locate or build a 9 volt DC power supply in order to power the DTMF encoder/decoder circuits. You can elect to build a power supply or try to locate a "wall wart" power supply from a surplus electronic supplier, to provide 9-V to the regulator at U3.

The Touch-Tone encoder or decoder circuit board can be mounted in a small plastic enclosure box to protect it. Use four plastic standoffs to mount the circuit board to the base plate of the plastic or metal enclosure box. Install a coaxial power jack for your 9-V DC power supply. Install an audio output jack if you are constructing the Touch-Tone Generator circuit. If you are building the Touch-Tone Decoder circuit, you will want to mount the LCD on the top or front of the enclosure, along with the power switch at S1. You will also need to install a four position jack on the enclosure for the programming cable connections.

Now you are ready to test out your Touch-Tone Encoder or Decoder circuit. Programming the STAMP 2 controller is achieved using the TX, RX, ATN, and ground pins 1 though 4. Connect up the programming cable between your personal computer and the 4-pin programming pins on the BASIC STAMP 2 microcontroller. Now connect your 9-volt "wall-wart" power supply to the circuit, and you are ready to begin programming and testing the Touch-Tone Generator/Decoder circuits.

Make a directory for the Touch-Tone Encoder and Touch-Tone Decoder programs. Type in or download the two programs into the directory. Go to www.paralax.com and locate the Windows BASIC STAMP Editor 2.5 or above from the DOWNLOADS section of the website. Install the STAMP 2 editor program. From the STAMP 2 editor program, open up either the Touch-Tone Generator (**dial.bs2**) or Touch-Tone Decoder (**dtmf_rcv.bs2**) program. In the Touch-Tone Generator program you will notice that there is a simulated telephone number listed. You will need to change the *sample* phone number to the *desired* phone number. With power applied to your circuit and the programming cable connected, between the programming computer and BASIC STAMP 2, go to the RUN tab on the toolbar in the editor program and press RUN and download either the Touch-Tone Encoder or Decoder program into your circuit board depending upon which circuit you built. Once the program is installed into the BASIC STAMP 2, you can begin using your new Touch-Tone Generator/Decoder circuits.

Building the Touch-Tone encoder or generator, will allow you to dial a phone number or generate control tones which can be sent through your telephone line or radio equipment. In operation you would use the STAMP 2 frequency output or Freqout command to send DTMF tones out of the circuit.

Building the Touch-Tone decoder will permit you to decode phone numbers dialed on your phone line, so you could monitor the phone line or you could use the Touch-Tone decoder to "read" tone from a radio receiver which ham operators use to control objects or repeater systems.

The Radio DTMF Encoder/Decoder circuits can be used for many different applications; we only illustrated two but your imagination can think up many more. Once you become more familiar with programming the BASIC STAMP 2, the encoder and decoder programs will lend themselves to many user modifications!

Touch-Tone Generator/dialing program

`DIAL.BS2`

```
' Program: DIAL.BS2 (BS2 sends DTMF tones via the 8880)
' This program demonstrates how to use the CM8880 as a DTMF tone
' generator. All that is required is to initialize the 8880 properly,
' then write the number of the desired DTMF tone to the 8880s
' 4-bit bus.
' The symbols below are the pin numbers to which the 8880s
' control inputs are connected, and one variable used to read
' digits out of a lookup table.
RS       con 4          ' Register-select pin (0 = data).
RW       con 5          ' Read/Write pin (0 = write).
CS       con 6          ' Chip-select pin (0 = active).
digit    var nib        ' Index of digits to dial, 1-15.
' This code initializes the 8880 for dialing by writing to its
' internal control registers CRA and CRB. The write occurs when
' CS (pin 6) is taken low, then returned high. See the accompanying
' article for an explanation of the 8880s registers.
OUTL = 127                  ' Pins 0-6 high to deselect 8880.
DIRL = 127                  ' Set up to write to 8880 (pins 0-6 outputs).
OUTL = %00011011            ' Set up register A, next write to register B.
high CS
OUTL = %00010000            ' Clear register B; ready to send DTMF.
high CS
' This for/next loop dials the seven digits of my fax number. For
' simplicity, it writes the digit to be dialed directly to the output
' pins. Since valid digits are between 0 and 15, this also takes RS,
' RW, and CS low—perfect for writing data to the 8880. To complete
' the write, the CS line is returned high. The initialization above
' sets the 8880 for tone bursts of 200 ms duration, so we pause
' 250 ms between digits. Note: in the DTMF code as used by the phone
' system, zero is represented by ten (1010 binary) not 0. That is why
' the phone number 459-0623 is coded 4,5,9,10,6,2,3.
for digit = 0 to 6
lookup digit,[ 4,5,9,10,6,2,3], OUTL     ' Get digit from table.
high CS                                   ' Done with write.
pause 250                                 ' Wait to dial next digit.
next
end
```

Touch-Tone decoding/display program

DTMF_RCV.BS2

```
' Program: DTMF_RCV.BS2 (Receives/displays DTMF using 8880 with BS2)
' This program demonstrates how to use the 8880 as a DTMF decoder. As
```

```
' each new DTMF digit is received, it is displayed on an LCD Serial
' Backpack screen. If no tones are received within a period of time
' set by sp_time, the program prints a space (or other selected character)
' to the LCD to record the delay. When the display reaches the right-hand
' edge of the screen, it clears the LCD and starts over at the left edge.
RS          con 4          ' Register-select pin (0 = data).
RW          con 5          ' Read/Write pin (0 = write).
CS          con 6          ' Chip-select pin (0 = active).
dtmf        var   byte     ' Received DTMF digit.
dt_Flag     var   bit      ' DTMF-received flag.
dt_det      var   INL.bit2 ' DTMF detected status bit.
home_Flag   var   bit      ' Flag: 0 = cursor at left edge of LCD.
polls       var   word     ' Number of unsuccessful polls of DTMF.
LCDw        con 16         ' Width of LCD screen.
LCDcol      var   byte     ' Current column of LCD screen for wrap.
LCDcls      con 1          ' LCD clear-screen command.
I           con 254        ' LCD instruction toggle.
sp_time     con 1500       ' Print space this # of polls w/o DTMF.
n24n        con $418D      ' Serout constant: 2400 baud inverted.
' This code initializes the 8880 for receiving by writing to its
' internal control registers CRA and CRB. The write occurs when
' CS (pin 6) is taken low, then returned high.
OUTL = %01111111            ' Pin 7 (LCD) low, pins 0 through 6 high.
DIRL = %11111111            ' Set up to write to 8880 (all outputs).
OUTL = %00011000            ' Set up register A, next write to register B.
high CS
OUTL = %00010000            ' Clear register B; ready to send DTMF.
high CS
DIRL = %11110000            ' Now set the 4-bit bus to input.
high RW                     ' And set RW to "read."
serout 7,n24n,[ I,LCDcls,I] ' Clear the LCD screen.
' In the loop below, the program checks the 8880's status register
' to determine whether a DTMF tone has been received (indicated by
' a '1' in bit 2). If no tone, the program loops back and checks
' again. If a tone is present, the program switches from status to
' data (RS low) and gets the value (0-15) of the tone. This
' automatically resets the 8880's status flag.
again:
high RS                     ' Read status register.
low CS                      ' Activate the 8880.
dt_flag = dt_det ' Store DTMF-detected bit into flag.
high CS                     ' End the read.
if dt_Flag = 1 then skip1   ' If tone detected, continue.
polls = polls+1             ' Another poll without DTMF tone.
if polls < sp_time then again  ' If not time to print a space, poll.
if LCDcol = LCDw then skip2 ' Don't erase the screen to print spaces.
dtmf = 16                   ' Tell display routine to print a space.
gosub Display               ' Print space to LCD.
```

```
skip2:
polls = 0                             ' Clear the counter.
goto again                            ' Poll some more.
skip1:                                ' Tone detected:
polls = 0                             ' Clear the poll counter.
low RS                                ' Get the DTMF data.
low CS                                ' Activate 8880.
dtmf = INL & %00001111                ' Strip off upper 4 bits using AND.
high CS                               ' Deactivate 8880.
gosub display                         ' Display the data.
goto again                            ' Do it all again.
Display:
if LCDcol <LCDw then skip3            ' If not at end of LCD, do not clear screen.
serout 7,N24N,[I,LCDcls,I]            ' Clear the LCD screen.
LCDcol = 0                            ' And reset the column counter.
skip3:' Look up the symbol for
   the digit.
if LCDcol=0 AND dtmf=16 then ret      ' No spaces at first column.
lookup dtmf,["D1234567890*#ABC-"] ,dtmf
serout 7,N24N,[dtmf]                  ' Write it to the Backpack display.
LCDcol = LCDcol + 1                   ' Increment the column counter.
ret:
return
```

Chapter 29

Caller-ID/Blocker Project

Parts list

Parts Bin

Caller-ID/Blocker

R1, R2, R3 10 K ohm
$\frac{1}{2}$ watt resistor

R4 18 K ohm $\frac{1}{4}$ W
resistor

R5 15 K ohm $\frac{1}{4}$ W
resistor

R6 270 K ohm $\frac{1}{4}$ W
resistor

R7 4.7 megohm $\frac{1}{4}$ W
resistor

R8 10 megohm $\frac{1}{4}$ W
resistor

R9 22 K ohm $\frac{1}{4}$ W
resistor

R10 470 ohm $\frac{1}{4}$ W
resistor

R11, R12, R13, R14
1 K ohm $\frac{1}{4}$ W
resistor

R15, R16 10 K ohm
$\frac{1}{4}$ W resistor

R17, R18, R19 10 K ohm
$\frac{1}{4}$ W resistor

C1, C2 .2 µF 400 V
mylar capacitor

C3, C4 470 pF 400 V
mylar capacitor

C5 .2 µF 50 V disc
capacitor

C6 1 µF 50 V disc
capacitor

C7, C11 1 µF 50 V
electrolytic capacitor

C8, C9 30 pF 35 V
disc capacitor

C10, C12 1 µF 50 V
disc capacitor

MOV 300 V metal
oxide varistor

D1, D2, D4 1N4004
silicon diodes

D5 1N4001 silicon diode

DS1, DS2, DS3 LED

DS4, DS5 LED

Q1 2N2222 NPN transistor

U1 Motorola MC 145447
Caller-ID chip

U2 BASIC STAMP 2
(original)

U3 LM7805 5 V
regulator

XTL 3.579 color burst
crystal

K1 5 volt reed relay
SPST

LCD serial LCD display
www.seetron.com

SW1, SW2, SW3 momentary
pushbutton switch
(normally open)

SW4, SW5 momentary
pushbutton switch
(normally open)

Misc circuit board,
power supply, wire,
connectors, etc.

Unwanted calls are highly interruptive and very annoying. Why not screen your incoming calls? Everyone knows those pesky unsolicited callers always phone you when you are indisposed or just sitting down for dinner. Screen your calls from unwanted phone callers or bill collectors, or your ex-wife/husband. The Caller-ID/Blocker project, shown in Figure 29-1, will allow you to identify up to 10 incoming telephone calls and display the phone numbers on an LCD display. The Caller-ID unit will also allow you to block incoming calls.

The Caller-ID project centers around the infamous BASIC STAMP 2 microprocessor, a Motorola Caller-ID chip, and a handful of external components, shown in Figure 29-2.

The Caller-ID connects across your phone line with the *ring* and *tip* connections. A Metal Oxide Varistor (MOV) is placed across the phone line to limit high voltage "spikes" from damaging your circuit. Capacitors C2 and C4 are the phone line inputs to the Caller-ID chip shown in the block diagram in Figure 29-3.

The *ring* and *tip* phone line inputs are fed to an op-amp and then to a band pass filter, followed by a demodulator and valid data detector. The output of the valid data detector is the Caller-ID data information. An additional two capacitors, C1 and C2, are placed across the line ahead of a diode bridge, which is used to rectify the ringing voltage. This ringing detect signal is then inputted to the ring detect pins on the Motorola MC145447 Caller-ID chip, pins 3 an 4. The ring detect network is formed by resistors R3, R4, R5, and R6 along with C5. The ubiquitous 3.579 Mhz color burst crystal forms the basic reference oscillator. The output at pin 7 of the Caller-ID chip drives an NPN transistor, which in turn drives the data indicator LED at DS1. The Caller-ID data is next coupled to the P19 pin on the

BASIC STAMP 2. The Ring detect output of the Caller-ID chip is fed to P18 of STAMP 2. The ring detect signal determines that the circuit should "pick up" or initiate. Once the ring signal is detected, the STAMP 2 pulls in the relay at K1. Indicator DS5 confirms that the relay has pulled in. The indicator at DS4 is a monitor LED; it stays on normally and then toggles off when the Caller-ID program is running. The indicator at DS3 is a troubleshooting LED; it goes high after the data has been read out from the LCD display unit. The last indicator at DS2 is driven by P0. This LED is also for troubleshooting; it goes high after the data has been successfully read into from the MC145447.

There are four function or operational input switches on STAMP 2 pins P9 through P12; we will discuss the functional operations later. Function A is activated by a normally open pushbutton switch at SW4. Function B is operated through SW3 at P11. Function C is implemented by SW1 at STAMP 2 pin P9. The RUN or START pushbutton is started by SW2 when the circuit is first energized. Note, all pushbutton switches are tied to high to 5 V and when pressed the switches go to ground upon activation. The Caller-ID display is connected to the STAMP 2 via pin P15. The LCD display is a single wire serial "backpack" style from Scott Edwards Electronics at www.seetron.com.

Power to the Caller-ID chip and the STAMP 2 is a supplied by a 5 V regulator at U3. On the MC145447, power is applied to pin 16, while pin 21 is used on the STAMP 2. A system reset switch is shown at SW5, a normally open pushbutton switch. A 9 to 12 V "wall-wart" power supply can be used to power the Caller-ID circuit ahead of the regulator at U3. Alternatively, a 9-V transistor radio battery could be used but a power supply is recommended.

Circuit assembly

Before we begin constructing the Telephone Caller-ID project, you will need to locate a clean, well-lit worktable or workbench. Next we will gather a small 25–30-W pencil tipped soldering iron, a length of 60/40 tin–lead solder, and a small canister of Tip Tinner, available at your local RadioShack store; Tip Tinner is used to condition the soldering tip between solder joints. You drive the soldering tip into the compound and it cleans and prepares the soldering tip. Next grab a

Figure 29-1 *Caller ID front panel. Courtesy of J Gary Sparks*

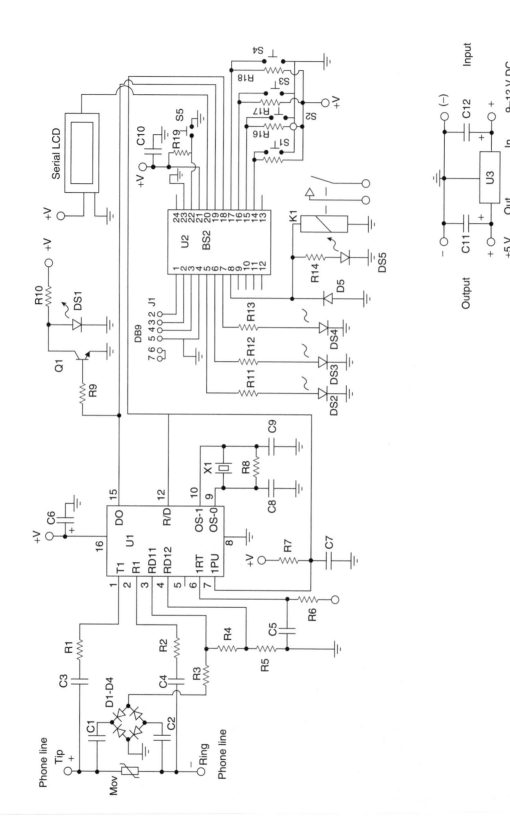

Figure 29-2 *Caller ID project circuit*

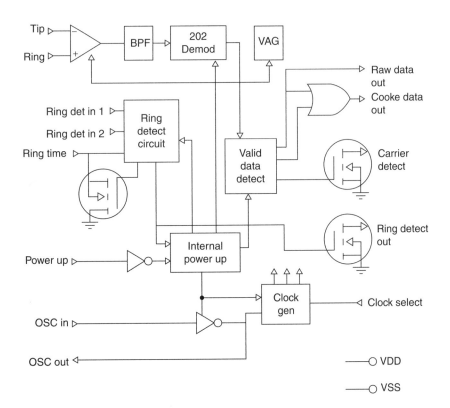

Figure 29-3 *MC145447 Caller-ID chip*

few small hand tools; try to locate small flat blade and Phillips screwdrivers. A pair of needle-nose pliers, a pair of tweezers along with a magnifying glass, and a pair of end-cutters and we will begin constructing the project. Now locate your schematic diagram and parts layout diagram along with all of the components needed to build the project.

The Caller-ID circuit was built on a small single sided glass epoxy circuit board. You can elect to construct you own PC board or you could assemble the circuit on a STAMP 2 carrier board if desired. The circuit board was then placed in a 4 × 6″ plastic box.

Once all the components are in front of you, you can check them off against the project parts list to make sure you are ready to start building the project. Now, locate Tables 29-1 and 29-2, and place them in front of you.

Table 29-1 illustrates the resistor code chart and how to read resistors, while Table 29-2 illustrates the capacitor code chart which will aid you in constructing the project.

Refer now to the resistor color code chart in Table 29-1, which will help you identify the resistor values.

Note that each resistor will have either three or four color bands, which begin at one end of the resistor's body. The first colored band represents the resistor's first digit, while the second color represents the second digit of the resistor's value. The third colored band represents the resistor's multiplier value, and the fourth color band represents the resistor's tolerance value. If there is no fourth colored band then the resistor has a 20% tolerance value. If the resistor's fourth band is silver then the resistor has a 10% tolerance value and if the fourth band is gold then the resistor's tolerance value will be 5%.

Finally, we are ready to begin assembling the project, so let's get started. As mentioned earlier, the prototype project was constructed on a single-sided printed circuit board or PCB; with your PC board in front of you, we can now begin populating the circuit board. We are going to place and solder the lowest height components first, the resistors and diodes. First, locate our first resistor R1. Resistors R1, R2, and R3 are 10,000 ohms, ½ W with a 5% tolerance. Now look for three resistors whose first color band is brown, black, orange, and gold. From the chart you will notice that brown is represented by the digit (1), the second color band is

Table 29-1

Resistor color code chart

Color Band	1st Digit	2nd Digit	Multiplier	Tolerance
Black	0	0	1	
Brown	1	1	10	1%
Red	2	2	100	2%
Orange	3	3	1,000 (K)	3%
Yellow	4	4	10,000	4%
Green	5	5	100,000	
Blue	6	6	1,000,000 (M)	
Violet	7	7	10,000,000	
Gray	8	8	100,000,000	
White	9	9	1,000,000,000	
Gold			0.1	5%
Silver			0.01	10%
No color				20%

Table 29-2

Three-digit capacitor codes

pF	nF	μF	Code	pF	nF	μF	Code
1.0			1R0	3,900	3.9	.0039	392
1.2			1R2	4,700	4.7	.0047	472
1.5			1R5	5,600	5.6	.0056	562
1.8			1R8	6,800	6.8	.0068	682
2.2			2R2	8,200	8.2	.0082	822
2.7			2R7	10,000	10	.01	103
3.3			3R3	12,000	12	.012	123
3.9			3R9	15,000	15	.015	153
4.7			4R7	18,000	18	.018	183
5.6			5R6	22,000	22	.022	223
6.8			6R8	27,000	27	.027	273
8.2			8R2	33,000	33	.033	333
10			100	39,000	39	.039	393
12			120	47,000	47	.047	473
15			150	56,000	56	.056	563
18			180	68,000	68	.068	683
22			220	82,000	82	.082	823
27			270	100,000	100	.1	104
33			330	120,000	120	.12	124
39			390	150,000	150	.15	154
47			470	180,000	180	.18	184
56			560	220,000	220	.22	224

Table 29-2—cont'd
Three-digit capacitor codes

pF	nF	μF	Code	pF	nF	μF	Code
68			**680**	270,000	270	.27	**274**
82			**820**	330,000	330	.33	**334**
100		.0001	**101**	390,000	390	.39	**394**
120		.00012	**121**	470,000	470	.47	**474**
150		.00015	**151**	560,000	560	.56	**564**
180		.00018	**181**	680,000	680	.68	**684**
220		.00022	**221**	820,000	820	.82	**824**
270		.00027	**271**		1,000	1	**105**
330		.00033	**331**		1,500	1.5	**155**
390		.00039	**391**		2,200	2.2	**225**
470		.00047	**471**		2,700	2.7	**275**
560		.00056	**561**		3,300	3.3	**335**
680		.00068	**681**		4,700	4.7	**475**
820		.00082	**821**			6.8	**685**
1,000	1.0	.001	**102**			10	**106**
1,200	1.2	.0012	**122**			22	**226**
1,500	1.5	.0015	**152**			33	**336**
1,800	1.8	.0018	**182**			47	**476**
2,200	2.2	.0022	**222**			68	**686**
2,700	2.7	.0027	**272**			100	**107**
3,300	3.3	.0033	**332**			157	**157**

Code = 2 significant digits of the capacitor value + number of zeros to follow

For values below 10 pF, use "R" in place of decimal e.g. 8.2 pF = 8R2

10 pF = 100

100 pF = 101

1,000 pF = 102

22,000 pF = 223

330,000 pF = 334

1 μF = 105

black, which is represented by (0). Notice that the third color is orange and the multiplier is (1,000), so (1)(0) × 1,000 = 10,000 or 10 K ohms.

Go ahead and locate where resistors R1, R2, and R3 are located on the PC and install them, next solder them in place on the PC board, and then with a pair of end-cutters trim the excess component leads flush to the edge of the circuit board. Next locate the remaining resistors and install them in their respective locations on the main controller PC board. Solder the resistors to the board, and remember to cut the extra lead with your end-cutters.

Note that capacitors generally come from two major groups or types. Non-polar types, such as the ones in this project, have no polarity and can be placed in either direction on the circuit board. The other type of capacitors are called polarized types, these capacitors have polarity and have to be inserted on the board with respect to their polarity marking in order for the circuit to operate correctly. On these types of capacitors, you will observe that the components actually have a plus or minus marking or a black or white band with a plus or minus marking on the body of the capacitors. Capacitors must be installed properly with respect to polarity in order for the circuit to work properly. Failure to install electrolytic capacitors correctly could cause damage to the capacitors as well as to other components in the circuit when power is first applied.

Many capacitors are quite small in physical size, so manufacturers have devised a three-digit code which can be placed on the capacitor instead of the actual value. The code occupies a much smaller space on the capacitor and can usually fit nicely on the capacitor body, but without the chart the builder would have a difficult time determining the capacitor's value.

Now look for the capacitors from the component stack and locate capacitors C6, C10, and C11, small capacitors labeled (104). Refer to Table 29-2 to find code (104) and you will see that it represents a capacitor value of .1 μF. Go ahead and place C6, C10, and C11 on the circuit board and solder them in place. Next locate the 30 pF disc capacitors, they will most likely be marked (33). Install these capacitors in their respective locations. Now take your end-cutters and cut the excess component leads flush to the edge of the circuit board. Once you have installed these components, you can move on to installing the remaining low profile capacitors.

Next locate the larger capacitors and electrolytic capacitors. Note that C1, C2, C3, and C4 are low value capacitors but large in size since they are high voltage capacitors placed across the phone line. Install these four capacitors and then solder them in place on the PC board. Finally locate the electrolytic capacitors. Make note of their polarity and install them with respect to their proper polarity, this is importance if the circuit is to work correctly. Solder the electrolytic capacitors in place and then trim the excess lead lengths with your end-cutters.

The Caller-ID project contains five silicon diodes. Diodes, as you will remember, will always have some type of polarity markings, which must be observed if the circuit is going to work correctly. You will notice that each diode will have a black or white colored band at one edge of the diode's body. The colored band represents the diode's cathode lead. Check your schematic and parts layout diagrams when installing the diodes on the circuit board. The diode bridge consisting of diodes D1 through D4 is at the input of the circuit, located on its left side. Diode D5 is located at pin 8 of U2. Place the diodes in their correct locations and then solder them in place on the PC board. Remember to cut the excess component leads with your end-cutters, flush to the edge of the circuit board.

The Caller-ID circuit utilizes a single conventional transistor at Q1, which is connected to the D0 or pin 15 of U1. Transistors are generally three lead devices, which must be installed correctly if the circuit is to work properly. Careful handling and proper orientation in the circuit is critical. A transistor will usually have a Base lead, a Collector lead, and an Emitter lead. The base lead is generally a vertical line inside a transistor circle symbol. The collector lead is usually a slanted line facing towards the base lead and the emitter lead is also a slanted line facing the base lead but it will have an arrow pointing towards or away from the base lead. Now, identify the two transistors, refer to the manufacturer's specification sheets as well as the schematic and parts layout diagrams when installing the two transistors. If you have trouble installing the transistors, get the help of an electronics enthusiast with some experience. Install the transistors in their respective locations and then solder them in place. Remember to trim the excess component leads with your end-cutters.

The Caller-ID circuit contains ICs. ICs are often static sensitive, so they must be handled with care. Use a grounded anti-static wrist strap and stay seated in one location when handling the ICs. Take out a cheap insurance policy by installing IC sockets for each of the ICs. In the event of a possible circuit failure, it is much easier to simply unplug a defective IC, than trying to unsolder 14 or 16 pins from a PC board without damaging the board. ICs have to be installed correctly if the circuit is going to work properly. IC packages will have some sort of markings which will help you orient them on the PC board. An IC will have either a small indented circle, a cut-out, or notch at the top end of the IC package. Pin 1 of the IC will be just to the left of the notch or cut-out. Refer to the manufacturer's pin-out diagram, as well as the schematic when installing these parts. If you doubt you ability to correctly orient the ICs, then seek the help of a knowledgeable electronics enthusiast.

The Caller-ID project also contains an IC regulator at U3 as well as the BASIC STAMP 2 microprocessor at U1. Take a look at the BASIC STAMP 2 module.

On the top-middle of the STAMP 2 module you will see a copper dot. Also note that the largest IC on the carrier board will be at the bottom; this will help you orient the BASIC STAMP module on the circuit board. Be sure to use a large 24 pin IC socket when installing the BASIC STAMP 2 module. When installing the regulator, be sure to properly identify the input and output before installing the component on the PC board. Check the PC layout or manufacturer's specifications before installing U3, to avoid damage to the circuit.

The Caller-ID circuit utilizes five LEDs, which also must be oriented correctly if the circuit is to function properly. An LED has two leads, a cathode and an anode lead. The anode lead will usually be the longer of the two leads and the cathode lead will usually be the shorter lead, just under the flat side edge of the LED; this should help orient the LED on the PC board. You may want to lead about ½ lead length on the LED, in order to have the LED protrude through the enclosure box. Solder the LEDs to the PC board and then trim the excess leads.

Now, locate the crystal at XTL and install it in its proper location between pins 9 and 10 of the Caller-ID chip at U1. Next locate and install the SPST, normally open reed relay at K1, which is connected to pin 8 of the STAMP 2. Finally, locate the metal oxide varistor (MOV) protector at the input of the telephone line.

Once the Caller-ID circuit board has been completed, take a short break and then we will inspect the PC board for possible "cold" solder joints and "short" circuits. Pick up the PC board with the foil side facing upwards toward you. Carefully inspect the solder joints, these should look smooth, clean, and bright. If any of the solder joints look dull, dark, or blobby, then you should remove the solder with a solder sucker or a solder wick and then resolder the joint all over again. Now we will inspect the PC board for possible short circuits. Short circuits can be caused from small solder blobs bridging the circuit traces or from cut component leads which can often stick to the circuit board once they have been cut. A sticky residue from solder will often trap component leads across the circuit traces. Look carefully for any bridging wires across circuit traces.

Once the inspection is complete, we can move on to installing the circuit board into an aluminum enclosure (see Figure 29-4).

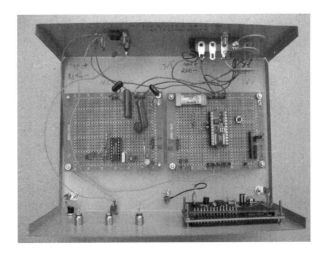

Figure 29-4 *Caller-ID inside view. Courtesy of J Gary Sparks*

Locate a suitable enclosure to house the Caller-ID circuit board within the enclosure. Use four plastic standoffs to mount the circuit board to the bottom plate of the enclosure. If you selected a convention relay you will have to mount it on the chassis and wire it to the PC board.

Mount a coaxial power jack on the rear panel of the chassis box, in order to plug in a small "wall-wart" 9-V DC power supply at input of the regulator. Mount the five status LEDs on the front panel of the enclosure box. Next you will have to install an RJ11 phone jack or a 2-position terminal strip on the front panel of the chassis to bring the phone into the chassis box. Locate all four momentary pushbutton function switches and arrange to mount them on the front panel of the enclosure. Mount a DB-9 serial connector on the rear panel which will be used to connect the programming cable between the Caller-ID circuit and your personal computer or laptop. Finally, install the LCD panel on the front panel of the enclosure box. Handle the LCD panel very carefully and do not press too hard on the LCD front while handling the LCD in order to avoid damaging it. The LCD will have three wires which will need to be connected to the Caller-ID circuit. You will have to connect the 5-V power lead to a +5-V bus on the circuit board. The LCD also must have a ground connection and finally the serial signal wire from the LCD must be connected to pin 20 of the BASIC STAMP 2 module.

Now you will need to make a directory on your computer labeled CallerID. Next you will need to download the Windows BASIC STAMP Editor version 2.5 or above from www.parallax.com. The program can be found under the DOWNLOADS pull-down button. Install the Windows editor program on your computer and then open the program and type in or load the **CallerID.bs2** program into the editor program. Now connect the programming cable between the Caller-ID circuit and your personal computer. Apply power to the Caller-ID circuit and we are ready to download the **CallerID.bs2** program into the BASIC STAMP 2 module.

Operation of the Caller-ID project begins when the phone starts ringing. The ring detector tells the STAMP 2 to wake-up and begin accepting data from the Caller-ID chip. Upon power-up the LCD reads "Waiting," then the STAMP 2 program reads in the Caller-ID or CID data, in multiple data message format (MDMF). This format presents the number and name of the originating call. Note, the SDMF formats send only the telephone numbers and will not work with this program in its present form. The program for this project displays the originating number on the LCD screen and stores up to 10 numbers in the EEPROM. The system will overwrite the oldest numbers displayed as new ones are received. In order to see past numbers, you will need to press the Function A button once, or several times during a delay loop, to cycle back through the past read-ins. Table 29-3 illustrates the functions of the pushbuttons used on the Caller-ID unit.

The program will store up to 10 numbers in to the EEPROM as numbers which are considered "blocked". To store a number in the EEPROM simply press Function C and hold the button for about a second until the LED pin 2 is out. To see what numbers are blocked cycle through the blocked numbers by depressing

Table 29-3

Momentary pushbutton switch functions

Function A	Function B	Function C	
Read	Blocked	Store	press once
Options		X	press C then A
	Erase	X	press C then B

Function B as you did for Function A and note the blocked numbers run from #11–20. The blocked number will be stored at the memory location of the last read-out number as cycled from Function B. The number which is stored into the blocked memory will be the last one read out as cycled by Function A.

In order to erase a number which has been blocked, you will need to hold down the Function C button until the LED on pin 2 goes out, then momentarily press the Function B button. Notice that zeros will replace the blocked number.

To turn on the call blocking feature; you will need to hold down the Function C button until the LED on pin 2 goes out. Next momentarily press the Function A button. The LCD display will cycle through "Call Blocking Off," "Inhibit Blocked Calls," "Connect Not Blocked Calls" for each push of the Function C and the Function A buttons. The last two options are nearly the same except: (1) when there is an error in data read, the first will allow the call to go through, the second will also allow the call to go through but only (2) when there is a call from a blocked number, only the first ring will get heard (until the CID data has been processed) while the second will not ring at all; (3) when there is a call from a not blocked number, the second will silence the first ring.

'CallerID.BS2 by J. Gary Sparks

```
'THIS PROGRAM READS IN AND STORES CALLER-ID DATA FROM A MC145447, STORES
'NUMBERS WHICH ARE DESIGNATED 'BLOCKED', READS OUT NUMBERS ON A SERIAL
'BACKPACK LCD, DECIDES TO BLOCK THE CALL IF A 'BLOCKED NUMBER' HAS CALLED
'*************************NOTES******************************
'read in first world length; if #, read in #;if not # store an OTH(er)
'    and fill rest of ram with ascii spaces
'I did a test and the first (hex) word length was dec 39 for a #
'    and dec 16 for out of area; I DISCARDED THIS OPTION SINCE
'    SOMETIMES ONLY THE NUMBER IS SENT AND THIS CONFUSES THE PROGRAM
'    PLUS THE STAMP SEEMED TOO SLOW TO DECIDE ON THE WORD LENGTH AND
'    THEN WAIT FOR AN (02)
'read direct from MC145447
'relay is on at first
'fctc 0 (delay) -> fcta 0 -> fcta 1 -> fctc 1 =
'    cycle through blocking options
'fctc 0 (delay) store cid # in blked # register
'fctc 0 (delay) -> fctb 0 -> fctb 1 -> fctc 1 = store 00s in blked numbers
'    to erase a number
'fctc must be held down for a delay to prevent accidents
'start storing info with mem addr 12
'pin 15 is lcd output line
'pin 14 is data input line
'pin 13 is here we will come, data to be sent, line
'IF PIN 13 = 0, NRD, a ring has been detected, not ring detect
'IF PINS 12 OR 11 OR 9 = 0 = FCTA or FCTB or FCTC
'pin 12 is functiona input line, look at data, normally 1
'pin 11 is functionb input line, look at blocked #s, normally 1
'if pin 10 = 0 reset mem loc 0 (stored data), mem loc 1 (blocked #s)
'    and come up in blocking off via nb,bi
'tie pin 10 (skip) high with resistor and be able to jump to ground
'    the very first time only to set address counters to initial values
'pin 9 is functionc input line, store a blocked #, normally 1
'if you shut down or you reload the program you can save stored info
'THIS HAS FOR...NEXT EEPROM OUTPUT LOOP, SHOWS #: ######### via FUNCTION
'    LIMIT 20 STORED NUMBERS
'SEE CONSTANT DEFINITION BELOW: DTLMT IS FOR 10 STORED #S, NUMBERED 1-10
'              BLLMT IS FOR 10 BLOCKED #S, NUMBERED 11-20
'READ OUT BLOCKED NUMBERS VIA FUNCTIONB, PIN 11
'READ OUT CID NUMBERS VIA FUNCIONA, PIN 12
'STORE NUMBERS TO BLOCK VIA FUNCTIONC, PIN 9

'pin 0 is led for read in complete—test purposes only
'pin 1 is led for read out complete—test purposes only
'pin 2 is led for function subroutine in operation, use a led
'pin 3 is for relay
'output "WAITING" on power up, if error, after set blocking
'******************************************************************
```

```
'***********************SET IN/OUT*****************************
input 14: input 13: input 12
input 11: input 10: input 9          'in lines
low 0: low 1                         'out lines; set 2, 3 later
'*************************************************************

'***********************DECLARE*******************************
'these are the ram spots that are read into
a       var     byte
b       var     byte
c       var     byte
d       var     byte
e       var     byte
f       var     byte
g       var     byte
h       var     byte
i       var     byte
j       var     byte
k       var     byte
'l      var     byte

'input lines
nrd     var     in13    'not Ring Detect from MC145447
fcta    var     in12    'function a
fctb    var     in11    'function b
skp     var     in10    'ground on very first power up to initialize
fctc    var     in9     'function c

'constants
dtst    con     12      'this is the start address for data
dtlmt   con     111     'the add. of the final data byte, 10 #s, 12-111
bllmt   con     211     'the add. of the final blk no byte, 10#s, 112-211
cidad   con     0       'this is the adr. of CID reception and storage
                        'and first read out
blnad   con     1       'this is the adr. of the blocked number
                        'read in and read out
cidro   con     2       'this is the adr. of the CID rdout when sequencing
bi      con     3       'this is the adr. of 1 inhibits blked call
nb      con     4       'this is the adr. of 1 connects line when not blked
                        'call

'more variables
x       var     byte    'scratch pad
y       var     byte    'hold addresses, etc.
z       var     byte    'scratch pad

qa      var     byte    'used in "test if # is blocked"
qb      var     byte
qc      var     byte
'*************************************************************

'THIS WILL RESET DATA COUNTERS AND BLOCKING IF PIN 10 IS LOW
```

```
        if skp = 1 then start1          'if pin 12 is high don't reset
        write cidad,dtst                'will be adr. of CID rcption
        write blnad,dtlmt+1             'will be adr. of blkd nos.
        write bi,0: write nb,0          'come up with blocking off


'THIS IS THE SET-RESET SEGMENT FOR THE LCD
start1: pause 1000                      'let lcd settle
        serout 15,396+$4000,[254,1]     'cls
        pause 1000
        serout 15,396+$4000,["WAITING"] 'at start, if error,
                                        'after set blocking


'THIS IS THE MAIN HOLD LOOP
'GO TO THE BEGINNING OF SERIN-ING THE DATA IF PIN13 = 0
'OR GO TO A FUNCTION SUBROUTINE IF PINS 12 OR 11 OR 9 = 0
start2: high 2: high 3                  'set led, relay each time
start2a:if nrd = 0 then start3x         'nrd = 0, data soon
        if fcta = 0 then fnctn          'function switch pressed
        if fctb = 0 then fnctn
        if fctc = 0 then zz             'put in delay to prevents oops
        goto start2a                    'loop
zz:     pause 2000                      'to prevent accidents
        if fctc = 1 then start2a        'oops has been prevented
        goto fnctn


'COME HERE FROM ABOVE IF NRD HAS BEEN LOW
'IF "CONNECT NOT BLOCKED" IS SET TURN OFF RELAY
'NOTE HOW THIS WORKS IN CONJUNCTION WITH
' "TEST TO SEE IF THE NUMBER IS A BLOCKED ONE AND HOW BLOCKING IS SET"
' LATER
start3x:read nb,z                       'check if "connect not blocked" is set
        if z =1 then goon               'if it is on, low 3
        goto start3                     'if "connect not blocked" not set, do nil
goon:   low 3


'THIS IS THE MAIN READ-IN SECTION OF THE PROGRAM
'the first CID info is hex 80 to denote MDMF, the format this program
'is written for; wait for that
'1. then read in the next byte which will be large if full data is sent
'2. or small if is "out of area" or "private"
'after waiting hex 80, read in word-length and then decide whether
'3. to wait for hex 02 = number parameter or to go to reason for absence
'4. see "notes to myself" for how I got 28, half-way between 16 and 39
'1-4 HAS BEEN CHANGED; THE STAMP APPEARS NOT TO BE FAST ENOUGH FOR RELIABLE
'OPERATION OF THIS FEATURE OF THE EARLIER PROGRAM CID23.BS2
start3: low 2                           'indicator led
        serin 14,813,6000,start1,[WAIT (128)]
                                        'wait hex 80
```

```
serin 14,813,1000,nonum,[WAIT (02),a,b,c,d,e,f,g,h,i,j,k]
                        'read in word-length, then the number
                        'the word-length will be dumped
goto anum

nonum: b = "O":c = "T":d = "H"       'OTH for other
       e = " ":f = " ":g = " ":h = " ":i = " ":j = " ":k = " "
         'set spaces in other variables

anum:  high 0                                  'got this far

'THIS IS THE MAIN OUTPUT SECTION OF THE PROGRAM
     serout 15,396+$4000,[254,1]                       'cls
     serout 15,396+$4000,[b,c,d," ",e,f,g," ",h,i,j,k] 'output
     high 1                                       'got this far


'HERE WE STORE DATA IN EEPROM
       read cidad,y      'define y
       gosub st          'store via subroutine
       if y = dtlmt+1 then far1 'dtlmt-dtst+1/10 #s, start with dtst again
       goto far2         'ADDR. COUNTER ALWAYS NEXT INPUT
far1:  write cidad,dtst 'new cycle
       goto test
far2:  write cidad,y     'save how far the memory has been used+1
       goto test


'*****************************************************************************
'SUBROUTINE
st:  write y,b    'STORE THE DATA IN EEPROM
     y = y+1
     write y,c
     y = y+1
     write y,d
     y = y+1
     write y,e
     y = y+1
     write y,f
     y = y+1
     write y,g
     y = y+1
     write y,h
     y = y+1
     write y,i
     y = y+1
     write y,j
     y = y+1
     write y,k
     y = y+1          'ready for next one
return
'*****************************************************************************
```

```
'TEST TO SEE IF THE NUMBER IS A BLOCKED ONE AND HOW BLOCKING IS SET
test:     qc = bllmt-dtlmt-10       'amount of blocked mem. - 10
          for qa = 0 to qc step 10  'IS # BLOCKED?
          for qb = 1 to 10 step 1   'compare
          read dtlmt+qa+qb,y         'compare
          read cidad,x               'compare
          read x+qb-11,z             'compare
          if y = z then here1        'compare
          goto here2                 'compare
here1:    next                       'compare
          goto test0                 'was a blocked #
here2:    next                       'compare
          goto fnctaQ                'it wasn't a blocked #, go on
test0:    read bi,y
          if y = 0 then fnctnQ       'INHIBIT BLOCKED off, go on
          low 3                      'if bi is 1 inhibit
          goto fnctnQ                'go on
'NOTE HOW THESE DECISIONS WORK IN CONJUNCTION WITH EARLIER:
'          "IF 'CONNECT NOT BLOCKED' IS SET TURN OFF RELAY"


'*********************************************************************
'STAY AS LONG AS RINGING
fnctaQ:   high 3                     'turn on relay
fnctnQ:   x = 0                      'scratchpad = 0
          goto stay2                 'avoid a reset
stay1:    x = 0
stay2:    if nrd = 0 then   stay1    'do nothing as long as ringing
          pause 50                   'stopped ringing
          x = x+1                    'add 1 every 50 ms
          if x < 100 then stay2      'if less than 5 sec keep testing
          low 0: low 1               'reset 0, 1
          goto start2
'*********************************************************************


'THE FUNCTION SWITCH COMES HERE
'fncta:    high 3                    'turn on 3 if need be
fnctn:    low 2                      'COME HERE FROM FUNCTION, blip led
          pause 250                  'debounce
          if fcta = 0 then so1       'go to stored output
          if fctb = 0 then so3
          if fctc = 0 then mem1      'goto holding for function c
          pause 250                  'debounce more
'         low 0                      'reset LEDs that confirmed earlier work
'         low 1
          goto start2


'HERE IS THE STORED OUTPUT
```

```
        'STORED NUMBERS
so1:    read cidad,y              'y is addr. of next store, in cidad
        if y <> dtst then so2     'reset cidad for highest data addr.
        y = dtlmt+1               'if recycling
so2:    gosub socid               'OUTPUT DATA STORE IN EEPROM
        write cidro,y             'THIS WILL BE USED IN STORING READ
                                  'WHERE-LEFT-OFF INTO BLKED # MEM
        if y <= dtst then resty1  'reset y when first # is done
        goto hold1
resty1: y = dtlmt+1
        goto hold1
        'BLOCKED NUMBERS
so3:    read blnad,y
        goto so5
so4:    y = y+10
        if y<=bllmt then so5
        y = dtlmt+1
so5:    gosub sobln
        write blnad,y
        goto hold2

'************************************************************************
'THIS SUBROUTINE OUTPUTS THE DATA STORED IN EEPROM
socid:  y = y-10
sobln:  serout 15,396+$4000,[254,192]  'down a line; come here for
        z = y-2                        'consecutive stored output
        z = z/10                       'get entry # digit(s)
        serout 15,396+$4000,[dec z, ": "] 'this is entry #
        for z = y to y+9               'OUTPUT EEPROMED DATA HERE
        read z,x
        serout 15,396+$4000,[x]
        next
        serout 15,396+$4000,[" "]      'get rid of past
return
'************************************************************************

'WHAT TO DO NEXT IN FUNCTION MODE
hold1:  if fcta = 0 then hold1
hold2:  if fctb = 0 then hold2
        pause 100                      'debounce
        z = 0
hold4:  z = z+1                        'do again if another fcta = 0
        pause 10                       'within one second
        if fcta = 0 then so2
        if fctb = 0 then so4
        if z = 100 then fnctn          'return to function to reset leds and go
        goto hold4                     'back to main wait loop


'STORE A BLOCKED # OR STORE 0S
```

```
mem1:   pause 100                   'debounce
mem2:   if fctb = 0 then do1        'store 0s if fctb = 0
        if fcta = 0 then do2
        if fctc = 0 then mem2       'to erase
        goto on1
do1:    b = 48:c = 48:d = 48:e = 48:f = 48:g = 48:h = 48:i = 48:j = 48:k = 48
        goto on2                    'set zeros

on1:    read cidro,y               'from last read out
        read y,b                   'GET #S INTO b, c, d, etc.
        y = y+1
        read y,c
        y = y+1
        read y,d
        y = y+1
        read y,e
        y = y+1
        read y,f
        y = y+1
        read y,g
        y = y+1
        read y,h
        y = y+1
        read y,i
        y = y+1
        read y,j
        y = y+1
        read y,k
'       y = y+1                     'unnecessary
on2:    read blnad,y               'to blocked data at point last looked at
        gosub st                   'store it
        pause 100                  'debounce
sit1:   if fctc = 0 then sit1
sit2:   if fctb = 0 then sit2
        goto fnctn


'THIS SETS WHETHER THE PROGRAM SHOULD BLOCK CALLS OR NOT
do2:    serout 15,396+$4000,[254,1] 'cls
        pause 100                   'debounce
        read bi,z                   'read setting
        if z = 0 then tgl1
        write bi,0: write nb,1      'settings
        serout 15,396+$4000,["connect not",254,192,"blocked calls"]
        goto sit3
tgl1:   read nb,z
        if z = 0 then tgl2
        write bi,0: write nb,0      'settings
        serout 15,396+$4000,["call blocking", 254,192,"is off"]
```

```
        goto sit3
tgl2:   write bi,1: write nb,0        'settings
        serout 15,396+$4000,["inhibit blocked", 254,192,"calls"]
sit3:   if fcta = 0 then sit3
sit4:   if fctc = 0 then sit4
        goto start1
```

Chapter 30

Page-Alert

Parts list

Page-Alert

R1, R2   240 ohm $\frac{1}{2}$ W resistor

R3, R4, R5, R6, R7 10 K ohm $\frac{1}{4}$ W resistor

R8, R9   330 ohm $\frac{1}{4}$ W resistor

R10   1 K ohm $\frac{1}{4}$ W, 5% resistor

C1, C2   .1 µF 35 V electrolytic capacitor

C3   35 µF 35 V electrolytic capacitor

C4   5 µF, 35 V electrolytic capacitor

D1   red LED

V1   Sidactor P3002AB-ND

U1   Cermetek CH1817D-DAA module

U2   BASIC STAMP 2 microcontroller (original)

U3   LM7805 regulator (5 V)

S1   On-Off power switch - toggle switch

S2   On-Off power switch - toggle switch

RY1   On-Off power switch - toggle switch

J1   RJ11 PC mount telephone jack

J2   4 pin programming header

J3   10 pin female header

J4   10 position screw terminal

Misc   PC board, wire, standoffs, chassis box, etc.

Switch/sensor Module

R1, R3   20 K ohm $\frac{1}{4}$ W resistor

R2, R4   10 K ohm $\frac{1}{4}$ W resistor

C1   .01 µF 25 volt disc capacitor

D1, D2   1N914 silicon diode

Q1   2N2222 NPN transistor

Misc   Sensors, PC board, wire, hardware, header, etc.

Temperature/voltage Level Module

R1, R3   2 K ohm $\frac{1}{4}$ W resistor

R2   50 K ohm potentiometer

R5   1 megohm $\frac{1}{4}$ W resistor

R4, R6   10 K ohm $\frac{1}{4}$ W resistor

R7   10 megohm $\frac{1}{4}$ W resistor

R8   3.3 K ohm $\frac{1}{4}$ W resistor

C1 .1 μF 25 V disc
    capacitor

C2 .01 μF 25 V disc
    capacitor

U1 LM393 comparator IC

S1 Resistive sensor,
    thermistor, etc.

Misc PC board, header,
      wire, hardware, etc.

Figure 30-1 *Page-Alert circuit board*

Page-Alert

The Page-Alert is a low cost microprocessor controlled alarm reporting system, which can monitor up to four different alarm conditions and immediately report them to your numerical pager (see Figure 30-1).

The Page-Alert can be used to protect your home, office, shop, or vacation home while you are away. The Page-Alert can be configured to monitor voltage levels, temperature, movement, doors, windows, and electronic equipment such as computer failures (see block diagram in Figure 30-2). The Page-Alert can free you to be two places at once!

The optional enhancement modules such as the Pyro-electric Motion Module can be used to sense "body heat" up to 30′ away and provide an output which will activate the Page-Alert. The Temperature/Voltage level module can be utilized to monitor temperature fluctuations or voltage level changes depending upon the configuration, and report the problem to your pager.

The Page-Alert utilizes your existing telephone line, so it incurs no additional monthly phone bills, which are generally required by alarm companies. Simply plug in the power supply and connect it to a regular phone line and connect up at least one sensor and you are ready to remotely monitor just about any alarm condition. Upon receiving a call from your Page-Alert, you can elect to respond yourself, or call a neighbor, friend, or co-worker to solve the problem. You could also elect to notify the police if desired.

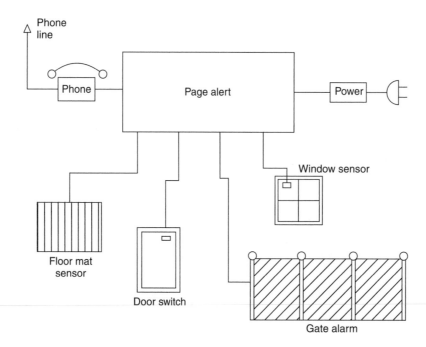

Figure 30-2 *Page-Alert block diagram*

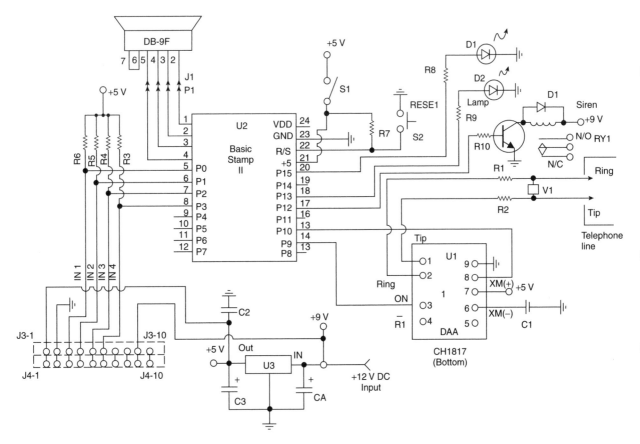

Figure 30-3 *Page-Alert circuit diagram*

The Page-Alert multichannel alarm reporting system centers around the BASIC STAMP 2 (BS2) microcomputer at U2, as shown in Figure 30-3.

The BS2 microcomputer is a small but powerful computer capable of up to 4,000 instructions per second. The BS2 micro-computer consists of the main processor, memory, reset, regulators, and touch tone/X-10 generator circuits all combined in a 24 pin chip carrier. The BS2 computer runs on an interpreted "BASIC" language. The BS2 has 16 input/output lines which can be seen in Table 30-1, which describes the pinouts of the BS2 microcontroller, including the serial port connections to your programming computer. The BS2 microcomputer is generally used for specific or dedicated control application.

The Page-Alert scans up to four alarm input channels simultaneously. All four channels are configured as normally open inputs, with 10 K ohm resistors across

Table 30-1

Basic STAMP 2 microprocessor pinouts

Pin	Name	Function	Description
1	SOU	Serial out	temporarily connects to PC's (Rx)
2	SIN	Serial out	temporarily connects to PC's (Tx)
3	ATN	Attention	temporarily connects to PC's (DTR)
4	VSS	Ground	temporarily connects to PC's ground
5	P0	User I/O 0	user ports which can be used for inputs
6	P1	User I/O 1	or outputs:
7	P2	User I/O 2	

(Continued)

Table 30-1—cont'd

Basic STAMP 2 microprocessor pinouts

Pin	Name	Function	Description
8	P3	User I/O 3	Output mode: Pins will source from VDD. Pins should
9	P4	User I/O 4	not be allowed to source more than 20 ma, or sink more than
10	P5	User I/O 5	25 ma. As groups, P0–P7 and P8–P15 should not be
11	P6	User I/O 6	allowed to source more than 40 ma, or sink 50 ma.
12	P7	User I/O 7	
13	P8	User I/O 8	Input mode: Pins are floating(less than 1 μa. leakage)
14	P9	User I/O 9	The logic threshold is approx. 1.4 V.
15	P10	User I/O 10	
16	P11	User I/O 11	Note: to realize low power during sleep, make sure that no
17	P12	User I/O 12	pins are floating, causing erratic power drain. Either drive
18	P13	User I/O 13	them to VSS or VDD, or program them as outputs that do
19	P14	User I/O 14	not have to source current.
20	P15	User I/O 15	
21	VDD	Regulator out	Output from 5 V regulator (VIN powered). Should not be allowed to source more than 50 ma, including P0–P15 loads.
22	RES	Power in	Power input (VIN not powered). Accepts 4.5–5.5 V. Current consumption is dependent on run/sleep mode and I/O.
23	VSS	Reset I/O	When low, all I/Os are inputs and program execution is suspended. When high, program executes from start. Goes low when VDD is less than 4 V or ATN is greater than 1.4 V. Pulled to VDD by a 4.7 K resistor. May be monitored as a brown-out/reset indicator. Can be pulled low externally (i.e. button to VSS) to force a reset. Do not drive high.
24	VIN	Ground	Ground. Located next to Vin for easy battery backup.
		Regulator In	Input to 5 V regulator. Accepts 5.5–15 V. If power is applied directly to VDD, pin may be left unconnected.

the inputs at pins P0 through P3. The inputs to the Page-Alert are brought out to the screw terminals at J4 for easy connections to the outside world. Inputs P0 through P3 are used as input channels; P4 through P7 are not used in this project. The output of the microcontroller at P9 is utilized to drive the Data Access Arrangement module (DAA) at U1. The Cermetek DAA is a Telco approved telephone interface, which should be used to couple and isolate electronic circuits to the phone line. It provides the correct input/output level interfacing, relay, and protection circuits needed for interfacing (see Figure 30-4).

Once an alarm input is triggered, the Page-Alert activates pin 14, which in turn drives the OH pin in the DAA module, allowing the phone to go "off hook." The microprocessor now begins the dialing sequence to call your pager.

The BS2 microprocessor contains a touch-tone generator which is utilized to dial your pager and also generate the identifier codes. Pin 15 or P10 on the BS2 is used to drive the audio signal from the touch-tone output to the TX (+) pin on the DAA, which dials the phone. The TX (–) pin on the DAA is connected to ground via a .1 disc capacitor. Pin 7 of the DAA is the 5 V power connection, while pin 9 is the ground connection. The RCV and RI pins on the DAA are not used here. Pins 1 and 2 of the DAA are connected to the phone line's ring and tip lines via the Sidactor

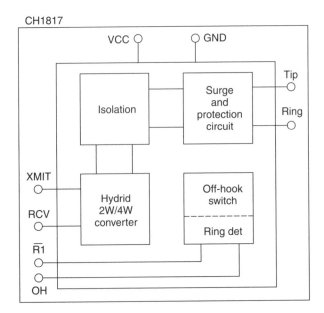

CH1817

Figure 30-4 *DAA Telephone interface module*

protection device, followed by two 100 ohm resistors which are used couple the DAA to the RJ11 phone plug.

A reset function is provided at pin 22 of U2, and is connected via R5 to pin 21(VDD). This pin is a "brown out" detector and reset device. A reset push button is connected between pin 22 and ground. A bypass capacitor is coupled across the power leads at pins 23 and 24. The power input on pin 24 can accept 5 to 15 V DC, which powers the internal regulators in the BS2. Pin 21 is the 5 V system power pin from the regulator.

The regulator at U3 is used to power the Page-Alert system. A 12 DC "wall wart" power cube can be used to provide input power to the Page-Alert board at J4-2. The regulator provides 5 V to pin 21 of the BS2 controller at U2. The 5 V source is also utilized to provide power to the sensor daughter boards which can be plugged into the female header at J3.

Circuit assembly

Before we begin constructing the Page-Alert, you will need to find a clean, well-lit worktable or workbench. Next we will gather a small 25–30 W pencil tipped soldering iron, a length of 60/40 tin–lead solder, and a small canister of Tip Tinner, available at your local RadioShack store; Tip Tinner is used to condition the soldering tip between solder joints. You drive the

soldering tip into the compound and it cleans and prepares the soldering tip. Next grab a few small hand tools; try to locate small flat blade and Phillips screwdrivers. A pair of needle-nose pliers, a pair of tweezers along with a magnifying glass, and a pair of end-cutters and we will begin constructing the project. Now locate your schematic diagram and parts layout diagram along with all of the components needed to build the project.

Once all the components are in front of you, you can check them off against the project parts list to make sure you are ready to start building the project. Now, locate Tables 30-2 and 30-3, and place them in front of you. Table 30-2 illustrates the resistor code chart and how to read resistors, while Table 30-3 illustrates the capacitor code chart which will aid you in constructing the project.

Refer now to the resistor color code chart shown in Table 30-2, which will help you identify the resistor values. Note that each resistor will have either three or four color bands, which begin at one end of the resistor's body. The first colored band represents the resistor's first digit, while the second color represents the second digit of the resistor's value. The third colored band represents the resistor's multiplier value, and the fourth color band represents the resistor's tolerance value. If there is no fourth colored band then the resistor has a 20% tolerance value. If the resistor's fourth band is silver then the resistor has a 10% tolerance value and if the fourth band is gold then the resistor's tolerance value will be 5%.

Finally, we are ready to begin assembling the project, so let's get started. The prototype project was constructed on a $2\frac{1}{2} \times 4\frac{1}{4}''$ single-sided printed circuit board or PCB. With your PC board in front of you, we can now begin populating the circuit board. We are going to place and solder the lowest height components first, the resistors and diodes. Let's locate our first two resistors R1 and R2. Resistors R1 and R2 each have a value of 100 ohms and are a 5% tolerance type. Now take a look for a resistor whose first color band is brown, black, brown, and gold. From the chart you will notice that brown is represented by the digit (1), the second color band is black, which is represented by (0). Notice that the third color is brown and the multiplier is (10), so $(1)(0) \times 10 = 100$ ohms.

Table 30-2
Resistor color code chart

Color Band	1st Digit	2nd Digit	Multiplier	Tolerance
Black	0	0	1	
Brown	1	1	10	1%
Red	2	2	100	2%
Orange	3	3	1,000 (K)	3%
Yellow	4	4	10,000	4%
Green	5	5	100,000	
Blue	6	6	1,000,000 (M)	
Violet	7	7	10,000,000	
Gray	8	8	100,000,000	
White	9	9	1,000,000,000	
Gold			0.1	5%
Silver			0.01	10%
No color				20%

Table 30-3
Three-digit capacitor codes

pF	nF	µF	Code	pF	nF	µF	Code
1.0			**1R0**	3,900	3.9	.0039	**392**
1.2			**1R2**	4,700	4.7	.0047	**472**
1.5			**1R5**	5,600	5.6	.0056	**562**
1.8			**1R8**	6,800	6.8	.0068	**682**
2.2			**2R2**	8,200	8.2	.0082	**822**
2.7			**2R7**	10,000	10	.01	**103**
3.3			**3R3**	12,000	12	.012	**123**
3.9			**3R9**	15,000	15	.015	**153**
4.7			**4R7**	18,000	18	.018	**183**
5.6			**5R6**	22,000	22	.022	**223**
6.8			**6R8**	27,000	27	.027	**273**
8.2			**8R2**	33,000	33	.033	**333**
10			**100**	39,000	39	.039	**393**
12			**120**	47,000	47	.047	**473**
15			**150**	56,000	56	.056	**563**
18			**180**	68,000	68	.068	**683**
22			**220**	82,000	82	.082	**823**
27			**270**	100,000	100	.1	**104**
33			**330**	120,000	120	.12	**124**
39			**390**	150,000	150	.15	**154**
47			**470**	180,000	180	.18	**184**
56			**560**	220,000	220	.22	**224**

Table 30-3—cont'd
Three-digit capacitor codes

pF	nF	μF	Code	pF	nF	μF	Code
68			**680**	270,000	270	.27	**274**
82			**820**	330,000	330	.33	**334**
100		.0001	**101**	390,000	390	.39	**394**
120		.00012	**121**	470,000	470	.47	**474**
150		.00015	**151**	560,000	560	.56	**564**
180		.00018	**181**	680,000	680	.68	**684**
220		.00022	**221**	820,000	820	.82	**824**
270		.00027	**271**		1,000	1	**105**
330		.00033	**331**		1,500	1.5	**155**
390		.00039	**391**		2,200	2.2	**225**
470		.00047	**471**		2,700	2.7	**275**
560		.00056	**561**		3,300	3.3	**335**
680		.00068	**681**		4,700	4.7	**475**
820		.00082	**821**			6.8	**685**
1,000	1.0	.001	**102**			10	**106**
1,200	1.2	.0012	**122**			22	**226**
1,500	1.5	.0015	**152**			33	**336**
1,800	1.8	.0018	**182**			47	**476**
2,200	2.2	.0022	**222**			68	**686**
2,700	2.7	.0027	**272**			100	**107**
3,300	3.3	.0033	**332**			157	**157**

Code = 2 significant digits of the capacitor value + number of zeros to follow

For values below 10 pF, use "R" in place of decimal e.g. 8.2 pF = 8R2

10 pF = 100
100 pF = 101
1,000 pF = 102
22,000 pF = 223
330,000 pF = 334
1 μF = 105

Go ahead and locate where resistors R1 and R2 are placed on the PC and install them, next solder them in place on the PC board, and then with a pair of end-cutters trim the excess component leads flush to the edge of the circuit board. Next locate the remaining resistors and install them in their respective locations on the main controller PC board. Solder the resistors to the board, and remember to cut the extra lead with your end-cutters.

Note that capacitors generally come from two major groups or types. Non-polar types, such as the ones in this project, have no polarity and can be placed in either direction on the circuit board. The other type of capacitors are called polarized types, these capacitors have polarity and have to be inserted on the board with respect to their polarity marking in order for the circuit to operate correctly. On these types of capacitors, you will observe that the components actually have a plus or minus marking or a black or white band with a plus or minus marking on the body of the capacitors. Capacitors must be installed properly with respect to polarity in order for the circuit to work properly. Failure to install electrolytic capacitors correctly could cause damage to the capacitors as well as to other components in the circuit when power is first applied.

Next refer to Table 30-3; this chart helps to determine the value of small capacitors. Many capacitors are quite small in physical size, so manufacturers have devised a three-digit code which can be placed on the capacitor instead of the actual value. The code occupies a much smaller space on the capacitor and can usually fit nicely on the capacitor body, but without the chart the builder would have a difficult time determining the capacitor's value.

Now look for the capacitors from the component stack and locate capacitors C1 and C2, they are small capacitors labeled (104). Refer to Table 30-3 to find code (104) and you will see that it represents a capacitor value of .1 μF. Go ahead and place C1 and C2 on the circuit board and solder them in place. Now take your end-cutters and cut the excess component leads flush to the edge of the circuit board. Once you have installed these components, you can move on to installing the remaining capacitors. Finally locate the two electrolytic capacitors at C3 and C4 and go ahead and install them on the PC board, remembering these capacitors have polarity and it must be observed if the circuit is to work correctly. The minus terminals of the capacitors are connected to ground.

The Page-Alert project contains a single silicon diode, placed across the relay RY1. Diodes, as you will remember, will always have some type of polarity markings, which must be observed if the circuit is going to work correctly. You will notice that each diode will have a black or white colored band at one edge of the diode's body. The colored band represents the diode's cathode lead. Check your schematic and parts layout diagrams when installing these diodes on the circuit board. Place the diode on the circuit board in its respective location and solder it in place on the PC board. After soldering in the diodes and the diode bridge, remember to cut the excess component leads with your end-cutters, flush to the edge of the circuit board.

The Page-Alert utilizes three conventional transistors. Transistors are generally three lead devices, which must be installed correctly if the circuit is to work properly. Careful handling and proper orientation in the circuit is critical. A transistor will usually have a Base lead, a Collector lead, and an Emitter lead. The base lead is generally a vertical line inside a transistor circle symbol. The collector lead is usually a slanted line facing towards the base lead and the emitter lead is also a slanted line facing the base lead but it will have an arrow pointing towards or away from the base lead. Now, identify the two transistors, and refer to the manufacturer's specification sheets as well as the schematic and parts layout diagrams when installing the two transistors. If you have trouble installing the transistors, get the help of an electronics enthusiast with some experience. Install the transistors in their respective locations and then solder them in place. Remember to trim the excess component leads with your end-cutters.

The Page-Alert circuit also contains two ICs. ICs are often static sensitive, so they must be handled with care. Use a grounded anti-static wrist strap and stay seated in one location when handling the integrated circuits. Take out a cheap insurance policy by installing IC sockets for each of the ICs. In the event of a possible circuit failure, it is much easier to simply unplug a defective IC, than trying to unsolder 14 or 16 pins from a PC board without damaging the board. ICs have to be installed correctly if the circuit is going to work properly. IC packages will have some sort of markings which will help you orient them on the PC board. An IC will have either a small indented circle, a cutout, or notch at the top end of the IC package. Pin 1 of the IC will be just to the left of the notch, or cutout. Refer to the manufacturer's pin-out diagram, as well as the schematic, when installing these parts. If you doubt your ability to correctly orient the ICs, then seek the help of a knowledgeable electronics enthusiast.

The DAA module has 9 pins which protrude through the bottom of the module. You can solder the module directly to the PC, but be careful to install it correctly. Note, pins 1 and 9 are across from each other and pins 4 and 5 are across from each other and there is a gap between pins 2 and 3 on the same side of the module.

The Page-Alert employs two LEDs at D1 and D2, which also must be oriented correctly if the they are to function properly. An LED has two leads, a cathode and an anode lead. The anode lead will usually be the longer of the two leads and the cathode lead will usually be the shorter lead, just under the flat side edge of the LED; this should help orient the LED on the PC board. You may want to lead about ½ lead length on the LED, in order to have the LED protrude through the enclosure box.

Solder the LEDs to the PC board and then trim the excess leads.

The prototype utilized a mini PC board SPDT relay at RY1. Study the relay pin-outs carefully before attempting to install the relay on the PC board. You will want to make sure that you can properly identify the relay coil leads as well as the relay pole outputs before installing the relay on the PC board and soldering it in place.

Once the main Page-Alert board has been completed, take a short break and then we will inspect the PC board for possible "cold" solder joints and "short" circuits. Pick up the PC board with the foil side facing upwards toward you. Carefully inspect the solder joints, these should look smooth, clean, and bright. If any of the solder joints look dull, dark, or blobby, then you should remove the solder with a solder sucker or a solder wick and then resolder the joint all over again. Now we will inspect the PC board for possible short circuits. Short circuits can be caused from small solder blobs bridging the circuit traces or from cut component leads which can often stick to the circuit board once they have been cut. A sticky residue from solder will often trap component leads across the circuit traces. Look carefully for any bridging wires across circuit traces.

Using the Page-Alert

Note the 10 position screw terminal strip at J4 which is connected to a 10 pin female header at J3. The 10 pin header J3 is used to accept the Motion Module, or the TVM add-on modules, or other expansion devices. Table 30-4 illustrates the pinouts for the screw terminal strip and the 10 pin header connections.

Once the Page-Alert has been completed, recheck the component placement and your solder connections. You are now ready to power your Page-Alert unit. Locate a 12 V DC power supply which will be used to power your Page-Alert. You can readily elect to utilize the ubiquitous 12 V DC "wall wart" to power your Page-Alert; usually, these power supplies are quite economical. The Page-Alert prototype was housed in an economical 4 × 5″ plastic Pactec enclosure.

The Page-Alert is initially programmed via pins 1 through 4. These pins are the serial input/output connections used to program the BS2. The RX line is

Table 30-4

Accessory daughter board I/O connections

J3 – 10 Pin header – power and I/O
J4 – 10 Pin header – sensor daughter boards

Pin	Pinout Description
1	+5 volt power pin
2	System ground
3	IN1 input pin
4	IN2 input pin
5	IN3 input pin
6	IN4 input pin
7	No connection
8	No connection
9	+9 volt power
10	No connection

shown at pin1, while the TX line is at pin2. A DTR line is provided at pin 3, and the ground is at pin 4.

To program your Page-Alert, you simply apply power to the screw terminals at J3/J4 and connect the Page-Alert via the serial cable to your personal computer. Next, go to www.parallax.com and download the Windows BASIC STAMP Editor version 2.5 or above. Open up the editor program and type in or download the **palert.bs2** program into the editor. Once the **palert.bs2** program appears on your computer screen, you will have to enter your pager's number into the first phone number position and then you must enter the second number or identifier into the program. With the power connected up to the Page-Alert board and the programming cable connected, you can RUN the **palert.bs2** program from the editor and it will load the program into the BASIC STAMP computer on the Page-Alert board.

Now you are ready to utilize the Page-Alert. You can connect a normally open alarm switch/sensor between each input terminal and the 5 V terminal on the screw terminal, for up to four channels. You can use any type of normally open alarm switch or sensor. Door or window switches could be used as well as any other sensor with a normally open set of contacts. You could also elect to build the Switch/Sensor Module, which will allow you to connect normally open and normally closed switches or sensors to the Page-Alert. Additional modules are described in Chapter 31.

In order to test activate the Page-Alert, you could also substitute a normally open pushbutton to start the Page-Alert unit. Once activated, the Page-Alert should come to life! First, the status lamp will begin to flicker to indicate the program has started. Next, the OH line on the DAA is activated and the phone line goes "off hook." Next the microprocessor begins touch-tone dialing your pager's phone number. The microprocessor then waits for a short interval, and the triggered alarm channel's ID or identifier is sent to your numerical pager.

Once alerted via your pager, you can respond yourself or you can elect to call a friend, neighbor, co-worker, or even the police depending upon the severity of the problem. The Page-Alert uses your existing phone line, so there is no additional phone cost to use it. The Page-Alert can be used as a self contained silent alarm using the Switch/Sensor Module or the Motion Sensor Module described in the next chapter. The Page-Alert can also be used with a multi-channel alarm control box; in this way the Page-Alert becomes the phone dialer for the alarm control box.

The Page-Alert is a low cost means to free you and alert you to intruders, or equipment failures of impending doom. Why not have some fun? Build one!

PAGE-ALERT PROGRAM

```
'Palert.BS2
'identify variables

flash           VAR byte
new_io_state    VAR byte
old_io_state    VAR byte
call_state      VAR byte
io              VAR byte
timer           VAR byte

'identify constants
OHPin           CON 9
TxPin           CON 10
Siren           Con 12
Lamp            Con 13
LEDPin          CON 15
recall_delay    CON 120

'initialize variables and program
init:
DIRL = %00000000
```

```
DIRH = %11111111
new_io_state = %00000000
old_io_state = %00000000

for flash = 1 to 3
   high LEDPin
   pause 1500
   low LEDpin
   pause 1500
next
main:
high LEDPin
new_io_state = INL
pause 100
low LEDPin
pause 100
if new_io_state <> %00000000 then
    dial_pager
after_io1:
old_io_state = new_io_state
goto main
delay_and_scan:
old_io_state = new_io_state
for timer = 1 to recall_delay
   high LEDPin
   new_io_state = INL
   pause 500
   low LEDPin
   pause 500
   if new_io_state <> old_io_state
     then dial_pager
   old_io_state = new_io_state
next
goto after_io1

dial_pager:
if new_io_state = %00000000 then main
high LEDPin
high OHPin
high Siren
high Lamp
pause 500

dtmfout TxPin,500, 500, [8]
pause 200
dtmfout TxPin,400, 200, [7,9,9,6,
    6,5,8]
sleep 6
gosub send_msg
```

```
goto delay_and_scan
send_msg:
io = new_io_state
'debug ibin8 io, cr
dtmfout TxPin, 300, 200, [5,5,io.bit0,
    io.bit1,io.bit2,io.bit3,11]
pause 1000
low OHPin
low LEDPin
pause 30000
low Siren
low Lamp
pause 100
return
end
```

Alarm Switch/Sensor Module

Normally open/normally closed input sensor board

The Switch/Sensor Module can be used to connect any number of alarm sensors to the input of the Tele-Alert system. Using the Alarm Sensor Module you can easily turn your Tele-Alert into a portable burglar alarm system, which can be used to call you and let you know someone has violated a particular space that you have protected.

The diagram in Figure 30-5 depicts a clever method of utilizing both normally open and normally closed input sensors or alarm switches, using a simple circuit.

Normally open sensors or switch contacts are shown at S1, S2, and S3. Normally closed switches or alarm sensors are shown S4, S5, S6, and S7. Any number of normally closed or normally open switches can be used in this input circuit. Two 1N914 diodes, an NPN transistor, and a few resistors form the heart of this detector. You could elect to build one of these input circuits for each of the BASIC STAMP 2 input pins if desired. The output of this input converter circuit provides a 5 volt signal to the input of the STAMP 2 upon an alarm signal. A few of these detector circuits could be built on a single circuit board to create a multi-channel alarm system using your Tele-Alert system. The alarm sensor daughter board is simply plugged into the main Tele-Alert board through the set of header pins on each of the boards.

Circuit assembly

Before we begin constructing the Switch/Sensor board, you will need to locate a clean, well-lit worktable or workbench. Next we will gather a small 25–30 W pencil tipped soldering iron, a length of 60/40 tin–lead solder, and a small canister of Tip Tinner, available at your local RadioShack store; Tip Tinner is used to condition the soldering tip between solder joints. You drive the soldering tip into the compound and it cleans and prepares the soldering tip. Next grab a few small hand tools; try to locate small flat blade and Phillips screwdrivers. A pair of needle-nose pliers, a pair of tweezers along with a magnifying glass, and a pair of end-cutters and we will begin constructing the project. Now locate your schematic diagram and parts layout diagram along with all of the components needed to build the project.

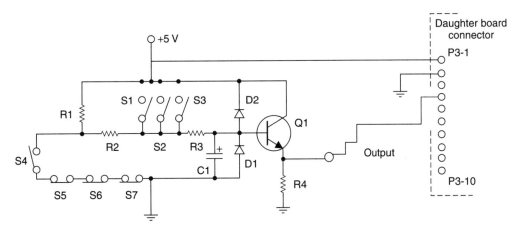

Figure 30-5 *Switch/sensor module*

The Switch/Sensor Board is built on single sided PC board. Once all the components are in front of you, you can check them off against the project parts list to make sure you are ready to start building the project. Now, locate Tables 30-2 and 30-3, and place them in front of you. Table 30-2 illustrates the resistor code chart and how to read resistors, while Table 30-3 illustrates the capacitor code chart which will aid you in constructing the project.

Refer now to the resistor color code chart shown in Table 30-2, which will help you identify the resistor values for the project. Note that each resistor will have either three or four color bands, which begin at one end of the resistor's body. The first colored band represents the resistor's first digit, while the second color represents the second digit of the resistor's value. The third colored band represents the resistor's multiplier value, and the fourth color band represents the resistor's tolerance value. If there is no fourth colored band then the resistor has a 20% tolerance value. If the resistor's fourth band is silver then the resistor has a 10% tolerance value and if the fourth band is gold then the resistor's tolerance value will be 5%.

Finally, we are ready to begin assembling the project, so let's get started. The prototype project was constructed on a single-sided printed circuit board or PCB. With your PC board in front of you, we can now begin populating the circuit board. We are going to place and solder the lowest height components first, the resistors and the diodes. First, locate our first resistor R1. Resistor R1 has a value of 20,000 ohms and is a 5% tolerance type. Now take a look for a resistor whose first color band is red, black, orange, and gold. From the chart you will notice that red is represented by the digit (2), the second color band is black, which is represented by (0). Notice that the third color is orange and the multiplier is (1,000), so (2)(0) × 1,000 = 20,000 or 20 K ohms.

Go ahead and locate where resistor R1 goes on the PC and install it, next solder it in place on the PC board, and then with a pair of end-cutters trim the excess component leads flush to the edge of the circuit board. Next, locate the remaining resistors and install them in their respective locations on the main controller PC board. Solder the resistors to the PC board, and remember to cut the extra component leads with your end-cutters.

Note that capacitors generally come from two major groups or types. Non-polar types, such as the ones in this project, have no polarity and can be placed in either direction on the circuit board. The other type of capacitors are called polarized types, these capacitors have polarity and have to be inserted on the board with respect to their polarity marking in order for the circuit to operate correctly. On these types of capacitors, you will observe that the components actually have a plus or minus marking or a black or white band with a plus or minus marking on the body of the capacitors. Capacitors must be installed properly with respect to polarity in order for the circuit to work properly. Failure to install electrolytic capacitors correctly could cause damage to the capacitors as well as to other components in the circuit when power is first applied.

Next refer to Table 30-3, this chart helps to determine value of small capacitors. Many capacitors are quite small in physical size, so manufacturers have devised a three-digit code which can be placed on the capacitor instead of the actual value. The code occupies a much smaller space on the capacitor and can usually fit nicely on the capacitor body, but without the chart the builder would have a difficult time determining the capacitor's value.

Now look for the capacitors from the component stack and locate capacitor C1; it is a small capacitor labeled (103). Refer to the Table 30-3 to find code (103) and you will see that it represents a capacitor value of .1 µF. Go ahead and place these capacitors on the circuit board and solder them in place. Now take your end-cutters and cut the excess component leads flush to the edge of the circuit board. Once you have installed these components, you can move on to installing the remaining capacitors. Save the larger electrolytic capacitors for mounting last. The electrolytic capacitors will have a plus or minus sign or a black or white color band. Be sure to orient the electrolytic capacitors correctly when placing them on the PC board.

The Alarm Switch/Sensor board contains two silicon diodes. Diodes, as you will remember, will always have some type of polarity markings, which must be observed if the circuit is going to work correctly. You will notice that each diode will have a black or white colored band at one edge of the diode's body. The colored band represents the diode's cathode lead. Check your schematic and parts layout diagrams when installing these diodes on the circuit board. Place the diodes on the circuit board in their respective locations and solder them in place on the PC board. After soldering in the diodes and the diode bridge, remember to cut the excess component leads with your end-cutters, flush to the edge of the circuit board.

The Switch/Sensor Board uses a single transistor at circuit's output which interfaces with the main Tele-Alert board. Transistors generally have three leads, a Base lead, a Collector lead and an Emitter lead. Refer to the schematic and you will notice that there will be two diagonal leads pointing to a vertical line. The vertical line is the transistor's base lead and the two diagonal lines represent the collector and the emitter. The diagonal lead with the arrow on it is the emitter lead, and this should help you install the transistor on to the circuit board correctly. Go ahead and install the transistor on the PC board and solder it in place, then follow-up by cutting the excess components leads.

If your Switch/Sensor board only has a single channel as shown in the schematic, then the output header only needs a signal output and ground. The single channel output from the Alarm Sensor board can be sent to the BS2 at input P0. If you elected to build a multi-channel Switch/ Sensor board then the additional channels would be sent to STAMP inputs P1, P2, P3, and P4.

Once the Alarm Sensor circuit board has been completed, take a short break and then we will inspect the PC board for possible "cold" solder joints and "short" circuits. Pick up the PC board with the foil side facing upwards toward you. Carefully inspect the solder joints, these should look smooth, clean, and bright. If any of the solder joints look dull, dark, or blobby, then you should remove the solder with a solder sucker or a solder wick and then resolder the joint all over again. Now we will inspect the PC board for possible "short" circuits. Short circuits can be caused from small solder blobs bridging the circuit traces or from cut component leads which can often stick to the circuit board once they have been cut. A sticky residue from solder will often trap component leads across the circuit traces. Look carefully for any bridging wires across circuit traces.

Once your Switch/Sensor board is complete and ready to go, simply attach your normally open sensors or switches in parallel at S1, S2, and S3. You can also connect up normally closed sensors and switches in series at S4, S5, S6, and S7. If you installed a 10-position header on the bottom of the Switch/Sensor Module then you can simply plug in the module's P3 connector into the Page-Alert's J3 connector on the Page-Alert board. You can add as many switches of either type, however, with too many switches configured in series and parallel in this fashion you will not be able to trace the actual alarm sensor very easily. In order to pinpoint the sensors more easily, you may want to consider building a four channel output Switch/Sensor Module instead of a single channel board as shown.

The Alarm Switch/Sensor board is a great addition to the Page-Alert or Tele-Alert system.

Temperature/Voltage level module

The Temperature/Voltage Level Module board is a versatile addition to the Tele-Alert microprocessor alarm controller (see Figure 30-6).

The Temperature/Voltage Level Module is a single channel module that will permit you to monitor temperature level changes either upwards or downwards, or allow you to measure voltage level changes by setting a user threshold control.

The Temperature/Voltage Level or TVM board consists of a single LM393 comparator integrated circuit and a handful of components, which is illustrated in the schematic of Figure 30-7.

Figure 30-6 *Page-Alert TVM temperature board*

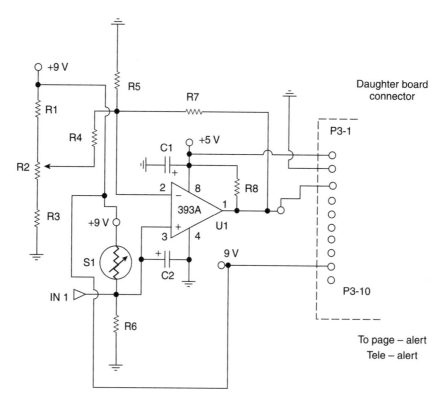

Figure 30-7 *Temperature-voltage module (TVM)*

The circuit begins with S1, a resistive sensor and R6. The thermistor or temperature sensors at S1 and R6 form a voltage divider to ground. In this configuration you are able to monitor upward temperature changes. If the position of S1 and R6 are reversed, then you are able to monitor decreasing temperature changes. The thermistor outputs at pin 3 of U1(A) represents the plus (+) inputs to the comparator circuits. The resistor network at R12 acts as the threshold control on the minus (−) inputs of the comparator. The output of the comparator at pin 1 produces a (high) output when the threshold is sensed.

The TVM Module can also be used to monitor voltage level changes by eliminating the Thermistor S1 and resistor R6 and replacing them with "scaling" resistors instead. Scaling resistors will now represent voltage dividers consisting of two resistors at each input channel. Ratios such as 100 K to 1 K for R1/R2 or 1 K to 100 ohms for R3/R4 would be used as input scalers. The higher value resistor would replace the thermistors while the lower value resistor would go to ground. Note, capacitors C1 and C3 provide noise reduction, when using external inputs. To measure voltage level changes, you simply connect the input wire IN1 to the circuit or

system being monitored, adjust the threshold control, and you are ready to go.

Circuit Assembly

Before we begin constructing the TVM board, you will need to locate a clean, well lit worktable or workbench. Next we will gather a small 25–30 W pencil tipped soldering iron, a length of 60/40 tin–lead solder, and a small canister of Tip Tinner, available at your local RadioShack store; Tip Tinner is used to condition the soldering tip between solder joints. You drive the soldering tip into the compound and it cleans and prepares the soldering tip. Next grab a few small hand tools; try to locate small flat blade and Phillips screwdrivers. A pair of needle-nose pliers, a pair of tweezers along with a magnifying glass, and a pair of end-cutters and we will begin constructing the project. Now locate your schematic diagram and parts layout diagram along with all of the components needed to build the project.

Once all the components are in front of you, you can check them off against the project parts list to make sure you are ready to start building the project. Now, locate Tables 30-2 and 30-3, and place them in front of you. Table 30-2 illustrates the resistor code chart and how to read resistors, while Table 30-3 illustrates the capacitor code chart which will aid you in constructing the project.

Refer now to the resistor color code chart shown in Table 30-2, which will assist you in identifying the resistor values for the project. Note that each resistor will have either three or four color bands, which begin at one end of the resistor's body. The first colored band represents the resistor's first digit, while the second color represents the second digit of the resistor's value. The third colored band represents the resistor's multiplier value, and the fourth color band represents the resistor's tolerance value. If there is no fourth colored band then the resistor has a 20% tolerance value. If the resistor's fourth band is silver then the resistor has a 10% tolerance value and if the fourth band is gold then the resistor's tolerance value will be 5%.

Finally, we are ready to begin assembling the project, so let's get started. The prototype project was constructed on a single-sided printed circuit board or PCB. With your PC board in front of you, we can now begin populating the circuit board. We are going to place and solder the lowest height components first, the resistors and diodes. First, locate our first resistors R1 and R3. Both resistors have a value of 2,000 ohms and are a 5% tolerance type. Now take a look for a resistor whose first color band is red, black, red, and gold. From the chart you will notice that red is represented by the digit (2), the second color band is black, which is represented by (0). Notice that the third color is red and the multiplier is (100), so $(2)(0) \times 100 = 2,000$ or 2 K ohms.

Go ahead and locate where resistor R1 goes on the PC and install it, next solder it in place on the PC board, and then with a pair of end-cutters trim the excess component leads flush to the edge of the circuit board. Note, for initial testing S1/R6 should not be inserted. When later installing the thermistors or scaling resistors, be sure to observe carefully the placement of S1/R6, depending upon the temperature direction you desire to measure or if you want to sense voltage levels. Trim Potentiometer R2 is used to set the threshold values for either temperature or voltage level sensing. The trim-pots are located at the top of the circuit board.

Remember that when voltage level detection is desired, scaling resistors are use in place of the thermistors as mentioned earlier. Next locate the remaining resistors and install them in their respective locations on the main controller PC board. Solder the resistors to the PC board, remembering to cut the extra component leads with your end-cutters.

Note that capacitors generally come from two major groups or types. Non-polar types, such as the ones in this project, have no polarity and can be placed in either direction on the circuit board. The other type of capacitors are called polarized types, these capacitors have polarity and have to be inserted on the board with respect to their polarity marking in order for the circuit to operate correctly. On these types of capacitors, you will observe that the components actually have a plus or minus marking or a black or white band with a plus or minus marking on the body of the capacitors. Capacitors must be installed properly with respect to polarity in order for the circuit to work properly. Failure to install electrolytic capacitors correctly could cause damage to the capacitors as well as to other components in the circuit when power is first applied.

Many capacitors are quite small in physical size, so manufacturers have devised a three-digit code which can be placed on the capacitor instead of the actual value. The code occupies a much smaller space on the capacitor and can usually fit nicely on the capacitor body, but without the chart the builder would have a difficult time determining the capacitor's value.

Now look for the capacitors from the component stack and locate capacitor C1; it is a small capacitor labeled (104). Refer to Table 30-3 to find code (104) and you will see that it represents a capacitor value of .1 µF. Now locate capacitor C2; it may be labeled (103) or .01 µF. Go ahead and place these capacitors on the circuit board and solder them in place. Now take your end-cutters and cut the excess component leads flush to the edge of the circuit board. Once you have installed these components, you can move on to installing the remaining capacitors. Save the larger electrolytic capacitors for mounting last. The electrolytic capacitors will have a plus or minus sign or a black or white color band. Be sure to orient the electrolytic capacitors correctly when placing them on the PC board.

The resistive temperature or photocell can be placed at S1 for sensing or you could elect to use the circuit to detect voltage level changes by connecting a low voltage input to the terminal labeled IN1. The LM393A is a dual comparator, so you could easily fabricate a dual channel TVM board with two outputs. Simply build two of the circuits shown on a single daughter board.

The pinouts at the edge of the TVM board are described below and are shown in Table 30-4.

At the top of the board the first solder pad used is pin #2 which is for the 9 V reference connection for the inputs at IN1 and trim potentiometer R2. Pin #3 is used for the system ground connection. Pins #4 and #5 are used for the two TVM outputs to the Tele-Alert board. Solder pads 6 and 7 are not used, while solder pads 8 and 9 are reserved for inputs IN1 and IN2 to the comparators. The last pin at position 10 is used to supply 5 V to the comparator.

Once the board has been completed, you will need to place the 10 pin header on the component side of the PC board and solder it on the foil side of the PC board. The 10 pin header allows connection with the Page-Alert or Tele-Alert boards. The 10 pin header on the TVM board then can be inserted into the 10 pin female header (second inside header) on the main Page-Alert or Tele-Alert board once you are ready. Note that pins #8 and #9 are inputs to the comparator and these pins should be either clipped or bent so as not to be inserted into the Tele-Alert board. The voltage level inputs from the circuit being remotely monitored are connected to these pins #8 and #9, i.e. inputs IN1 and IN2 respectively.

Once the TVM circuit board has been completed, take a short break and then we will inspect the PC board for possible "cold" solder joints and "short" circuits. Pick up the PC board with the foil side facing upwards toward you. Carefully inspect the solder joints, these should look smooth, clean, and bright. If any of the solder joints look dull, dark, or blobby, then you should remove the solder with a solder sucker or a solder wick and then resolder the joint all over again. Now we will inspect the PC board for possible short circuits. Short circuits can be caused from small solder blobs bridging the circuit traces or from cut component leads which can often stick to the circuit board once they have been cut. A sticky residue from solder will often trap component leads across the circuit traces. Look carefully for any bridging wires across circuit traces.

Your Temperature/Voltage Level Module is now complete and ready to go. It is advised that you test your new TVM board before you connect it to the main Tele-Alert board. The TVM board was designed to incorporate a 10-position header pin set which can be used to simply plug the TVM board into the Page-Alert or Tele-Alert Board at J3.

Testing the TVM board

You could elect to construct a multi-channel TVM board by incorporating up to four of the circuits shown onto a single TVM PC board. Connect a scope or multi-meter to the channel 1 output on the TVM board. Next, connect the plus (+) lead from a 5 volt power supply to pins #2 and #10, then connect the minus (−) lead from the power supply to the system ground at pin #3. For this initial testing, resistors R1/R2 and R3/R4 are not inserted. Now, locate an adjustable low voltage power supply. Connect the plus lead (+) from the adjustable supply to the input at IN1 or pin #8, and connect the minus (−) lead to the system ground at pin #3. Be sure the adjustable power supply is initially adjusted for 0 V before applying terminals at the edge of the TVM board. Now, turn on the first 5 V power supply and slowly advance the adjustable power supply voltage from

Table 30-4

Accessory daughter board I/O connections

J3 – 10 Pin header – power and I/O
J4 – 10 Pin header – sensor daughter boards

Pin	Pinout description
1	+ 5 volt power pin
2	System ground
3	IN1 input pin
4	IN2 input pin
5	IN3 input pin
6	IN4 input pin
7	No connection
8	No connection
9	+ 9 volt power
10	No connection

zero to one or two volts. You may have to set the threshold to set the trip point at which you will begin to see an output on the scope. Once the comparator is tripped the scope reading should change from 0 to 5 V. Once channel 1 has been tested, you can move on to testing channel 2 in the same manner. Once the TVM board has been tested and you have decided if you wish to have temperature or voltage level sensing you can insert the input resistors S1/R2 or the scalers.

Chapter 31

Tele-Alert Project

Parts list

Tele-Alert

R1, R2, R3, R12
  4.7 K ohm $\frac{1}{4}$ W resistor

R4, R5, R6, R7 10 K ohm
  $\frac{1}{4}$ W resistor

R8, R9, R10, R11
  10 K ohm $\frac{1}{4}$ W resistor

R13 330 ohm $\frac{1}{4}$ W
  resistor

R14 620 ohm $\frac{1}{4}$ W
  resistor

R15 1 K ohm $\frac{1}{4}$ W
  resistor

R16 150 ohm $\frac{1}{4}$ W
  resistor

R17 130 V rms MOV

C1 10 µF 35 V
  electrolytic capacitor

C2, C7, C8, C9 .1 µF
  35 V capacitor

C3, C4, C5, C6 1 µF
  35 V electrolytic
  capacitor

C10 .001 µF 35 V
  capacitor

C11 .1 µF 250 V mylar
  capacitor

C12 47 µF 35 V
  electrolytic capacitor

D1 1N914 silicon diode

D2, D3 3.9 V zener
  diode

D4 LED

XTL 20 MHz ceramic
  resonator

T1 600-600 ohm audio
  transformer

J1, J2 10 position male
  header jacks

J3 3 position male
  header jacks

J4, J5 4 position male
  header jack

U1 PIC 16C57
  microprocessor

U2 24LC16B EEPROM memory

U3 MAX232 Serial com
  chip

U4 PVT412L MOS relay

U5 LM7805 5 volt DC
  regulator

Misc PC board, wire,
  IC sockets, standoffs,
  chassis box, etc.

PIR Motion Module

R1 47 K ohms $\frac{1}{4}$ W
  resistor

R2 200 K ohm trim
  potentiometer

R3 3.9 K ohm $\frac{1}{4}$ W
  resistor

R4 56 K ohm $\frac{1}{4}$ W
  resistor

R5 10 K ohm $\frac{1}{4}$ W
  resistor

R6 1 megohm trim
  potentiometer

R7 10 K ohm ¼ W
   resistor

C1, C4, C7 100 nF 35 V
   electrolytic capacitor

C2 220 pF 35 V mylar
   capacitor

C3, C5 10 μF 35 V
   electrolytic capacitor

C6 .1 μF 35 V disc
   capacitor

C8 4.7 nF 35 V disc
   capacitor

Q1 2N2222 transistor

LED red LED

U1 78L05 5 V
   regulator

U2 PIR controller chip
   KC778B

PIR PIR sensor RE200B

F1, F2 PIR Fresnel
   lens-wide/narrow beam

Misc PC board, header,
   IC socket, wire,
   hardware, etc.

kit available:
   www.kitsrus.com

Listen-in Module

R1 2.2 K ohm ¼ W
   resistor

R2, R7 10 K ohm ¼ W
   resistor

R3, R5 100 K ohm ¼ W
   resistor

R4 5.6 K ohm ¼ W
   resistor

R6 5 megohm ¼ W
   resistor

C1 2.2 μF 35 V
   electrolytic capacitor

C2 .1 μF 35 V disc
   capacitor

C3, C4 10 μF 35 V
   electrolytic capacitor

U1 LM358 dual op-amp

MIC electret microphone

Misc PC board, 10 pin
   header, IC sockets,
   wire, hardware, etc.

Tele-Alert

The Tele-Alert is a unique new low cost multi-channel microprocessor controlled remote event/alarm reporting system which will immediately notify you anywhere in the country of any event via your cell phone; see Figure 31-1.

The Tele-Alert can monitor up to four different alarm or event conditions from a host of different types of

Figure 31-1 *Tele-alert main board*

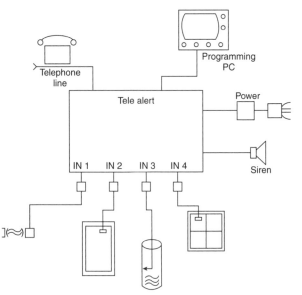

Figure 31-2 *Tele-alert block diagram*

sensors as shown in Figure 31-2, and report the particular channel which was activated.

The Tele-Alert can be configured to monitor voltage levels, temperature changes, movement, contact closures, windows, doors, safes, and perimeters, as well as computer equipment. The Tele-Alert can be utilized to protect your home, office, shop, or vacation home.

The Tele-Alert is used with your existing telephone line, so no additional charges are added to your monthly phone bill. The Tele-Alert can be programmed to call you on your cell phone, or to call a friend, neighbor, co-worker, or a relative if desired. The Tele-Alert project is flexible and the software programs can be expandable for future applications. The Tele-Alert can monitor four different input events or alarm conditions.

The Tele-Alert is compact and easy to use; simply plug in a 9-V power supply, connect up your telephone line and the input connections and you are ready to go. The optional Motion Module allows the Tele-Alert to detect moving persons and will trigger the Tele-Alert; in addition the Motion Module also contains two normally open and two normally closed alarm loop channels, so a number of alarm sensors/switches can be used. The optional Temperature/Voltage Level module allows your Tele-Alert to monitor up to four temperature or voltage level presets, and report the changes via your cell phone or pager. The optional Listen-in Module will allow you to listen in for up to two minutes to the area being protected.

The Tele-Alert multi-channel alarm reporting system begins with the Alternative BASIC STAMP 2 circuit diagram shown in Figure 31-3.

The heart of the Tele-Alert is the PIC 16C57 chip which emulates the BASIC STAMP II computer. The PIC 16C57 is preloaded with a BASIC interpreter much like the Parallax BASIC STAMP 2 (BS2). The PIC 16C57, however, is a much less expensive approach to solving a problem, but requires a few more external parts than the BS2. The PIC 16C57 requires very little in the way of support chips to make it function. The PIC 16C57 basically requires only two support chips, a 24LC16B EEPROM for storage and a MAX232 for communication. Only a few extra components are needed to form a functional microprocessor with serial input/outputs. A 20 MHz ceramic resonator and a few resistors and a diode are all that are needed to use the processor in its most simple form. The 28 pin

PIC 16C57 is a very capable and versatile little microprocessor which can perform many tasks.

In order to turn the BS2 "look alike" into a Tele-Alert project, a few more parts must be added to the basic configuration. First, you need to configure pins 10 through 17 as inputs for a total of 8 inputs which can be used to sense events or alarm conditions. The first four input pins are brought out to the terminal block for easy access. These first four inputs IN1 through IN4 are used as inputs to the microprocessor. Pins 18 through 25 are configured as output pins. Pin 18 is used to activate or enable the microphone in the Listen-in Module, if utilized. Pin 19 is used to drive the solid state relay at U4. The microprocessor activates the LED in U4, which in turn closes the relay contacts at pins 4 and 6 of the relay, which essentially shorts out, blocking capacitor C11 and allows the phone to go "on hook" and begins dialing your cell phone or pager. Pins 20 and 21 of U1 are audio output pins. Pin 20 of U1 outputs the touch-tone signals needed to dial a phone when an alarm condition is sensed, while Pin 21 is used to output the tone sequences which are used to indicate via your cell phone which event/alarm channel has been activated. The capacitors at C9 and C10 are used to couple the touch tones and the alarm sequence tones to the coupling transformer network formed by T1 and the associated components. Capacitor C10 is placed across T1 while Zener diodes D2 and D3 are used as voltage clamps diodes at the input of the transformer.

Transformer T1 is used to couple the audio from the Tele-Alert to the phone line, once the unit has been triggered. Resistor R9 is used to couple one side of the transformer to the phone line at J6, and also acts to hold the phone line. The other secondary transformer lead is coupled to the phone line via C11. C11 is a blocking capacitor which keeps the phone off hook until an alarm condition occurs. The solid state relay at U4 acts to short out C11 when an event or alarm condition occurs, thus coupling the T1 to the phone line. Resistor R16 is a metal oxide varistor (MOV) and is used to protect the circuit from high voltage spike. Pins 22 and 23 may be utilized as auxiliary outputs to drive local alarm sirens or outdoor lamps if desired, while pin 24 is left for further expansion. Pin 25 of U1 is used as a status indicator, when the microprocessor is configured as a Page-Alert. The 20 MHz ceramic resonator is connected between pins 26 and 27 to establish the clock reference

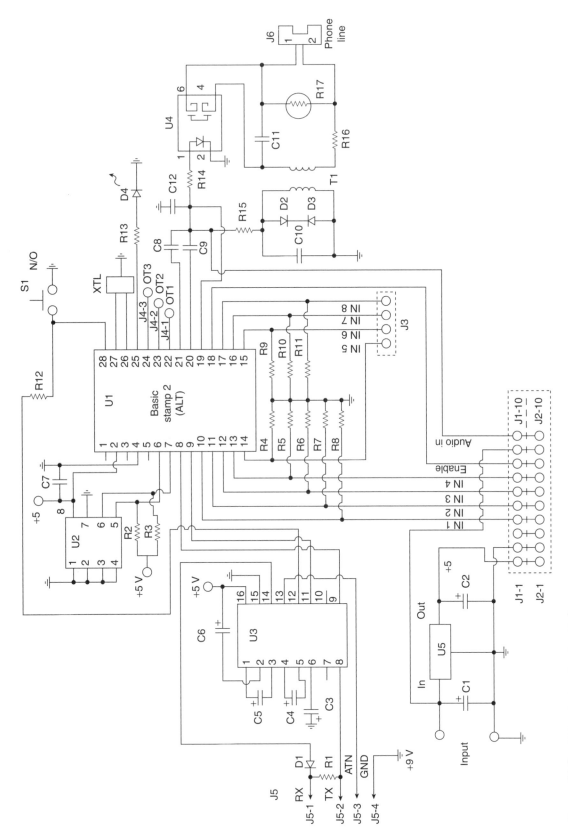

Figure 31-3 *Tele-alert main circuit diagram*

for the BASIC interpreter chip at U1. Pin 28 is utilized to reset the micro-processor, via S1 if the system "locks up". The serial 16 K EEPROM memory is coupled to the microprocessor via pins 6 and 7.

The microprocessor communicates via U3, a MAX232 serial communication chip. The MAX232 is coupled to the microprocessor through pins 8 and 9, very simply an input and output pin respectively. Four capacitors are all that are required to animate the MAX232, serial communication chip. These capacitors are required as a charge pump to create a minus voltage for the serial chip. The MAX232 is coupled to a 9-pin serial connector for serial communication with a laptop or personal computer for programming purposes.

The Tele-Alert circuit is powered via the regulator at U5 which provides 5 V to U1, U2, and U3. A nine volt DC "wall wart" or wall cube power supply is used to provide power to the Tele-Alert circuit. The 9 volt source is also used to provide power to optional enhancement modules.

Circuit assembly

Before we begin constructing the Tele-Alert, you will need to locate a clean, well-lit worktable or workbench. Next we will gather a small 25–30-W pencil tipped

soldering iron, a length of 60/40 tin–lead solder, and a small canister of Tip Tinner, available at your local RadioShack store; Tip Tinner is used to condition the soldering tip between solder joints. You drive the soldering tip into the compound and it cleans and prepares the soldering tip. Next grab a few small hand tools; try to locate small flat blade and Phillips screwdrivers. A pair of needle-nose pliers, a pair of tweezers along with a magnifying glass, and a pair of end-cutters and we will begin constructing the project. Now locate your schematic diagram and parts layout diagram along with all of the components needed to build the project.

The Home Phone Intercom controller is built on double-sided PC board. The main Tele-Alert circuit board measures $4 \times 2\,\frac{1}{2}''$ and can be housed in a suitable plastic enclosure. Note, the basic Tele-Alert is only about $\frac{1}{2}''$ high, but if you intend to add optional modules at a later time you should consider an enclosure which has more height to accommodate the optional circuit boards. Once all the components are in front of you, you can check them off against the project parts list to make sure you are ready to start building the project. Now, locate Tables 31-1 and 31-2, and place them in front of you. Table 31-1 illustrates the resistor code chart and how to read resistors, while Table 31-2 illustrates the capacitor code chart, which will aid you in constructing the project.

Table 31-1

Resistor color code chart

Color Band	1st Digit	2nd Digit	Multiplier	Tolerance
Black	0	0	1	
Brown	1	1	10	1%
Red	2	2	100	2%
Orange	3	3	1,000 (K)	3%
Yellow	4	4	10,000	4%
Green	5	5	100,000	
Blue	6	6	1,000,000 (M)	
Violet	7	7	10,000,000	
Gray	8	8	100,000,000	
White	9	9	1,000,000,000	
Gold			0.1	5%
Silver			0.01	10%
No color				20%

Table 31-2

Three-digit capacitor codes

pF	nF	µF	Code	pF	nF	µF	Code
1.0			**1R0**	3,900	3.9	.0039	**392**
1.2			**1R2**	4,700	4.7	.0047	**472**
1.5			**1R5**	5,600	5.6	.0056	**562**
1.8			**1R8**	6,800	6.8	.0068	**682**
2.2			**2R2**	8,200	8.2	.0082	**822**
2.7			**2R7**	10,000	10	.01	**103**
3.3			**3R3**	12,000	12	.012	**123**
3.9			**3R9**	15,000	15	.015	**153**
4.7			**4R7**	18,000	18	.018	**183**
5.6			**5R6**	22,000	22	.022	**223**
6.8			**6R8**	27,000	27	.027	**273**
8.2			**8R2**	33,000	33	.033	**333**
10			**100**	39,000	39	.039	**393**
12			**120**	47,000	47	.047	**473**
15			**150**	56,000	56	.056	**563**
18			**180**	68,000	68	.068	**683**
22			**220**	82,000	82	.082	**823**
27			**270**	100,000	100	.1	**104**
33			**330**	120,000	120	.12	**124**
39			**390**	150,000	150	.15	**154**
47			**470**	180,000	180	.18	**184**
56			**560**	220,000	220	.22	**224**
68			**680**	270,000	270	.27	**274**
82			**820**	330,000	330	.33	**334**
100		.0001	**101**	390,000	390	.39	**394**
120		.00012	**121**	470,000	470	.47	**474**
150		.00015	**151**	560,000	560	.56	**564**
180		.00018	**181**	680,000	680	.68	**684**
220		.00022	**221**	820,000	820	.82	**824**
270		.00027	**271**		1,000	1	**105**
330		.00033	**331**		1,500	1.5	**155**
390		.00039	**391**		2,200	2.2	**225**
470		.00047	**471**		2,700	2.7	**275**
560		.00056	**561**		3,300	3.3	**335**
680		.00068	**681**		4,700	4.7	**475**
820		.00082	**821**			6.8	**685**
1,000	1.0	.001	**102**			10	**106**
1,200	1.2	.0012	**122**			22	**226**
1,500	1.5	.0015	**152**			33	**336**
1,800	1.8	.0018	**182**			47	**476**

Table 31-2—cont'd
Three-digit capacitor codes

pF	nF	μF	Code	pF	nF	μF	Code
2,200	2.2	.0022	**222**			68	**686**
2,700	2.7	.0027	**272**			100	**107**
3,300	3.3	.0033	**332**			157	**157**

Code = 2 significant digits of the capacitor value + number of zeros to follow

For values below 10 pF, use "R" in place of decimal e.g. 8.2 pF = 8R2

10 pF = 100

100 pF = 101

1,000 pF = 102

22,000 pF = 223

330,000 pF = 334

1 μF = 105

Refer now to the resistor color code chart in Table 31-1, which will help you identify the resistor values before installing them on the PC board. Note that each resistor will have either three or four color bands, which begin at one end of the resistor's body. The first colored band represents the resistor's first digit, while the second color represents the second digit of the resistor's value. The third colored band represents the resistor's multiplier value, and the fourth color band represents the resistor's tolerance value. If there is no fourth colored band then the resistor has a 20% tolerance value. If the resistor's fourth band is silver then the resistor has a 10% tolerance value and if the fourth band is gold then the resistor's tolerance value will be 5%.

We are ready to begin assembling the project, so let's get started. The prototype project was constructed on a single-sided printed circuit board or PCB. With your PC board in front of you, we can now begin populating the circuit board. We are going to place and solder the lowest height components first, the resistors and diodes. First, locate our first resistor R1. Resistor R1 has a value of 4,700 ohms and is a 5% tolerance type. Now take a look for a resistor whose first color band is yellow, violet, red, and gold. From the chart you will notice that yellow is represented by the digit (4), the second color band is violet, which is represented by (7). Notice that the third color is red and the multiplier is (100), so (4)(7) × 100 = 4,700 or 4.7 K ohms.

Go ahead and locate where resistor R1 goes on the PC and install it, next solder it in place on the PC board,

and then with a pair of end-cutters trim the excess component leads flush to the edge of the circuit board. Next locate the remaining resistors and install them in their respective locations on the main controller PC board. Solder the resistors to the PC board, and remember to cut the extra component leads with your end-cutters.

Note that capacitors generally come from two major groups or types. Non-polar types, such as the ones in this project, have no polarity and can be placed in either direction on the circuit board. The other type of capacitors are called polarized types, these capacitors have polarity and have to be inserted on the board with respect to their polarity marking in order for the circuit to operate correctly. On these types of capacitors, you will observe that the components actually have a plus or minus marking or a black or white band with a plus or minus marking on the body of the capacitors. Capacitors must be installed properly with respect to polarity in order for the circuit to work properly. Failure to install electrolytic capacitors correctly could cause damage to the capacitors as well as to other components in the circuit when power is first applied.

Many capacitors are quite small in physical size, so manufacturers have devised a three-digit code which can be placed on the capacitor instead of the actual value. The code occupies a much smaller space on the capacitor and can usually fit nicely on the capacitor body, but without the chart the builder would have a difficult time determining the capacitor's value. Now look for the capacitors from the component stack and locate capacitors C2, C7, C8, and C9; they are small capacitors labeled (104). Refer to Table 31-2 to find code (104) and you will see that it represents a capacitor value of .1 μF. Go ahead and place these capacitors on the circuit board and solder them in place. Now take your end-cutters and cut the excess component leads flush to the edge of the circuit board. Once you have installed these components, you can move on to installing the remaining capacitors. Save the larger electrolytic capacitors for mounting last. The electrolytic capacitors will have a plus or minus sign or a black or white color band. Be sure to orient the electrolytic capacitors correctly when placing them on the PC board.

The Tele-Alert project contains three silicon diodes. Diodes, as you will remember, will always have some type of polarity markings, which must be observed if the circuit is going to work correctly. You will notice that each diode will have a black or white colored band at one edge of the diode's body. The colored band represents the diode's cathode lead. Check your schematic and parts layout diagrams when installing these diodes on the circuit board. Place all the diodes on the circuit board in their respective locations and solder them in place on the PC board. Diode bridge BR1 on the right side of the schematic can be installed now. After soldering in the diodes and the diode bridge, remember to cut the excess component leads with your end-cutters, flush to the edge of the circuit board.

The main Tele-Alert controller circuit also contains five ICs. ICs are often static sensitive, so they must be handled with care. Use a grounded anti-static wrist strap and stay seated in one location when handling the ICs. Take out a cheap insurance policy by installing IC sockets for each of the ICs. In the event of a possible circuit failure, it is much easier to simply unplug a defective IC than trying to unsolder 14 or 16 pins from a PC board without damaging the board. ICs have to be installed correctly if the circuit is going to work properly. IC packages will have some sort of markings which will help you orient them on the PC board. An IC will have either a small indented circle, a cutout, or notch at the top end of the IC package. Pin 1 of the IC will be just to the left of the notch, or cutout. Refer to the manufacturer's pin-out diagram, as well as the schematic, when installing these parts. If you

doubt you ability to correctly orient the ICs, then seek the help of a knowledgeable electronics enthusiast. IC U4 is an opto-isolator and has six pins but the procedure for installation is the same as for the other ICs.

The Tele-Alert controller circuit utilizes a single LED at D4, which also must be oriented correctly if the circuit is to function properly. An LED has two leads, a cathode and an anode lead. The anode lead will usually be the longer of the two leads and the cathode lead will usually be the shorter lead, just under the flat side edge of the LED; this should help orient the LED on the PC board. You may want to leave about ½ lead length on the LED, in order to have the LED protrude through the enclosure box. Solder the LED to the PC board and then trim the excess leads.

Now, locate the crystal (XTL) and install it in its proper location between pins 26 and 27 of U1 and ground as shown on the schematic. Next locate the MOV protection device which is placed across the telephone line at R17. This device has no polarity so it can be installed in either direction. Solder it in place and then trim the excess leads.

Next, install the coupling transformer at T1. The transformer is used to isolate the telephone line from the BS2. The transformer is a 600 to 600 ohm matching transformer so it can be mounted in either direction, since both windings are the same. Use the two transformer tabs to secure the transformer to the PC board. Solder the transformer in place now.

Go ahead and install the normally open RESET switch at S1. This PC board push-button switch mounts directly to the circuit board. Locate and install the 10-position header jack J1–J10 and install it at the bottom edge of the circuit board. Finally, mount the three position header jack OT1–OT3 on pins 22, 23, and 24 of U1, these are the relay output jacks.

Once the main Tele-Alert circuit board has been completed, take a short break and then we will inspect the PC board for possible "cold" solder joints and "short" circuits. Pick up the PC board with the foil side facing upwards toward you. Carefully inspect the solder joints, these should look smooth, clean, and bright. If any of the solder joints look dull, dark, or blobby, then you should remove the solder with a solder sucker or a solder wick and then resolder the joint all over again. Now we will inspect the PC board for possible short circuits. Short circuits can be caused from small solder blobs bridging the circuit traces or from cut component leads which can often stick to the circuit board once they have been cut. A sticky residue from solder will often trap component leads across the circuit traces. Look carefully for any bridging wires across circuit traces.

The main Tele-Alert circuit is powered via the regulator at U5. Power to the regulator is from a 9 V DC "wall wart" power supply. Notice that the 9 V are brought out to pin 9 on the header to power the add-on modules, if used. Five volts from the regulator output are brought out on pin1 of the header, while the ground is on pin 2 of the header.

Operation of the Tele-Alert is simple. First, connect up a 9 V wall wart power supply to the circuit via the two power input pins at the top of the board. Next you will need to make up a serial communication or programming cable which connects the Tele-Alert header J5 to the serial port of your programming computer; see Figure 31-4. Now you will need to "fire-up" your PC and load the supplied disk into your computer.

Make a directory called **Tele-Alert** and dump the contents of the disk into that directory. Download a copy of the BASIC STAMP Windows Editor, PBASIC 2.5 or above from www.parallax.com and place it on your PC. Bring up the Windows STAMP 2 editor program and type in or load the **talert.bs2** program and save it. Attach the programming cable from your PC or laptop to the BS2 microcontroller and then connect up the power source to the Tele-Alert board. Scroll down the displayed program, and look for the simulated phone number and replace it with your own phone number.

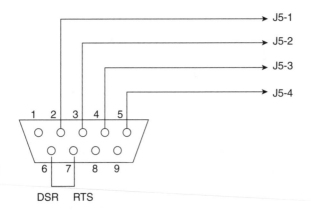

Figure 31-4 *STAMP 2 Programming cable*

Remember a digit 10 is programmed by a zero. If you want to dial a long distance number just add the full sequence of numbers. If you wish to dial out from a PBX telephone system, you will need to first put in the access number such as an 8 or 9 followed by a comma then the actual phone number.

Once the Tele-alert board is powered-up and the programming cable attached, then you can go to the tool bar in the STAMP 2 editor program and run the **talert.bs2** program.

Next, press ALT-R to load the program into the Tele-Alert circuit. Your Tele-Alert is now programmed and ready to operate. At last, you can connect your event or alarm inputs to the set of solder pads at the far edge of the circuit board, just below the phone line connection. Remember, you must use normally open circuit switches for the alarm inputs unless you are using the optional Switch/Sensor-module. The Switch/Sensor provides both normally open and normally closed inputs. Connect your phone line via the phone jack and your Tele-Alert is now ready to serve you.

Now you are ready to simulate an alarm condition to see if your Tele-Alert functions. You can test the circuit in one of two ways. A simulated approach is to connect a crystal headphone across the secondary of T1 at J6, with the Tele-Alert disconnected from the phone line. Apply power to the circuit, and connect a normally open switch across input IN1. Activate the switch at IN1 and you should begin hearing activity at the crystal headphone. First you should hear the touch-tone sequences followed by the alarm tone sequence. If this checks-out then you can move on to a "real" test by connecting the Tele-Alert to the phone line, and repeat the same test over again.

The main Tele-Alert circuit board has two ten position female header jacks, i.e. J1 and J2, near the input terminal solder pads. These two 10 pin headers allow you to add the optional enhancement modules, such as the Switch/Sensor Module, and the TVM Modules which are described in chapter 30. The Motion Module and the Listen-in Module are described below.

The Tele-Alert board provides two headers, the J1 and J2 headers. These headers are provided to plug in the various modules. Table 31-3 illustrates the header terminal pin-outs for the Tele-main Tele-Alert board.

Table 31-3

Input/output terminal connections

Alarm Input Terminals and Header Block – J1

J1-1	J2-1	+5 volt DC input	
J1-2	J2-2	Ground	
J1-3	J2-3	N/C	
J1-4	J2-4	IN1 input channel #1	
J1-5	J2-5	IN2 input channel #2	Tele-alert input pins
J1-6	J2-6	IN3 input channel #3	
J1-7	J2-7	IN4 input channel #4	
J1-8	J2-8	Microphone enable	
J1-9	J2-9	+9 volts DC output	
J1-10	J2-10	Audio input mic	

Auxiliary Inputs – Not implemented

J3-1	IN5	Future Expansion
J3-2	IN6	
J3-3	IN7	
J3-4	IN8	

Auxiliary Outputs

J4-1	OT1	Aux output channel #1	For external siren
J4-2	OT2	Aux output channel #2	For external lamp
J4-3	OT3	Aux output channel #3	Future use

Serial Communication

J5-1	RX	Pin 2	RS232	DB-9S
J5-2	TX	Pin 3	RS232	DB-9S
J5-3	ATN	Pin 4	RS232	DB-9S
J5-4	GND	Pin 5	RS232	DB-9S
J6-1	L1	Phone line connection		
J6-2	L2	Phone line connection		

The Tele-Alert system is very flexible and can be used in a number of different alarm configurations. The Tele-Alert could be used to monitor safes, doors, windows, floor mats, computers, movement, temperature and voltage changes, as well as smoke and fire sensors. The Tele-Alert can even monitor existing local alarms by using the Tele-Alert as a multi-channel dialer. With the Tele-Alert and the Motion Module you can easily protect an entire vacation house or cabin, using the existing phone line. You can even monitor an overheating computer or UPS failure. The Tele-Alert can be utilized in almost any alarm configuration from multi-zoned alarm systems to simple multi-event annunciators with a little imagination. The Tele-Alert can also drive a siren or flashing outdoor lights to create a local "noisy" alarm if desired. The diagram in Figure 31-5 illustrates optional local alarm connections.

Relay 1 can be used to drive small loads such as an electronic Sonalert, or with the addition of Relay 2 you can drive larger loads, such as motor sirens or flashing outdoor lamps. You can drive two external loads from OT-1 and OT-2. Output OT-3 is left for future expansion. The Tele-Alert can be used day or night with your existing phone line; no additional phone lines

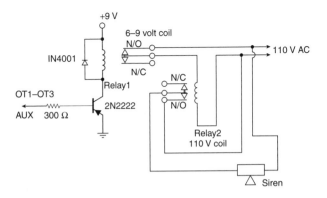

Figure 31-5 *AUX-output relay driver*

or phone bills are incurred. Now with the Tele-Alert you can be notified of an alarm condition when you are shopping, boating, driving, golfing, traveling, or working outdoors. The Tele-Alert can dial multiple phone numbers or repeatedly dial the same number, as desired. The Tele-Alert can be used to call a friend, a neighbor, a relative, or a co-worker in the event of an alarm condition. If a second party is notified, the condition can be verified and then the police could be called, depending upon the severity of the alarm condition.

The Tele-Alert is very flexible and can be tailored to many applications with a little imagination. Once you build the Tele-Alert you can experiment with different sounds and sound patterns which are sent to the remote phone. The Tele-Alert is always there to serve you, to free you, and to give you piece of mind! Why not build one for yourself!

Tele-Alert program

```
'Tele-Alert Program
'identify variables
flash          Var byte
new_io_state   Var byte
old_io_state   Var byte
call_state     Var byte
io             Var byte
timer          Var byte
i              Var byte
f              Var word

TXpin          Con 10
LedPin         Con 15
Lamp           Con 13
```

```
Siren          Con 12
Tele           Con 9
MIC            Con 8
recall_delay   Con 120
C              Con 2000
D              Con 1000
E              Con 500
G              Con 350
R              Con 0
```

```
init:
DIRL = %00000000
DIRH = %11111111
new_io_state = %00000000
old_io_state = %00000000

for flash = 1 to 3
  high LedPin
  pause 1500
  low LedPin
  pause 2500
next

main:
high LedPin
new_io_state = INL
pause 100
low LedPin
pause 100
if new_io_state <> %00000000 then
    dial_cell
after_io1:
old_io_state = new_io_state
goto main

delay_and_scan
old_io_state = new_io_state
  for timer = 1 to recall_delay
  high LEDPin
  new_io_state = INL
  pause 500
  low LedPin
  pause 500
  if new_io_state <> old_io_state
    then dial_cell
  old_io_state = new_io_state
  next
goto after_io1
```

```
dial_cell:
if new_io_state = %00000000 then main
high LedPin
high Tele
high Lamp
high Siren
pause 500
dtmfout TXPin, 600, 600, [8]
dtmfout TXPin, 500, 100, [7,9,9,4,
     9,9,0]
sleep 10
gosub send_msg

goto delay_and_scan

send_msg:
io = new_io_state

if io.bit0 = 1 then gosub_chan_1
if io.bit1 = 1 then gosub_chan_2
if io.bit2 = 1 then gosub_chan_3
if io.bit3 = 1 then gosub_chan_4
pause 100
gosub_chan_1:
for i = 1 to 15
lookup i,[E,D,E,D,E,D,E,D,E,D,E,D,
     E,D,E] ,f
freqout 11,750,f,(f-1) max 32768
next
high MIC
pause 10000
low Tele
low LEdPin
low Siren
low LAMP
low MIC
pause 100
return
end


gosub_chan_2:
for i = 1 to 20
lookup i,[E,D,C,E,D,C,E,D,C,E,D,C,E,
     D,C,E,D,C,E,D] ,f
freqout 11,500,f,(f-1) max 32768
next
'high MIC
'pause 10000
```

```
low Tele
low LedPin
low Siren
low Lamp
low MIC
pause 100
return

end


gosub_chan_3:
for i= 0 to 25
lookup i,[ G,E,D,C,G,E,D,C,G,E,D,
     C,G,E,D,C,G,E,D,C,G,E,D,C,G] ,f

freqout 11,350,f,(f-1) max 32768
next
'high MIC
'pause 10000
low Tele
low LedPin
low Siren
low Lamp
low MIC
pause 100
return
end
gosub_chan_4:

for i = 0 to 30
lookup i,[C,D,E,G,C,D,E,G,C,D,E,G,C,D,E,
     G,C,D,E,G,C,D,E,C,D,E,G,C,D,E] ,f
freqout 11,200,f,(f-1) max 32768
next
'high MIC
'pause 10000
low Tele
low LEDPin
low Siren
low Lamp
low MIC
pause 100
return
end
```

Motion module

The Pyro-electric Motion Module is a great addition to the Tele-Alert or Page-Alert microprocessor

Figure 31-6 *Tele-Alert infra-red sensor board*

alarm controller. The Pyro-electric Motion Module centers around a pyro-electric or "body heat" infra-red sensor and the controller chip, and is shown in Figure 31-6.

The PIR controller chip block diagram is depicted in Figure 31-7. The PIR controller is a complex chip, which provides amplification, filtering, clock, comparators, a daylight detector, and a

voltage regulator. Table 31-4 lists the pinouts of the 20 pin PIR controller chip.

The Motion Module circuit is illustrated in the system diagram in Figure 31-8. The circuit begins with the sensitive PIR detector. The pyro-electric infra-red sensor or PIR is a sensitive three-lead high impedance sensor, which is shown with a 47 K ohm output resistor at R1. A "sensitivity" or "range" control for the PIR sensors is shown at R2. The PIR sensor is coupled to a specially designed PIR controller chip which is optimized for PIR alarm sensors.

The PIR sensor requires a Fresnel beam focusing lens. The Fresnel lens is placed ahead of the PIR sensor to give the sensor a specific pattern of coverage. PIR Fresnel lenses come in two basic types. The most common beam pattern is the wide angle lens which looks out to about twelve feet with a 30–50 degree pattern; the second Fresnel lens is a narrow angle type which looks out to about fifty feet with

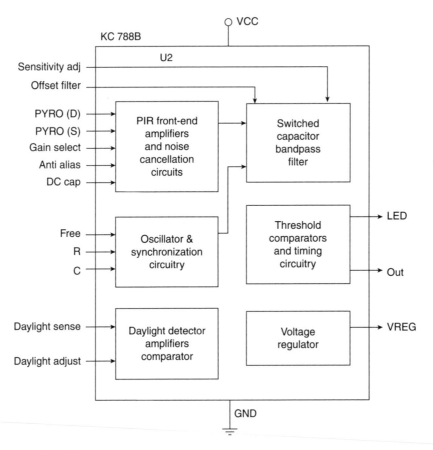

Figure 31-7 *PIR controller chip*

Table 31-4

KC778B PIR controller pinouts

Pin	Name	Description
1	Vcc	+5 volt supply
2	sensitivity adj	PIR sensitivity input
3	offset filter	PIR offset filter
4	anti alias	PIR anti alias filter
5	DC capacitor	PIR gain stabilization filter
6	V reg	voltage regulator output
7	pyro (D)	pyro drain reference
8	pyro (S)	pyro source input
9	ground (A)	analog ground
10	ground (D)	digital ground
11	daylight adj	daylight adj and and CDS
12	daylight sense	silicon photo dioed input
13	gain select	PIR gain select
14	on/auto/off	mode selecttri-state input
15	toggle	mode select toggle input
16	out	output
17	LED	indicator LED
18	C	off timer OSC
19	R	off timer OSC
20	Fref	frequency reference OSC

a narrow beam pattern. The PIR Motion Module kit comes with both types of Fresnel lens.

Capacitors C5, C6, and C7 are offset filters, anti-aliasing filters and a DC capacitor respectively.

Resistors R3 and R4 are utilized as a frequency reference between pin #20 and Vcc. The PIR sensor is placed across pins #2, #7, and #8 of the controller chip. The actual alarm output pin of the controller is located at pin #16, which is coupled to the output driver transistor at Q1. The R/C time constant for the output "on time" of the controller is between pins #18 and #19, and is controlled by R5, R6, and C8. A "movement" indicator LED is located between pin #17 and pins #9 and #10, the analog and digital ground pins. The chip employs a daylight detector which is not implemented in our application. Power to the controller chip is brought to pin 1 of U1 via a 78l05 regulator. The input to the regulator is a 9 V DC power supply which is used to power the Tele-Alert controller.

Circuit assembly

Before we begin constructing the PIR Motion Module board, you will need to locate a clean, well-lit worktable or workbench. Next we will gather a small 25–30-W pencil tipped soldering iron, a length of 60/40 tin–lead solder, and a small canister of Tip Tinner, available at your local RadioShack store; Tip Tinner is used to condition the soldering tip between solder joints. You drive the soldering tip into the compound and it cleans and prepares the soldering tip. Next grab a few small hand tools; try to locate small flat blade and Phillips screwdrivers. A pair of needle-nose pliers, a pair of tweezers along with a magnifying glass, and a pair of

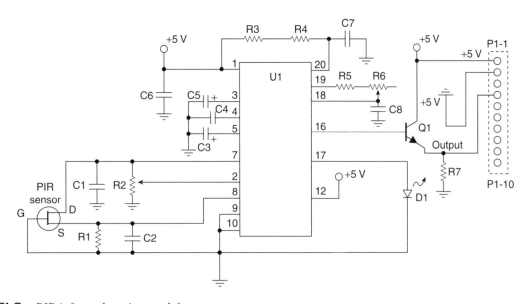

Figure 31-8 *PIR infra-red motion module*

end-cutters and we will begin constructing the project. Now locate your schematic diagram and parts layout diagram along with all of the components needed to build the project.

Once all the components are in front of you, you can check them off against the project parts list to make sure you are ready to start building the project. Now, locate Tables 31-1 and 31-2, and place them in front of you.

Table 31-1 illustrates the resistor code chart and how to read resistors, while Table 31-2 illustrates the capacitor code chart which will aid you in constructing the project. Refer now to the resistor color code chart shown in Table 31-1, which will assist you in identifying the resistor values used in the project.

Note that each resistor will have either three or four color bands, which begin at one end of the resistor's body. The first colored band represents the resistor's first digit, while the second color represents the second digit of the resistor's value. The third colored band represents the resistor's multiplier value, and the fourth color band represents the resistor's tolerance value. If there is no fourth colored band then the resistor has a 20% tolerance value. If the resistor's fourth band is silver then the resistor has a 10% tolerance value and if the fourth band is gold then the resistor's tolerance value will be 5%.

Finally, we are ready to begin assembling the project, so let's get started. The Motion Module is fabricated on a single sided 1¼ by 2½ inches circuit board. With your PC board in front of you, we can now begin populating the circuit board. We are going to place and solder the lowest height components first, the resistors and diodes. First, locate our first resistor R1. Resistor R1 has a value of 47,000 ohms and is a 5% tolerance type. Now take a look for a resistor whose first color band is yellow, violet, orange, and gold. From the chart you will notice that yellow is represented by the digit (4), the second color band is violet, which is represented by (7). Notice that the third color is orange and the multiplier is (1,000), so $(4)(7) \times 1,000 = 47,000$ or 47 K ohms.

Go ahead and locate where resistor R1 goes on the PC and install it, next solder it in place on the PC board, and then with a pair of end-cutters trim the excess component leads flush to the edge of the circuit board. Next locate the remaining resistors and install them

in their respective locations on the main controller PC board. Solder the resistors to the PC board, and remember to cut the extra component leads with your end-cutters.

Note that capacitors generally come from two major groups or types. Non-polar types, such as the ones in this project, have no polarity and can be placed in either direction on the circuit board. The other type of capacitors are called polarized types, these capacitors have polarity and have to be inserted on the board with respect to their polarity marking in order for the circuit to operate correctly. On these types of capacitors, you will observe that the components actually have a plus or minus marking or a black or white band with a plus or minus marking on the body of the capacitors. Capacitors must be installed properly with respect to polarity in order for the circuit to work properly. Failure to install electrolytic capacitors correctly could cause damage to the capacitors as well as to other components in the circuit when power is first applied.

Many capacitors are quite small in physical size, therefore manufacturers have devised a three-digit code which can be placed on the capacitor instead of the actual value. The code occupies a much smaller space on the capacitor and can usually fit nicely on the capacitor body, but without the chart the builder would have a difficult time determining the capacitor's value.

Now look for the capacitors from the component stack and locate capacitors C1, C4, and C7. These are all small capacitors labeled (104). Refer to the Table 31-2 to find code (104) and you will see that it represents a capacitor value of 100 nF. Go ahead and place these capacitors on the circuit board and solder them in place. Now take your end-cutters and cut the excess component leads flush to the edge of the circuit board. Once you have installed these components, you can move on to installing the remaining low profile capacitors. Save the larger electrolytic capacitors for mounting last. The electrolytic capacitors will have a plus or minus sign or a black or white color band. Be sure to orient the electrolytic capacitors correctly when placing them on the PC board. Locate capacitors C3 and C5, which are low leakage 10 µF electrolytic capacitors; go ahead and install on the PC board and solder them in place. Do not forget to trim the excess leads with your end-cutters.

The PIR Motion Module utilizes a single conventional transistor. Transistors are generally three lead devices, which must be installed correctly if the circuit is to work properly. Careful handling and proper orientation in the circuit is critical. A transistor will usually have a Base lead, a Collector lead, and an Emitter lead. The base lead is generally a vertical line inside a transistor circle symbol. The collector lead is usually a slanted line facing towards the base lead and the emitter lead is also a slanted line facing the base lead but it will have an arrow pointing towards or away from the base lead. Now, identify the two transistors, referring to the manufacturer's specification sheets as well as the schematic and parts layout diagrams when installing the two transistors. If you have trouble installing the transistors, get the help of an electronics enthusiast with some experience. Install the transistors in their respective locations and then solder them in place. Remember to trim the excess component leads with your end-cutters.

The Motion Module circuit contains a special purpose IC. ICs are often static sensitive, so they must be handled with care. Use a grounded anti-static wrist strap and stay seated in one location when handling the ICs. Take out a cheap insurance policy by installing IC sockets for each of the ICs. In the event of a possible circuit failure, it is much easier to simply unplug a defective IC, than trying to unsolder 14 or 16 pins from a PC board without damaging the board. ICs have to be installed correctly if the circuit is going to work properly. IC packages will have some sort of markings which will help you orient them on the PC board. An IC will have either a small indented circle, a cutout, or notch at the top end of the IC package. Pin 1 of the IC will be just to the left of the notch, or cutout. Refer to the manufacturer's pin-out diagram, as well as the schematic, when installing these parts. If you doubt your ability to correctly orient the ICs, then seek the help of a knowledgeable electronics enthusiast to help you.

The Motion Module circuit utilizes a single LED at D1, which also must be oriented correctly to function properly. An LED has two leads, a cathode and an anode lead. The anode lead will usually be the longer of the two leads and the cathode lead will usually be the shorter lead just under the flat side edge of the LED, this should help orient the LED on the PC board. You may want to leave about ½ lead length on the LED, in order to have

the LED protrude through the enclosure box. Solder the LED to the PC board and then trim the excess leads.

The Motion Module uses two printed circuit potentiometers, a 200 K ohm at R2 and a 1 megohm potentiometer at R6. Mount the two potentiometers on the PC board and solder them in place, trimming leads as necessary.

Finally, locate and carefully pickup the PIR motion sensor using your grounded anti-static wrist band and install it on the PC board. Note that the Gate lead connects to ground, while the Source lead connects to the junction of R1/C2 and the Drain lead is connected to capacitor C1. The components, especially the controller chip and PIR sensor, can be damaged if they are placed incorrectly. Note that the PIR sensor is placed on the foil side of the circuit board.

At one edge of the PC board, you will find the 10 pin solder tabs which are used to secure the 10 pin male header pins that couple the Motion Module to the Tele-Alert. The top pin #1 is reserved for the 9 volt power input power pin. Pin #2 is used as the system ground, while pin #3 is the output pin from the Motion Module to the Tele-Alert main board. The rest of the header pins are not used. The 10 pin header is placed on the component side of the PC board and soldered on the foil side of the Motion Module PC board.

Once the Motion Module circuit board has been completed, take a short break and then we will inspect the PC board for possible "cold" solder joints and "short" circuits. Pick up the PC board with the foil side facing upwards toward you. Carefully inspect the solder joints, these should look smooth, clean, and bright. If any of the solder joints look dull, dark, or blobby, then you should remove the solder with a solder sucker or a solder wick and then resolder the joint all over again. Now we will inspect the PC board for possible short circuits. Short circuits can be caused from small solder blobs bridging the circuit traces or from cut component leads which can often stick to the circuit board once they have been cut. A sticky residue from solder will often trap component leads across the circuit traces. Look carefully for any bridging wires across circuit traces.

Circuit testing

Once the Motion Module has been completed, you can easily test the board, by applying a 9 volt transistor battery to pins #1 and #2 of the header. Connect a voltmeter or oscilloscope to the output pin #4 and the ground pin #2 and you are ready to test the Motion Module board. Once power is applied, wait about five seconds, then wave your hand in front of the PIR sensor. The Fresnel lens does not have to be in front of the PIR for this test. The indicator LED should light once your hand is waved in front of the PIR sensor. At this point in time, you should also see the output pin #3 jump from zero volts to 5 volts on your meter or scope. You may have to adjust the PIR sensitivity control at R2 or the "time on" control at R6 for optimum, once the Fresnel lens is in place and the enclosure is shut and you become familiar with the detector's operation. This completes the testing of your new Motion Module.

Next, you will need to connect the Motion Module to the Tele-Alert controller. The Motion Module should be plugged into the 10 pin female header on the main Tele-Alert board labeled J1. On the Tele-Alert board, the first header next to the screw terminals is the 10 pin header which is used for the Listen-in board, while the second header (J2) is a 10 pin one which is used for the TVM or Motion Module board. Once the two boards are fastened together, you will need to place a hole in the Tele-Alert enclosure to allow the PIR to "see" the room. You will at this point have to place the Fresnel lens ahead of the PIR sensor before securing the top of the Tele-Alert enclosure.

Your Tele-Alert is now a complete motion detector alarm controller/dialer which can be used to protect a house, cottage, or work-shop area. If you are clever in the placement of the PIR sensor, you can easily protect large areas using a single PIR Motion sensor.

Listen-in module

The Tele-Alert Listen-in Module allows you to listen-in to the room or area being monitored by the Tele-Alert alarm controller. With the Listen-in board you can listen-in for up to two minutes after you have been called by the Tele-Alert unit. The Listen-in board is shown in Figure 31-9.

The Listen-in board begins at the sensitive electret microphone at MIC, shown in the schematic diagram at Figure 31-10.

Figure 31-9 *Tele-Alert listen-in board*

The sensitive electret microphone is first biased via R1, which is enabled via the microprocessor. The sound output from the electret microphone is coupled through C1 and R2 to the dual op-amp at U1(A) which acts as an audio pre-amplifier. The LM358 op-amp at U1 is a single supply device eliminating the need for the usual two power supplies used with most op-amps. The gain of U1 is essentially controlled by resistors R3 and R6. The output of the first pre-amp stage is next coupled to the second amplifier stage at U1(B) via C2 and R5. The output of the second audio amplifier stage at U1(B) is coupled directly to the telephone coupling transformer T1 via C3. The listen-in board kit is powered by the 9 volt system power source through the main Tele-Alert board.

Circuit assembly

Before we begin constructing the Listen-in module, you will need to locate a clean, well-lit worktable or workbench. Next we will gather a small 25–30-W pencil tipped soldering iron, a length of 60/40 tin–lead solder, and a small canister of Tip Tinner, available at your local RadioShack store; Tip Tinner is used to condition the soldering tip between solder joints. You drive the soldering tip into the compound and it cleans and prepares the soldering tip. Next grab a few small hand tools; try to locate small flat blade and Phillips screwdrivers. A pair of needle-nose pliers, a pair of tweezers along with a magnifying glass, and a pair of end-cutters and we will begin constructing the project. Now locate your schematic diagram and parts layout diagram along with all of the components needed to build the project.

Once all the components are in front of you, you can check them off against the project parts list to make sure

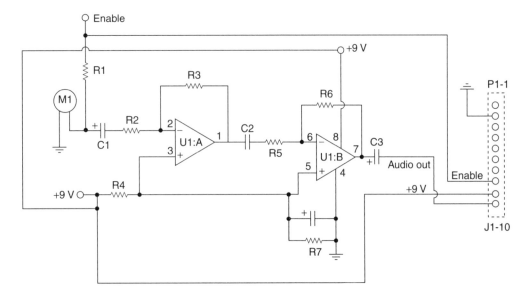

Figure 31-10 *Listen-in module*

you are ready to start building the project. Now, locate Tables 31-1 and 31-2, and place them in front of you.

Table 31-1 illustrates the resistor code chart and how to read resistors, while Table 31-2 illustrates the capacitor code chart which will aid you in constructing the project.

Refer now to the resistor color code chart shown in Table 31-1, which will help you identify the resistor values for the project. Each resistor will have either three or four color bands, which begin at one end of the resistor's body. The first colored band represents the resistor's first digit, while the second color represents the second digit of the resistor's value. The third colored band represents the resistor's multiplier value, and the fourth color band represents the resistor's tolerance value. If there is no fourth colored band then the resistor has a 20% tolerance value. If the resistor's fourth band is silver then the resistor has a 10% tolerance value and if the fourth band is gold then the resistor's tolerance value will be 5%.

So finally we are ready to begin assembling the project. The Listen-in board kit measures 2 by 1¼ inches, and was designed to plug-in easily to the main Tele-Alert board. The Listen-in board is simple to construct, and can be completed in about a half hour or so. So let's get started. With your PC board in front of you, we can now begin populating the circuit board. We are going to place and solder the lowest height

components first, the resistors and diodes. First, locate our first resistor R1. Resistor R1 has a value of 2,200 ohms and is a 5% tolerance type. Now take a look for a resistor whose first color band is red, red, red, and gold. From the chart you will notice that blue is represented by the digit (2), the second color band is black, which is represented by (2). Notice that the third color is red and the multiplier is (100), so $(2)(2) \times 100 = 2,200$ or 2.2 K ohms.

Go ahead and locate where resistor R1 goes on the PC and install it, next solder it in place on the PC board, and then with a pair of end-cutters trim the excess component leads flush to the edge of the circuit board. Next locate the remaining resistors and install them in their respective locations on the main controller PC board. Solder the resistors to the board, and remember to cut the extra lead with your end-cutters.

Note that capacitors generally come from two major groups or types. Non-polar types, such as the ones in this project, have no polarity and can be placed in either direction on the circuit board. The other type of capacitors are called polarized types, these capacitors have polarity and have to be inserted on the board with respect to their polarity marking in order for the circuit to operate correctly. On these types of capacitors, you will observe that the components actually have a plus or minus marking or a black or white band with a plus or minus marking on the body of the capacitors.

Capacitors must be installed properly with respect to polarity in order for the circuit to work properly. Failure to install electrolytic capacitors correctly could cause damage to the capacitors as well as to other components in the circuit when power is first applied.

Many capacitors are quite small in physical size, so manufacturers have devised a three-digit code which can be placed on the capacitor instead of the actual value. The code occupies a much smaller space on the capacitor and can usually fit nicely on the capacitor body, but without the chart the builder would have a difficult time determining the capacitor's value.

Now look for the capacitors from the component stack and locate capacitor C2, which is a small capacitor labeled (104). Refer to Table 31-2 to find code (104) and you will see that it represents a capacitor value of .1 µF. Go ahead and place C2 on the circuit board and solder it in place. Now take your end-cutters and cut the excess component leads flush to the edge of the circuit board. Now you can move on to installing the remaining capacitors. Next, we will install the remaining capacitors, which are all electrolytic types. Remember electrolytic capacitors have polarity and must be installed correctly if the circuit is to function properly. Go ahead and install these capacitors paying close attention to the polarity markings. Remember to trim the excess component leads from the PC with your end-cutters.

The Listen-in module circuit contains a single dual stage IC. ICs are often static sensitive, so they must be handled with care. Use a grounded anti-static wrist strap and stay seated in one location when handling the ICs. Take out a cheap insurance policy by installing an IC socket for the IC. In the event of a possible circuit failure, it is much easier to simply unplug a defective IC than trying to unsolder 14 or 16 pins from a PC board without damaging the board. ICs have to be installed correctly if the circuit is going to work properly. IC packages will have some sort of markings which will help you orient them on the PC board. An IC will have either a small indented circle, a cutout, or notch at the top end of the IC package. Pin 1 of the IC will be just to the left of the notch, or cutout. Refer to the manufacturer's pin-out diagram, as well as the schematic when installing these parts. If you doubt your ability to correctly orient the ICs, then seek the help of a knowledgeable electronics enthusiast to help you.

Next we will install the electret microphone. The electret microphone has two leads which first must be identified before placing it on the circuit board. One lead is fastened to the case or body of the microphone and this is the ground lead. The other microphone lead is the "hot" or active lead that is powered by the enable pin through resistor R1 which is connected to the header pins at the edge of the PC board.

The Listen-in module contains a single in-line row of 10 male header pins at one edge of the circuit board which allows the board to connect to the main Tele-Alert board. The first of the ten header pins at P1-1 is the 5-V supply, followed by the system ground connection at pin P1-2. Many of the solder pads are not used except for the power pins, the audio output pin at P1-10, and audio enable connection at pin P1-8. The ten pin male header is inserted and soldered on the foil side of the circuit board, once the board has been completed. Both circuit boards should have the components facing upwards when connected together and finished.

Once the main Listen-in module board has been completed, take a short break and then we will inspect the PC board for possible "cold" solder joints and "short" circuits. Pick up the PC board with the foil side facing upwards toward you. Carefully inspect the solder joints, these should look smooth, clean, and bright. If any of the solder joints look dull, dark, or blobby, then you should remove the solder with a solder sucker or a solder wick and then resolder the joint all over again. Now we will inspect the PC board for possible short circuits. Short circuits can be caused from small solder blobs bridging the circuit traces or from cut component leads which can often stick to the circuit board once they have been cut. A sticky residue from solder will often trap component leads across the circuit traces. Look carefully for any bridging wires across circuit traces.

Once completed the Listen-in board can be tested by applying a 9-V source to the two power pins, system ground at P1-2, and 9 volts at P1-9 (see pinouts in Table 31-5).

You can then connect a headphone to the output pin P1-10 and ground; you must also apply 5 V to the enable pin at P1-8, in order to activate the electret microphone. You should now be able to hear room sounds in the headphone, and the Listen-in board is complete and ready to use.

Table 31-5

Listen-in module pinouts

Pin	Name	Description
1	N/C	system ground
2	Gnd	microphone enable
3	N/C	audio ouput pin from module
4	N/C	9 volt system power for module
5	N/C	
6	N/C	
7	N/C	
8	enable	
9	Audio	
10	+9 V	

You are now ready to attach the Listen-in board to the main Tele-Alert board. Take a look at the Tele-Alert board; at one edge you will notice the input terminal connections at the outside edge of the board, followed by two rows of header sockets. The first row of female 10 pin header sockets at the outside edge are used to hold the Listen-in board while the second set of inside header pins are used to power the TVM board and the Motion module board. The TVM or Motion module "daughter" boards can be utilized at the same time as the Listen in board if desired. Grasp the Tele-Alert board in one hand and now plug the Listen-in module to the first or outside 10 pin female header.

Now you are now ready to utilize the Listen-in board with your main Tele-Alert system board. Connect a 9 V "wall wart" power supply, phone line, and alarm switch or sensor contacts to the Tele-Alert main board and you are ready to go! The Listen-in board can be used in conjunction with the Motion module to detect movement, then let you listen-in to room sounds once motion is detected, so your Tele-Alert can be used as a complete self contained alarm.

Chapter 32

Remote Temperature Monitor

Parts list

Parts Bin

Remote Temperature Monitor

R1, R6, R8 330 ohm,
$\frac{1}{4}$ W, 5% resistor

R2, R3, R4 10 K ohm,
$\frac{1}{4}$ W, 5% resistor

R5, R7, R9 10 K ohm,
$\frac{1}{4}$ W, 5% resistor

R10, R11 100 ohm, $\frac{1}{4}$ W,
5% resistor

R12 1 K ohm, $\frac{1}{4}$ W,
5% resistor

R14, R15 330 ohm, $\frac{1}{4}$ W,
5% resistor

C1, C2 47 μF, 35 V
electrolytic capacitor

C3, C4 0.1 μF, 35 V
ceramic capacitor

Q1, Q2 2N3906 PNP
transistor

M1 250 volt, metal
oxide varistor (MOV)

D1, D2, D3, D5 Red LEDs

D4 Yellow LED

U1 BASIC STAMP 2 -
microcontroller
(Parallax, Inc)

U2 CH1786 MODEM -
(Parallax, Inc)

U3 DS1620 - Temperature
Sensor (Dallas
Semiconductor)

U4 LM2940 regulator
(National)

S1 momentary pushbutton
(Reset)

J1 RJ-11 telephone jack

J2 DB-9F - RS-232
female - serial port
connector

Misc PC board, wire,
IC sockets, hardware,
enclosure, etc.

The Remote Temperature Monitor is a useful and fun microprocessor controlled remote reading temperature project. You can remotely read the high, low, and current temperature of your home, your office, or your cabin in the woods. You connect the Remote Temperature Monitor circuit to a standard telephone line in you remote location and then simply use your computer to call and remotely read the temperature using the Remote Temperature Monitor's modem (see Figure 32-1).

Figure 32-1 *Remote temperature monitor. Courtesy of Parallax, Inc*

The Remote Temperature Monitor centers around the BASIC STAMP 2 microprocessor which is illustrated in the block diagram shown in Figure 32-2.

The STAMP 2 is a powerful microprocessor with 15 general purpose input/output pins which can be used for sensing and control. The STAMP 2 is easily programmed in a small modified form of BASIC called PBASIC, and was originated by Parallax, Inc.

The Remote Temperature Monitor schematic is shown in Figure 32-3.

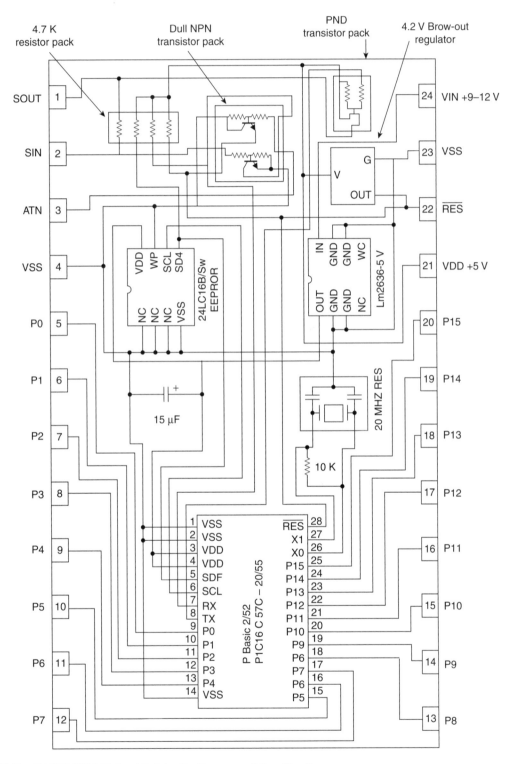

Figure 32-2 *BASIC STAMP 2 – (Original). Courtesy of Parallax Inc*

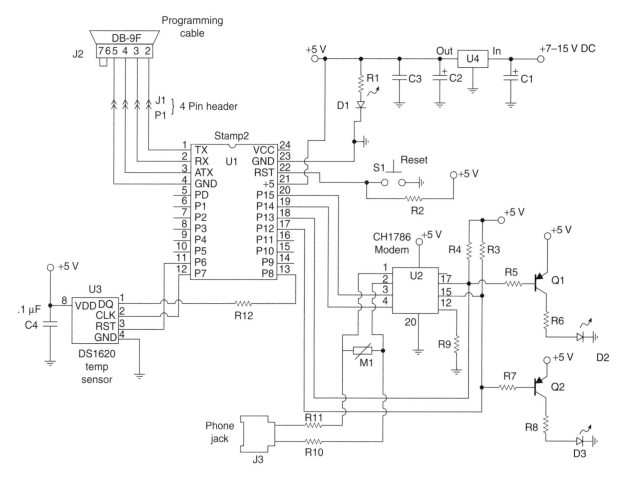

Figure 32-3 *Remote temperature monitor*

The Remote Temperature Monitor consists of four major building blocks which center around the BASIC STAMP 2 microprocessor, a CH1786 modem, a digital DS1620 temperature sensor, a power supply, and a few transistors and status LEDs. The heart of the circuit revolves around the BASIC STAMP 2 (BS2) microprocessor which animates the temperature sensor, the modem, and the status lamps. The incoming phone line is connected via the RJ11 telephone jack. The two telephone line wires are connected to the circuit via two 100-ohm resistors across the phone line. A 250 volt metal oxide varistor (MOV) is used to protect the modem from ringing voltage and voltage spikes. The ring and tip phone wires enter the mini-modem at pins 1 and 2 as shown. The data carrier detect (DCD) output pin on the modem goes low after the modem, goes "off-hook", or answers the call. The output of this pin drives a transistor at Q1 which lights LED D2. The ring indicator (RI) pin on the modem is used to drive transistor Q2 which in turn lights the LED at D3 when the phone is ringing. The data terminal ready (DTR)

pin on the modem is tied to ground via a 10 K ohm resistor. The modem is connected to the BS2 microprocessor through four pins. The TX and RX pins are the communication lines which allow the modem and STAMP to serially talk to each other. The DCD and RI lines on pins 17 and 15 of the modem connect to P13 and P12 of the BS2 microprocessor.

The Dallas Semiconductor DS1620 digital temperature sensor is a small 8-pin DIP style package, which communicates with the microprocessor through three pins (see the block diagram in Figure 32-4).

The chart in Table 32-1 describes the pin-out functions of the chip. The DS1620 is a digital thermometer and thermostat which provides a 9–bit temperature reading which is used indicate the ambient room temperature in our project. The DQ or input/output lead conveys the temperature signal data to BS2 through a 1 K ohm resistor. The clock pin (CLK) is used by the microprocessor to clock the temperature chip in order to send the

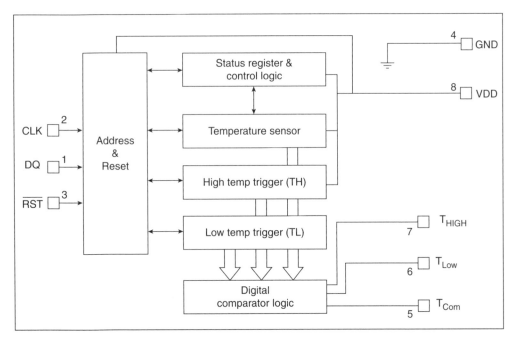

Figure 32-4 *DS1620 Digital temperature sensor. Courtesy of Dallas Semiconductor*

data readings to the microprocessor. The Reset pin (RST)on the temperature sensor is controlled by the microprocessor. The data line or DQ line on pin 1 of the DS1620 is coupled to pin 8 of BS2, while the CLK line at pin 2 is fed to the P7 I/O data line on the microprocessor. The RST line from the temperature sensor on pin 3 is sent to the P6 I/O line on the BS2 processor. A .1 µF capacitor is tied across the power or Vdd pin to ground.

The second major building block of the Remote Temperature Monitor is the mini-modem.

Table 32-1

DS 1620 Digital temperature sensor pinouts

Pin	Symbol	Description
1	DQ	Input/output pin for 3-wire communication
2	CLK/Conv	Clock input pin for 3-wire communication
3	RST	Reset input pin for 3-wire communication
4	GND	Ground pin
5	T_{COM}	High/low combination trigger – goes high with temp exceeds TH
6	T_{LOW}	Low temp trigger – goes high when temp falls below TL
7	T_{HIGH}	High temp trigger – goes high when temp exceeds TH
8	Vdd	+5 volt power supply pin

The mini-modem is a Cermetek CH1786 full function 2,400 bps modem. The modem measures a slight $1.01 \times 1.27 \times 0.52''$ and occupies a small footprint on the PC board; see Figure 32-5.

The block diagram of the mini-modem is shown in Figure 32-6. The CH1786 only requires two external interfaces, a serial interface that can be routed directly to a microprocessor, and a tip and

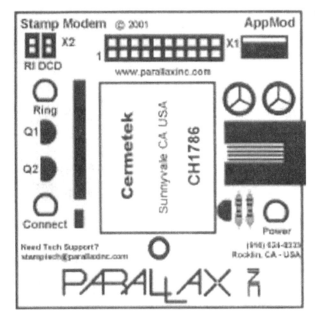

Figure 32-5 *Modem board layout. Courtesy of Parallax, Inc*

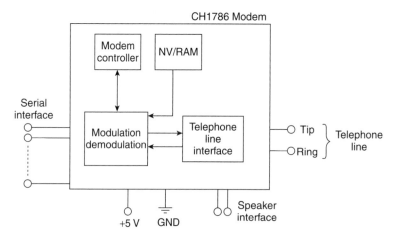

Figure 32-6 *CH1786 Modem block diagram. Courtesy of Parallax, Inc*

ring telephone line interface which goes directly to an RJ-11 jack for the PSTN line connection. The CH1786 modem can be controlled with industry standard AT commands. The CH1786 modem supports asynchronous operation at 2,400 bps, 1,200 bps, and 300 bps The CH1786 modem operates from a 5-V supply. A pinout chart for the mini-modem is shown in Table 32-2.

Table 32-2

CERMETEK CH1786 mini-modem pinouts

Pin	Name	TypeE	Function
1	RING	I/O	RING – connects to telephone line through RJ11 jack
2	TIP	I/O	TIP – connects to telephone line through RJ11 jack
3	RXA	O	analog voice injected – receive signal – let float if not used
4	TXA	I	analog voice injected – transmit signal – let float if not used
5	SPK	O	audio output for speaker if used
6	NC	—	no connection
7	NC	—	no connection
8	SLEEP	O	a low – indicates idle mode, used to control other devices
9	NC	—	no connection
10	TXD	I	serial transmit data input, marking or binary 1-is xmited when high
11	RXD	O	serial receive data output – rcvd marking or binary 1-indicated by high
12	V/D	O	voice data-used to switch between phone and modem use
13	DTR	I	data terminal ready – active low
14	DSR	O	data set ready – Low indicates handshaking with remote modem
15	RI	O	ring indication – follow frequency of ringing
16	CTS	O	clear to send – no connect if not using FAX option
17	DCD	O	data carrier detect – low indicates data carrier from a remote modem
18	HS	O	speed indication – low indicates 2,400 bps
19	VCC	—	power supply pin 5 V
20	GND	—	ground connection
21	RST	I	reset – active high – must be high for 10 ms to reset
22	NC	—	no connection

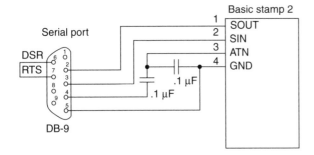

Figure 32-7 *Computer to STAMP programming cable*

The STAMP can be programmed via pins 1 through 4 connected to a DB-9F connector, as seen in Figure 32-7.

The chart shown in Table 32-3 depicts the BASIC STAMP 2 microprocessor pinout and functions.

Circuit assembly

Before we begin constructing the Remote Temperature Monitor, you will need to locate a clean, well-lit worktable or workbench. Next we will gather a small 25–30-W pencil tipped soldering iron, a length of 60/40 tin–lead solder, and a small canister of Tip Tinner, available at your local RadioShack store; Tip Tinner is used to condition the soldering tip between solder joints. You drive the soldering tip into the compound and it cleans and prepares the soldering tip. Next grab a few small hand tools; try to locate small flat blade and Phillips screwdrivers. A pair of needle-nose pliers,

a pair of tweezers along with a magnifying glass, and a pair of end-cutters and we will begin constructing the project. Now locate your schematic diagram and parts layout diagram along with all of the components needed to build the project.

Once all the components are in front of you, you can check them off against the project parts list to make sure you are ready to start building the project. Now, locate Tables 32-4 and 32-5, and place them in front of you.

Table 32-4 illustrates the resistor code chart and how to read resistors, while Table 32-5 illustrates the capacitor code chart which will aid you in constructing the project.

Refer now to the resistor color code chart shown in Table 32-4, which will assist you in identifying the resistor values for the project. Note that each resistor will have either three or four color bands, which begin at one end of the resistor's body. The first colored band represents the resistor's first digit, while the second color represents the second digit of the resistor's value. The third colored band represents the resistor's multiplier value, and the fourth color band represents the resistor's tolerance value. If there is no fourth colored band then the resistor has a 20% tolerance value. If the resistor's fourth band is silver then the resistor has a 10% tolerance value and if the fourth band is gold then the resistor's tolerance value will be 5%.

Finally, we are ready to begin assembling the project, so let's get started. The prototype was constructed on a

Table 32-3

Basic STAMP 2 – pinouts

Pin	Name	Description
1	Sout	Serial out, connects to PC serial port (DB-9, pin-2)
2	Sin	Serial in, connects to PC serial port (DB-9, pin-3)
3	ATN	Attention, connects to PC serial port -DTR pin (DB-9, pin-4)
4	VSS	System ground, connects to PC serial port (DB-9, pin-5)
5-20	P0-P15	General purpose input/output lines, source/sink 30 mA,
(21-36)	X0-X15	BS2P40-only – auxiliary bank of input/outputs source/sink 100 mA max
21(37)	VDD	5-volt DC input – from external regulator - recommended with loads
22(38)	RES	Reset pin, goes ow when power supply less than 4.2 V
23(39)	VSS	System ground same as pin-4, connects to power supply ground
24(40)	Vin	Unregulated power input accepts 5.5–12 V DC, 7.5 recommended

Table 32-4
Resistor color code chart

Color Band	1st Digit	2nd Digit	Multiplier	Tolerance
Black	0	0	1	
Brown	1	1	10	1%
Red	2	2	100	2%
Orange	3	3	1,000 (K)	3%
Yellow	4	4	10,000	4%
Green	5	5	100,000	
Blue	6	6	1,000,000 (M)	
Violet	7	7	10,000,000	
Gray	8	8	100,000,000	
White	9	9	1,000,000,000	
Gold			0.1	5%
Silver			0.01	10%
No color				20%

Table 32-5
Three-digit capacitor codes

pF	nF	µF	Code	pF	nF	µF	Code
1.0			1R0	3,900	3.9	.0039	392
1.2			1R2	4,700	4.7	.0047	472
1.5			1R5	5,600	5.6	.0056	562
1.8			1R8	6,800	6.8	.0068	682
2.2			2R2	8,200	8.2	.0082	822
2.7			2R7	10,000	10	.01	103
3.3			3R3	12,000	12	.012	123
3.9			3R9	15,000	15	.015	153
4.7			4R7	18,000	18	.018	183
5.6			5R6	22,000	22	.022	223
6.8			6R8	27,000	27	.027	273
8.2			8R2	33,000	33	.033	333
10			100	39,000	39	.039	393
12			120	47,000	47	.047	473
15			150	56,000	56	.056	563
18			180	68,000	68	.068	683
22			220	82,000	82	.082	823
27			270	100,000	100	.1	104
33			330	120,000	120	.12	124
39			390	150,000	150	.15	154
47			470	180,000	180	.18	184
56			560	220,000	220	.22	224

Table 32-5—cont'd
Three-digit capacitor codes

pF	nF	μF	Code	pF	nF	μF	Code
68			**680**	270,000	270	.27	**274**
82			**820**	330,000	330	.33	**334**
100		.0001	**101**	390,000	390	.39	**394**
120		.00012	**121**	470,000	470	.47	**474**
150		.00015	**151**	560,000	560	.56	**564**
180		.00018	**181**	680,000	680	.68	**684**
220		.00022	**221**	820,000	820	.82	**824**
270		.00027	**271**		1,000	1	**105**
330		.00033	**331**		1,500	1.5	**155**
390		.00039	**391**		2,200	2.2	**225**
470		.00047	**471**		2,700	2.7	**275**
560		.00056	**561**		3,300	3.3	**335**
680		.00068	**681**		4,700	4.7	**475**
820		.00082	**821**			6.8	**685**
1,000	1.0	.001	**102**			10	**106**
1,200	1.2	.0012	**122**			22	**226**
1,500	1.5	.0015	**152**			33	**336**
1,800	1.8	.0018	**182**			47	**476**
2,200	2.2	.0022	**222**			68	**686**
2,700	2.7	.0027	**272**			100	**107**
3,300	3.3	.0033	**332**		157		**157**

Code = 2 significant digits of the capacitor value + number of zeros to follow

For values below 10 pF, use "R" in place of decimal e.g. 8.2 pF = 8R2

10 pF = 100
100 pF = 101
1,000 pF = 102
22,000 pF = 223
330,000 pF = 334
1 μF = 105

single-sided printed circuit board or PCB. With your PC board in front of you, we can now begin populating the circuit board. We are going to place and solder the lowest height components first, the resistors and diodes. First, locate our first resistor R1. Resistor R1 has a value of 330 ohms and is a 5% tolerance type. Now take a look for a resistor whose first color band is brown, black, orange, and gold. From the chart you will notice that orange is represented by the digit (3), the second color band is orange, which is represented by (3). Notice that the third color is brown, so the multiplier is (10), then (3)(3) × 10 = 330 ohms.

Go ahead and locate where resistor R1 goes on the PC and install it, next solder it in place on the PC board, and then with a pair of end-cutters trim the excess component leads flush to the edge of the circuit board. Next locate the remaining resistors and install them in their respective locations on the main controller PC board. Solder the resistors to the board, and remember to cut the extra lead with your end-cutters.

Note that capacitors generally come from two major groups or types. Non-polar types, such as the ones in this project, have no polarity and can be placed in either direction on the circuit board. The other type of capacitors are called polarized types, these capacitors have polarity and have to be inserted on the board with respect to their polarity marking in order for the circuit to operate correctly. On these types of capacitors, you will observe that the components actually have a plus or minus marking or a black or white band with a plus or minus marking on the body of the capacitors. Capacitors must be installed properly with respect to polarity in order for the circuit to work properly. Failure to install electrolytic capacitors correctly could cause damage to the capacitors as well as to other components in the circuit when power is first applied.

Many capacitors are quite small in physical size, so manufacturers have devised a three-digit code which can be placed on the capacitor instead of the actual value. The code occupies a much smaller space on the capacitor and can usually fit nicely on the capacitor body, but without the chart the builder would have a difficult time determining the capacitor's value.

Now look for the lowest profile capacitors from the component stack and locate capacitors C3 and C4, labeled .1 µF or (104). Note that there are a number of small capacitors in the circuit that may not be labeled with their actual value, so you will have to check Table 32-5 to learn how to use the three-digit capacitor codes. Go ahead and place C3 and C4 on the circuit board and solder them in place. Now take your end-cutters and cut the excess component leads flush to the edge of the circuit board. Next, locate the remaining low profile capacitors and install them on the printed circuit board, then solder them in place. Remember to trim the extra component leads. There are a number of electrolytic capacitors in the circuit and placing them in the circuit with respect to their proper polarity is essential for proper operation of the circuit. Once you have located the larger electrolytic capacitors install them in their respective locations and solder them in place. Finally trim the excess component leads.

The Remote Temperature Monitor utilizes two conventional transistors. Transistors are generally three lead devices, which must be installed correctly if the circuit is to work properly. Careful handling and proper orientation in the circuit is critical. A transistor will usually have a Base lead, a Collector lead, and an Emitter lead. The base lead is generally a vertical line inside a transistor circle symbol. The collector lead is usually a slanted line facing towards the base lead and

the emitter lead is also a slanted line facing the base lead but it will have an arrow pointing towards or away from the base lead. Now, identify the two transistors, referring to the manufacturer's specification sheets as well as the schematic and parts layout diagrams when installing the two transistors. If you have trouble installing the transistors, get the help of an electronics enthusiast with some experience. Install the transistors in their respective locations and then solder them in place. Remember to trim the excess component leads with your end-cutters.

The Remote Temperature Monitor circuit contains four ICs. ICs are often static sensitive, so they must be handled with care. Use a grounded anti-static wrist strap and stay seated in one location when handling the ICs. Take out a cheap insurance policy by installing IC sockets for each of the ICs. In the event of a possible circuit failure, it is much easier to simply unplug a defective IC than trying to unsolder 14 or 16 pins from a PC board without damaging the board. ICs have to be installed correctly if the circuit is going to work properly. IC packages will have some sort of markings which will help you orient them on the PC board. An IC will have either a small indented circle, a cutout, or notch at the top end of the IC package. Pin 1 of the IC will be just to the left of the notch, or cutout. Refer to the manufacturer's pinout diagram, as well as the schematic when installing these parts. If you doubt you ability to correctly orient the ICs, then seek the help of a knowledgeable electronics enthusiast.

The intercom controller circuit employs three indicator or status LEDs at D1, D2, and D3 which also must be oriented correctly if the circuit is to function properly. An LED has two leads, a cathode and an anode lead. The anode lead will usually be the longer of the two leads and the cathode lead will usually be the shorter lead, just under the flat side edge of the LED; this should help orient the LED on the PC board. You may want to leave about ½ lead length on the LED, in order to have the LED protrude through the enclosure box. Solder the LEDs to the PC board and then trim the excess leads.

The Remote Temperature Monitor uses a single 250 V MOV at the input to the mini-modem to protect it from telephone ringing and surge voltages; this device has no polarity so it can be mounted in either direction in the circuit.

Next, go ahead and install the mini push-button switch at S1, which functions as a reset switch for the microprocessor. The reset switch is used in the event of a microprocessor lockup.

Once the Remote Temperature Monitor circuit board has been completed, take a short break and then we will inspect the PC board for possible "cold" solder joints and "short" circuits. Pick up the PC board with the foil side facing upwards toward you. Carefully inspect the solder joints, these should look smooth, clean, and bright. If any of the solder joints look dull, dark, or blobby, then you should remove the solder with a solder sucker or a solder wick and then resolder the joint all over again. Now we will inspect the PC board for possible short circuits. Short circuits can be caused from small solder blobs bridging the circuit traces or from cut component leads which can often stick to the circuit board once they have been cut. A sticky residue from solder will often trap component leads across the circuit traces. Look carefully for any bridging wires across circuit traces.

Once this inspection is complete, we can move on to installing the circuit board into a suitable plastic enclosure. Locate a suitable enclosure to house the circuit board. Arrange the circuit board and the transformer on the bottom of the chassis box. Now, locate four ¼″ standoffs, and mount the PC board between the standoffs and the bottom of the chassis box.

Finally, install a four pin header/jack connector at J1/P1 to allow connecting up the serial programming connector at J2. Next, go ahead and install the modular phone jack at J3 and a coaxial power jack for the power supply input. Solder these components to the circuit board.

You can power the Remote Temperature Monitor using a 12 V DC power cube or "wall-wart" type power supply which can be readily purchased from your local RadioShack or electronic supply house. These power supplies generally have coaxial plugs on them, which can mate to the coaxial power jack on the chassis of your Remote Temperature Monitor. Next, you will have to load the **Rtemp.bs2** program into the Remote Temperature Monitor circuit. Refer to the programming cable diagram, and connect the programming cable between the Remote Temperature Monitor and a personal computer or laptop. Call up the BASIC STAMP editor program that you downloaded from www.parallax.com. Copy the **rtemp.bs2** program into your laptop. Now call up the BS2 editor program and look for the **rtemp.bs2** program.

With power applied to the Remote Temperature Monitor circuit, download the **rtemp.bs2** program into the BS2 microcontroller via the programming cable and you are ready to begin remote monitoring. Your Remote Temperature Monitor can now be placed where you wish to monitor the remote location or cabin.

Now, when you return to your home or workshop you are ready to remotely interrogate the temperature sensor from your home or office location.

Operation

When you return to your home, turn on your computer and bring up the Windows Hyper-terminal modem Window or any other window modem program and configure it for 2,400 bps with settings of 8-N-1 data bits. From the Hyper-terminal Window dial the telephone number of your remote location, i.e. the Remote Temperature Monitor. The modems will begin communicating with each other and soon you will begin seeing the temperature readings appear in the Hyper-terminal window, as shown in Figure 32-8.

When you are finished reading the high, low and current temperature readings, you can ask the Hyper-terminal program to hang up; this will reset the modems for the next time you wish to call up the remote location and read the remote temperature.

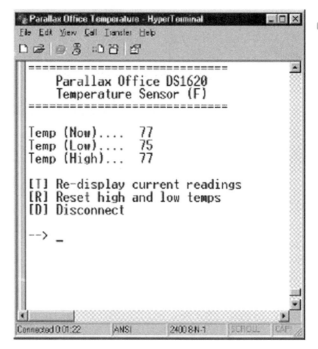

Figure 32-8 *Temperature readout screen*

```
'Temperature (CH1786Temperature.BS2) Parallax
' This program monitors a Dallas Seminconductor DS1620 digital thermometer
' while waiting for an incoming call. When a call is received, the STAMP
' causes the modem to answer the call then displays temperature data on
' the remote terminal.
' -----[I/O Definitions] ------------------------------------------------
'
' modem pins
'
TX1 CON 15 ' CH1786 "Tx" pin
RX1 CON 14 ' CH1786 "Rx" pin
RI_ VAR In12 ' ring indicator
DCD_ VAR In13 ' carrier detect
' DS1620 pins
'
Rst CON 6 ' DS1620.3
Clk CON 7 ' DS1620.2
DQ CON 8 ' DS1620.1
' -----[Constants] ------------------------------------------------------
'
True CON 1
False CON 0
No CON 1
Yes CON 0
T2400 CON 396 ' 2400 baud for modem
LF CON 10 ' line feed character
FF CON 12 ' form feed (clear remote screen)
' DS1620 commands
'
RTmp CON $AA ' read temperature
WTHi CON $01 ' write TH (high temp register)
WTLo CON $02 ' write TL (low temp register)
RTHi CON $A1 ' read TH
RTLo CON $A2 ' read TL
Strt CON $EE ' start conversion
StpC CON $22 ' stop conversion
WCfg CON $0C ' write configuration register
RCfg CON $AC ' read configuration register
NTasks CON 3 ' total number of tasks

' -----[Variables] ------------------------------------------------------
'
tmpIn VAR Word ' 9-bit temp input from DS1620
nFlag VAR tmpIn.Bit8 ' negative flag
hlfBit VAR tmpIn.Bit0 ' half degree C bit
```

```
tempF VAR Word ' converted fahrenheit value
tempC VAR Byte ' converted celcius value
tmpNow VAR Word ' current temperature
tmpLo VAR Word ' low temp
tmpHi VAR Word ' high temp
sign VAR Byte ' - for negative temps
sLo VAR Byte
sHi VAR Byte
inByte VAR Byte ' input from user terminal
cmd VAR Byte ' command pointer
answer VAR Byte ' user response to prompt
task VAR Byte ' task control variable
riFltr VAR Byte ' for ring indicator filter
' -----[EEPROM Data] -------------------------------------------------
'
' -----[Initialization] ----------------------------------------------
'
Init:
tmpLo = $FFFF ' start with opposite extremes
tmpHi = 0
I_Modm:
PAUSE 250 ' allow modem to power up
' train modem for speed
'
SEROUT TX1, T2400, 10, ["AT", CR]
SERIN RX1, T2400, 2500, Error, [WAIT ("OK")]
PAUSE 250
' auto answer on second ring (S0 = 2)
' set max time for carrier detect to 30 secs (S7 = 30)
'
SEROUT TX1, T2400, 10, ["ATS0 = 2 S7 = 30", CR]
SERIN RX1, T2400, 2500, Error, [WAIT ("OK")]
I_1620:
HIGH Rst ' alert the DS1620
SHIFTOUT DQ,Clk,LSBFIRST,[WCfg] ' write configuration
' use with CPU; free run mode
SHIFTOUT DQ,Clk,LSBFIRST,[%00000010]
LOW Rst
PAUSE 10 ' pause for DS1620 EE write cycle
HIGH Rst
SHIFTOUT DQ,Clk,LSBFIRST,[Strt] ' start temp conversions
LOW Rst
debug "Initializing DS1620",cr
NoDCD: IF DCD_ = Yes THEN NoDCD ' make sure DCD is clear
' -----[Main Code] ---------------------------------------------------
'
Main: GOSUB ScanT ' get current temperature
IF DCD_ = Yes THEN GetMdm ' call received
```

```
BRANCH task, [Task0, Task1, Task2]
GOTO Main
Task0: ' task code here
'
task = 1 ' select a specific task
GOTO NextT ' go do it
Task1: ' task code here
'
task = 2
GOTO NextT
Task2: ' task code here
'
task = 0
GOTO NextT
NextT: ' task = task + 1 // NTasks
' round-robin to next task
GOTO Main
END
' -----[Subroutines] -------------------------------------------------------
'
' ==========
' Modem Routines
' ==========
' error with modem
' - structured as separate routine to allow user indications/enhancements
'
Error: ' additional code here
PAUSE 1000
GOTO I_Modm ' try to initialize again
GetMdm: PAUSE 5000 ' let other end get ready
Modm1: GOSUB DoMenu ' show readings and menu
Get1: SERIN RX1, T2400, [inByte] ' wait for input
' process user input
cmd = 99
' convert letter to digit (0..5)
LOOKDOWN inByte, ["tTrRdD"], cmd
cmd = cmd/2 ' fix for BRANCH
' branch to handler
BRANCH cmd, [Cmd0, Cmd1, Cmd2]
GOTO Modm1
Cmd0: GOSUB ScanT ' get current temp
GOTO Modm1
Cmd1: GOSUB RstT ' reset high and low
GOTO Modm1
Cmd2: GOSUB Discon ' disconnect from user
IF answer = No THEN Modm1 ' stay with user
GOTO NoDCD ' back to the beginning
' clear remote terminal and display menu
'
```

```
DoMenu:
debug "DoMenu",cr
SEROUT TX1, T2400, [FF]
SEROUT TX1, T2400, ["===================", CR, LF]
SEROUT TX1, T2400, ["Remote STAMP - Station 001", CR, LF]
SEROUT TX1, T2400, ["===================", CR, LF]
SEROUT TX1, T2400, [LF]
SEROUT TX1, T2400, ["Temp (Now).... ", sign, DEC tmpNow, CR, LF]
SEROUT TX1, T2400, ["Temp (Low).... ", sLo, DEC tmpLo, CR, LF]
SEROUT TX1, T2400, ["Temp (High)... ", sHi, DEC tmpHi, CR, LF]
SEROUT TX1, T2400, [LF]
SEROUT TX1, T2400, ["[ T] Re-display current readings", CR, LF]
SEROUT TX1, T2400, ["[ R] Reset high and low temps", CR, LF]
SEROUT TX1, T2400, ["[ D] Disconnect", CR, LF]
SEROUT TX1, T2400, [LF, "--> "]
RETURN
' reset high and low temperatures
'
RstT: SEROUT TX1, T2400, [CR, LF, LF, "Reset? "]
GOSUB YesNo
IF answer = No THEN RstX
GOSUB ScanT
tmpLo = tmpNow
sLo = sign
tmpHi = tmpNow
sHi = sign
RstX: RETURN
' disconnect
'
Discon: SEROUT TX1, T2400, [CR, LF, LF, "Disconnect? "]
GOSUB YesNo
IF answer = No THEN DiscX
SEROUT TX1, T2400, [CR, LF, LF, "Disconnecting.", CR, LF]
' return modem to command state
' and hang up
'
PAUSE 2000
SEROUT TX1, T2400, ["+++"]
PAUSE 2000
SEROUT TX1, T2400, 10, ["ATH0", CR]
DiscX: RETURN
' confirm for [Y]es or [N]o
' and get user input (default = No)
'
YesNo: SEROUT TX1, T2400, ["Are you sure? (Y/N): "]
answer = No
' get answer
' - but only wait for 5 seconds
```

```
                    `
       SERIN RX1, T2400, 5000, YesNoX, [inByte]

       IF inByte = "y" THEN IsYes
       IF inByte = "Y" THEN IsYes
       GOTO YesNoX
       IsYes: answer = Yes
       YesNoX: RETURN
       ' process ring indicator
       ' - filters pulsing ring indicator
       ' - waits for about 0.25 second of no RI pulsing before returning
       '
       DoRing:
       ' your code here
       ' (i.e., count number of rings)
       '
       RIWait: riFltr = 0 ' clear the "no pulses" counter
       RIchk: IF RI_ = Yes THEN RIWait ' still pulsing
       riFltr = riFltr + 1 ' not pulsing, increment count
       IF riFltr > 50 THEN RIx ' RI clear now
       PAUSE 5 ' 5 ms between RI scans
       GOTO RIchk ' check again
       RIx: RETURN ' done - outta here
       ' ==========
       ' DS1620 Routines
       ' ==========
       ' get current temperature
       ' — update high and low readings
       '
       ScanT: HIGH Rst ' alert the DS1620
       SHIFTOUT DQ,Clk,LSBFIRST,[RTmp] ' read temperature
       SHIFTIN DQ,Clk,LSBPRE,[tmpIn\9] ' get the temperature
       LOW Rst
       GOSUB GetF ' convert to Farhenheit
       tmpNow = tempF
       IF (tmpLo < tmpNow) THEN THigh
       tmpLo = tmpNow ' set new low
       sLo = sign
       debug "Scan T",cr
       THigh: IF (tmpHi > tmpNow) THEN TDone
       tmpHi = tmpNow ' set new high
       sHi = sign
       TDone: RETURN
       ' convert reading from ½ degrees input (rounds up)
       '
       GetC: IF nFlag = 0 THEN CPos ' check negative bit (8)
       sign = "–" ' set sign
       tempC = –tmpIn/2 + hlfBit ' if neg, take 2's compliment
```

```
GOTO CDone
CPos: sign = " "
tempC = tmpIn/2 + hlfBit
CDone: RETURN
' convert ($\frac{1}{2}$ degrees C) to Fahenheit with rounding
' – general equation (for whole degrees): F = C * 9/5 + 32
'
GetF: sign = " "
IF nFlag = 0 THEN FPos1
tmpIn = -tmpIn & $FF ' convert from negative
IF tmpIn < 36 THEN FPos0
FNeg: sign = "–"
tempF = tmpIn * 9/10 + hlfBit – 32
GOTO FDone
FPos0: tempF = 32 – (tmpIn * 9/10 + hlfBit)
GOTO FDone
FPos1: tempF = tmpIn * 9/10 + 32 + hlfBit
FDone: RETURN
```

Appendix

Electronic Parts Suppliers

We have listed some electronic parts suppliers. We cannot guarantee any particular level of service from these companies and supply this list only as a convenience to our readers.

Alltronics — Electronic parts, over runs, tec
Box 730, Morgan Hill, CA 95038
Tel: 408 847 0033
Fax: 408 847 0133
http://www.alltronics.com

Carls Electronics — ElectroInc.s kits
484 Lakepark Ave, Suite 59, Oakland, CA 94610
Tel: 866 664 0627 (toll-free USA) or 510 451 4320
Fax: 510 903 9712
http://www.electronickits.com/

Cermetek Microelectronics, Inc. — Mini modems
1390 Borregas Avenue, Sunnyvale, CA 94089
Tel: 408 752 5000
Fx: 408 752 5004
www.cermetek.com
sales@cermetek.com

Circuit Specialists Inc. — Electronic parts, test
   equipment
Paul Thorpe PO Box 3047, Scottsdale, AZ 85271-3047
Tel: 800 528 1417
Fax: 602 464 5824
http://www.circuitspecialists.com/

Clare, Inc. – Teltone Products — Touch tone decoders
78 Cherry Hill Drive, Beverly, MA 01915,
Tel: 978 524 6700
Fax: 978 524 4700
http://www.clare.com/

Digi-Key Electronics — electronics parts
701 Brooks Avenue South, Thief River Falls,
   MN 56701
Tel: 800 344 4539 or 218 681 6674
Fax: 218 681 3380
http://www.digikey.com/

DIY Electronics (HK) Ltd. — Electronics kit
   manufacturer
Peter Crowcroft & Ladda, PO Box 88458,
   Sham Shui Po, Hong Kong
Tel: 852 2304 2250
Fax: 852 2729 1400

Elenco Electronics Inc. — Electronic kits,
   educational products
150 Carpenter Ave., Wheeling, IL 60090
Tel: 847 541 3800
Fax 847 520 0085
http://www.elenco.com/

Freescale Semiconductor, Inc. — Semiconductors
(formerly Motorola Semiconductor)
Tel: 800 521 6274
http://www.freescale.com/

Glolab — Electronic kits
307 Pine Ridge Drive, Wappingers Falls,
   New York (NY), 12590
Tel: 845 297 9772
Fax: 845 297 9772
http://www.glolab.com

Hobby Engineering — Electronics parts, kits, etc.
1405 Huntington Avenue, Suite 150, South
   San Francisco, CA 94080
Tel: 866 Robot 50 (866 762 6850) or 650 875 7987
Fax: 650 952 7629
http://www.hobbyengineering.com/

Intersil corp — Semiconductor
1001 Murphy Ranch Road, Milpitas, CA 95035
Tel: 408 432 8888 or 888 INTERSIL (888 468 3774)
Fax: 408 434 5351
http://www.intersil.com/cda/home/

Marlin P. Jones & Associates C — Electronic parts,
   over runs, power supplies, etc.
PO Box 12685, Lake Park FL 33403-0685
Tel: 407 848 8236
Fax: 407 844 8764
http://www.mpja.com/

Maxim Integrated Products — Semiconductors
(formerly Dallas Semiconductor)
120 San Gabriel Drive, Sunnyvale, CA 94086
Tel: 408 737 7600
Fax: 408 737 7194
http://www.maxim-ic.com/

Mouser Electronics, Inc. B — Large electronics
   parts house
1000 North Main Street, Mansfield, TX 76063
Tel: 800 346 6873 or 817 804 3888
Fax: 817 804 3899
http://www.mouser.com/

National Semiconductor B — Semiconductors
2900 Semiconductor Dr., P.O. Box 58090, Santa Clara,
   CA 95052-8090
Tel: 408 721 5000
http://www.national.com/

Newark Electronics — Extensive electronic parts
4801 N. Ravenswood, Chicago, IL 60640-4496
Tel: 773 784 5100
Fax: 888 551 4801
http://www.newark.com/

Ocean State Electronics — Electronic parts
PO Box 1458, 6 Industrial Drive, Westerly RI 02891
Tel: 401 596 3080
Fax: 401 596 3590
http://www.oselectronics.com/

Parallax, Inc. — BASIC STAMP, microcontrollers,
   accessories
599 Menlo Drive, Rocklin, CA 95765
Tel: 888 512 1024 (toll-free sales)
Tel: 888 997 8267 (toll-free tech support)
http://www.parallax.com

Phillips Semiconductor NXP B — Semiconductors
NXP Semiconductors
Tel: 800 447 1500
http://www.nxp.com/#[0]

Ramsey Electronics — Large kit manufacturer
590 Fishers Station Dr., Victor, NY 14564
Tel: 800 446 2295 or 585 924 4560
Fax: 585 924 4886
http://www.ramseyelectronics.com/

Samsung Electronics C —Semiconductors
3655 North First Street, San Jose, CA 95134
Tel: 1 408 954 7000
http://www.samsungsemi.com

Solder-it — Soldering accessories
P.O. Box 360, Chagrin Falls, OH 44022
Tel: 800 897 8989 or 440 247 6322
Fax: 440 247 4630
http://www.solder-it.com/

Talking Electronics B — Electronics kits, extensive
   web page projects
P.O. Box, Cheltenham, 3192 Victoria, Australia
Tel: 0417329788
http://www.talkingelectronics.com/

Texas Instruments, Inc. — Semiconductors
13532 N. Central Expressway, M/S 3807, Dallas,
   TX 75243-1108
http://www.ti.com/

Vellemen Inc. (USA) — Large electronic manufacturer
7354 Tower Street, Fort Worth, TX 76118
Tel: 817 284 7785
Fax: 817 284 7712
http://www.vellemanusa.com/us/

Winbond Electronics Corporation America —
  Memory storage devices
Information Storage Devices (ISD), 2727 North First
  Street, San Jose, CA 95134
Tel: 408 943 6666
Fax: 408 544 178

Index

Figures and Tables in **BOLD**
Projects in ***BOLD ITALIC***

Index